Innovative Approaches towards Ecological Coal Mining and Utilization

Innovative Approaches towards Ecological Coal Mining and Utilization

Jiuping Xu
Heping Xie
Chengwei Lv

Authors

Prof. Jiuping Xu
Sichuan University
School of Business
No. 24 South Section 1
Yihuan Road
610065 Chengdu
China

Prof. Heping Xie
Shenzhen University
518000 Shenzhen
China

Dr. Chengwei Lv
Sichuan University
School of Business
No. 24 South Section 1
Yihuan Road
610065 Chengdu
China

Cover
Frontcover: iStock / @ Evgeny Gromov

Library of Congress Card No.: applied for

British Library Cataloguing-in-Publication Data
A catalogue record for this book is available from the British Library.

Bibliographic information published by the Deutsche Nationalbibliothek
The Deutsche Nationalbibliothek lists this publication in the Deutsche Nationalbibliografie; detailed bibliographic data are available on the Internet at <http://dnb.d-nb.de>.

Print ISBN: 978-3-527-34692-9
ePDF ISBN: 978-3-527-82510-3
ePub ISBN: 978-3-527-82512-7
oBook ISBN: 978-3-527-82511-0

Cover Design Adam-Design, Weinheim, Germany
Typesetting Straive, Chennai, India
Printing and Binding CPI Group (UK) Ltd, Croydon, CR0 4YY

Printed on acid-free paper

C098646_111021

Contents

Preface

Coal has a very long and varied history which plays critical roles in the development of human beings, especially for the modern and contemporary society. Some historians believe that coal was first discovered and used around 1000 BCE and during the Industrial Revolution in the eighteenth and nineteenth centuries, coal experienced its most important expansion period of mining and utilization. During the development of human beings society, four epoches have been accessed: the age of farming, the age of industrial, the age of information, and the age of intelligence. In the agricultural age, coal might be the only energy resource that human beings are skilled, so its role is to act as a combustion-supporting agent which helps human beings to live on it. During the age of industrial, due to high energy supply ratio and the steady utilization technology, coal played as the most important catalyst to support the high developing speed of human society. Recently, in the information age, coal is the stabilizing agent around the world due to its most energy demand for people. As for the future, in the age of intelligence, coal will play the role of temper agent which will continue to help further development of human being society. In another word, the history of coal mining and utilization is inextricably linked with that of the Industrial Revolution. Along with the first practical coal-fired electric generating station, developed by Thomas Edison, went into operation in New York City in 1882, coal became the most important part in the energy structure and this situation lasts for over 100 years until the early twenty-first century. Even nowadays, coal still shares 27.6% of the primary energy consumption which ranked the second place among all the known energy resources according to the BP Statistical Review of World Energy 2018, and in some certain areas, such as electric power generation and steel industry, coal is still playing unmovable roles. As for electric power generation, coal is the world's dominant source, with a share of 38.1% in 2017 and in steel industry, similar situation takes place again. Another fact should be highlighted is that after several years of free fall, the coal market experienced a mini-revival in 2018, with both global consumption and production increasing and at the same time, with consideration of that the world proved coal reserves are currently sufficient to meet 134 years of global production, much higher than the R/P ratio for oil and gas. As coal is abundantly available, affordable, reliable, geographically well distributed and easy and safe to transport, its markets are well functioning and responsive to changes in supply and demand. It is convinced that coal will still be the most important and wide-used energy resource in the following decades.

However, everything in this world has their own two sides and when we evaluating the position and importance of coal along with the development of human beings, its positive influence should never be the only thing which should be focused on and the negative perspective should also be paid attention to. The major challenges facing coal are concerned with its environmental impacts. These include the release of pollutants, such as oxides of sulphur and nitrogen (SOx and NO_x), and particulate and trace elements, such as mercury and at the same time, greenhouse gas emission is another troublesome thing that must be highlighted right now. According to the BP Statistical Review of World Energy 2018 and the data from the World Bank, the total carbon emissions amount in 2017 reached 33 444.0 million tonnes which is 1.6% more compared to 2016 and in which the using of coal shared over 40%. Many researchers around the world noticed this critical problem and appealed to reduce or even forbid using of coal. However, this is not realistic mainly due to that for most regions in the world, coal shared over half of its primary energy consumption and the equipments for coal mining and utilization there are almost made up within the 30 years. For such regions, they cannot pay for the forbidding using of coal. Other researchers claim that some new technologies such as carbon capture and storage should be focused and put into practice. It is convinced that such hard technologies are quite efficient in reducing the carbon emissions and have already used in many developed countries and regions. However, the most challenges for such hard technologies are the economic efficiency and feasibility and they are too expensive for the most developing countries and regions which are the main users of coal.

Based on what was discussed above, it can be concluded that coal will still be critical for the development of human beings in the following decades and its harm on the environment should also be faced squarely. The performance of existing pathways and methods toward such issues are not satisfied and some more innovative approaches are needed. The book has 11 chapters and made up with 3 main parts. Chapters 1 and 2 are the first part of this book in which the development history, current status, and future possible pathways of ecological coal mining and utilization are introduced. The second part is from Chapter 3 to Chapter 6 and in this part, the typical environmental problems in coal mining and utilization which includes the groundwater damage and coal gangue excessive accumulation issues during coal mining and air pollutant emissions and greenhouse gas emissions reduction issues during coal utilization are discussed and solved in the ecological soft path ways. The third part is from Chapter 7 to Chapter 11 in which ecological problems in the coal involved integrated energy systems will be fully considered and discussed, at the same time, several practical soft paths also will be given to cope with the actual situation.

In the second part of this book, four typical ecological problems will be researched. The first one is the coal–water conflict under the multiple coal seam production system. An equilibrium strategy-based bilevel programming model under the co-production in multiple coal seams situation is proposed in which both the groundwater quality and quantity protection are focused, the equilibrium between environmental protection and economic development is analyzed, and the relationship and conflicts among the different stakeholders which includes the authority and collieries are taken into consideration at the same time. A possibility measure involved Karush–Kuhn–Tucker condition solution method is designed to solve the proposed model, and the successful application in the Luan coal field, China, shows the efficiency and feasibility of it. According to the second one, another innovative approach which makes full consideration of the impacts of seasonal changes on the groundwater level and production plans of collieries is proposed. A mining quota competition mechanism is built through this approach in which the authority allocates the initial mining quota to each sub-colliery and then, collieries can determine suitable production schemes for each season to improve the total environmental protection performance for the purpose of competing for as much mining quota as possible. Each colliery's mining plans are fed back to the authority, which adjusts the scheme based on the performance of each colliery. After a further adjustment, the mining quota allocation scheme is returned to collieries again, which changes their own mining plans based on this new quota. This process is repeated several times until a final scheme acceptable to both the authority and all collieries is agreed on. The proposed method is then used in the Yanzhou coal field, China, and achieves a satisfied objective. The above two research problems focus on the water environment problem along with the coal production and at the same time, the third research problem will focus on the solid pollution issues caused by coal production. To reduce the coal gangue accumulation amount and its damage on the soil and groundwater in the large-scale coal field, an innovative approach which integrated the equilibrium decision-making model and geographic information system (GIS) is proposed to identify and select the most suitable coal gangue facility construction site under the 3R principle (Reducing, Reusing, and Recycling). In this approach, the GIS technology is first employed to identify the possible candidate site of coal gangue facility and then an equilibrium strategy-oriented bilevel programming model is designed to make both the local authority and the sub-collieries as an integrated decision system to select the most suitable one. A real-world case study at Yanzhou coal field, China, is conducted and the results show that such approach has wonderful potential to reduce the coal gangue accumulation amount. The last research problem in the second part of this book will focus the air pollution problem as well as the greenhouse gas emissions issued in the coal utilization period. To improve the emission performance of coal-fired power plant (CPP) which is known as the most serious pollution source of the atmospheric environment, especially for reducing the carbon emissions amount from CPP, an integrated innovative decision-making system which includes the coal purchasing, blending, and distributing is conducted with the purpose of minimizing the total carbon emissions amount and the operational cost simultaneously.

The third part of this book consists of five chapters from Chapter 7 to Chapter 11, in which the ecological problems caused by the production and utilization of coal involved integrated energy system are researched. The first research topic in this part is the co-reducing multiple kinds of air pollutants from coal combustion. The total carbon emissions amount and the total PM_{10} emissions amount are taken into consideration at the same time, and the sufficient electric power supply and the economic benefit are also focused. A real-world case study at Sichuan Province, China, is employed and the results shows that such innovative coal blending method has a good performance in co-reducing carbon emissions and PM_{10} emissions. The second research topic is to improve the emissions performance and energy efficiency during coal mining. A real-world case study at Chaohua Colliery, China, is then discussed which would be the demonstration of the efficiency of the proposed approach. The third research topic in this part is with the purpose of improving the emissions performance of the coal involved electric power generation systems from the regional perspective, and an equilibrium strategy based on a hydro-wind-thermal complementary system with consideration of the cooperation of hydro power plants, wind power plants, and CPP is proposed. The randomness of seasonal wind speeds, the water flow uncertainty, and the CPP operational decisions are integrated into a whole decision-making model, and the steady power supply, minimizing the pollutants emissions amount, and the highest possible economic benefit are setting as the objective functions at the same time. Such decision-making system can make full utilization of the steady output of electric in CPP to offset the volatility of wind power and hydro power generation and at the same time, the cleanliness of the latter can just to neutralize the heavy pollution of the former. The Bijie City in China is employed as the real-world case to demonstrate the efficiency of the proposed approach.

With its emphasis on problem-solving and practical application, this book is ideal for researchers, practitioners, engineers, graduate students, and upper-level undergraduates with coal mining and utilization backgrounds in applied mathematics, management science, operations research, and engineering management.

September 2020 *Jiuping Xu, Heping Xie, Chengwei Lv*

Acknowledgments

This work is supported by the National Key Basic Research Development Plan (973 Program, Grant N0.2011CB201200), the Funds for Creative Research Groups of China (Grant No. 50221402), the National Social Sciences Foundation Monumental Projects (Grant No. 17ZDA286), and the National Funds for Distinguished Young Scientists of China (Grant No. 70425005). The authors want to take this opportunity to thank the researchers from Sichuan University, particularly, Liming Yao, Yi Lu, Ziqiang Zeng, Fengjuan Wang, Rui Qiu, Xiaoling Song, Jingqi Dai, Lurong Fan, Ning Ma, Wen Gao, Qian Huang, and Qing Feng. Authors would like to express a special acknowledgment to all editor boards of the Global Physical Sciences in Wiley, especially for Program Manager Dr. Lifen Yang, the senior managing editor Ms. Katherine Wong, the project editor Ms. Shirly Samuel and content refinement specialist, Abisheka Santhoshini for their wonderful cooperation and helpful comments. This book has benefited from the consultation of many references and the authors would like to thank all of these authors here. Finally, the authors express their deep gratitude to anonymous reviewer for their kind support and valuable insights and information from whom the authors have received significant enlightenment in the ecological coal mining and utilization.

1

Technical Developing Pathway of Ecological Coal Mining

It is believed that since BCE 1000, human beings had already begun to conduct coal mining and its relative activities and in these early years, coal mining was small-scale, nonstandard, experience-based, and inefficient. Such coal mining behavior was relative feasible at that time mainly due to that the demand of coal was very small. However, things began to change since the eighteenth century along with the flare-up of the Industrial Revolution which was first began in Britain and spread around all the Europe and North America soon after. As it is known, one of the most important changes that the Industrial Revolution brings to the human beings on the technology perspective was the availability of coal to power steam engines and this great invention promoted the rapid expansion of the international trade by building the coal-fed steam for the railways and steamships. As a result, the demand on coal experienced an extremely rapid rising and the former coal mining technologies were dead out and then in 1880s the coal cutting machines were introduced which was the milestone of the modern coal mining and later in 1912, the surface coal mining also welcomed its new chapter by the invention of the steam shovels. It can be concluded that coal mining technologies directly affect coal mining quantity and quality, which in turn affect global energy supplies and, on another hand, coal mining has already made a series of environmental problems which are gaining more and more attention around the world and the environmental-friendly-oriented ecological coal mining technologies and methods are needed in urgent. In this chapter, for the purpose of better understanding of the developing pathway of coal mining from the technological paradigmatic development perspective, a general data analysis was conducted. Through this analysis, the main coal mining technology developing stages can be summarized clearly and using the S-curve-oriented prediction method, the main development direction for coal mining technologies in the following years can be identified.

1.1 Background Introduction

Coal is one of the most abundant, affordable, and readily combustible energy resources all over the world and is consumed more than 53 million tonnes of oil equivalent (mtoe) per year, with a proportion of 28.1% on current global primary

Innovative Approaches towards Ecological Coal Mining and Utilization, First Edition.
Jiuping Xu, Heping Xie, and Chengwei Lv.

energy [Ye et al., 2013, British Petroleum, 2017, WCA, 2017b]. Although renewable energy is developing in an increasing speed, its consumption only accounts for 10% of total global primary consumption. As a result, power generated by coal is still expected to dominate in the global energy structure. While the use of coal causes severe environmental problems, concerns about global energy supplies have grown over the past 15 years with rapid urbanization and industrialization of economy in developing countries; therefore, it is predicted that the demand for coal will remain stable in the short and medium term [WCA, 2017a].

The global coal reserves are 11.139 billion tonnes, with the reserves to production (R/P) ratio of over 153 [British Petroleum, 2017], indicating the huge potential of coal resource exploitation. The energy return on investment (EROI) has been found to be a useful measure to assess resource availability [Hall et al., 2014, Court and Fizaine, 2017]. This term is the ratio of the amount of energy delivered by a given process to the amount of energy consumed, and obviously the higher the EROI, the greater the net energy delivered to society for economic growth [Hall et al., 2014]. The long-term EROI estimates for global coal production show a rising trend, indicating that the global coal production is expected to peak between 2025 and 2045, which means that coal exploitation remains significant EROI potential [Court and Fizaine, 2017]. To directly assess future global coal production capacity, a technological diffusion model was developed and used to simulate the prediction of coal product capacity. It has been found that coal is anticipated to continue to make up a considerable share of global prime energy to meet the energy demands related to technological development.

Since the demand for coal exploitation will remain high in the future, there is also the danger of exacerbating the accompanying environmental problems, especially as majority of the known coal reserves will have to be mined underground [Griffith and Clarke, 1979]. However, traditional underground coal mining practices have caused serious environmental problems such as water aquifer pollution and land surface subsidence in Australia and China, which has brought social and health problems to these regions and their surroundings [Kapusta and Stanczyk, 2011]. Due to the current limitations of coal mining technology, 85% of the world's coal resources cannot be mined by conventional methods [PricewaterhouseCoopers, 2011]. Therefore, it is necessary to increase the research on coal mining practices with high technical efficiency and environmental friendliness.

With the rapid development of science and technology, there are more and more studies on the future of coal mining and many processes and technological innovations have been developed [Scott et al., 2010, Bise, 2013]. Especially, significant progress has been made in coal seam mining methods and other difficulties associated with severe inclinations, instabilities and complex geological structures, and key problems related to deep mine mining pressure control, gas and thermal pollution governance, and tunnel arrangements have been partially solved [Saghafi, 2012, Atay et al., 2014]. However, despite the remarkable development of coal mining technology in recent decades, there has been few paradigm investigations due to the absence of a systematic analytical framework.

The technological paradigm was first proposed by Dosi based on Kuhn's scientific paradigm theory and it has proven to be a reliable method for studying past trends and predicting future possibilities [Kuusi and Meyer, 2007, Ivanova and Leydesdorff, 2015]. Dosi also proposed a technology trajectory to track technology progress within the paradigm and the economic and technological trade-offs required. Subsequently, many studies have applied this method to elucidate the operation and dynamic development within the paradigm, and some useful results have been obtained [Rashid et al., 2013, Chen et al., 2015]. Motivated by previous studies, this chapter uses the technological paradigm theory to determine the coal mining technology road map, puts forward a coal mining technology paradigm, and reveals the long-term technological development dynamics, and thus provides guidance for future mining technology development and coal policy management.

In order to guide the research development direction of the potential coal mining technology paradigm, two steps were taken. Firstly, based on bibliometrics methods, a generalized data analysis system was developed to qualitatively analyze the keyword trend of coal mining technology publications and map the knowledge network. The Web of Science™ core collection was selected as the main database to search relevant literature on coal mining technology, then CiteSpace was used to analyze the textual data of relevant literature in the database [Chen, 2016]. These procedures could establish a coal mining technological paradigm that identifies production development trends and coal mining technologies that should pay attention to, which formed the basis of the proposed integrated coal mining development system.

1.2 Coal Mining Technology Development

In this section, to fully understand the long-term coal mining technological development dynamics and future trends so as to comprehensively review the development history of coal mining technology and to provide guidance for the development of mining technology in the future, an analysis approach based on literature mining is developed, which uses the data from the Web of Science database and the CiteSpace software to conduct cluster and identification analysis.

1.2.1 Literature Analyses

Scientific literatures are always the most important carrier of the critical and frontier discovery of research for most topics and research area and as for coal mining technology, this situation is absolutely true. However, it is difficult to conveniently and effectively to access and summarize the coal mining technology developing history and trend through the traditional literature analysis method which mainly due to that the relative knowledge and information are always embodied in the large amount of the published literatures. Literature mining has proven to be a useful method for elucidating major trends across time in published scientific literature and for the building of topic maps [De Bruijn and Martin, 2002]. In this section, a generalized data analysis system is developed from previous research to reveal the coal mining technological development trends.

1.2.1.1 Data Analysis System

The literature on coal mining technology has a long history of nearly 90 years, meanwhile considering the range and depth of research in this field, it is difficult to identify knowledge gaps or explore future research possibilities, thus, an effective analytical method is necessary. The citation indexing of scientific literature suggested by Garfield has proved to be useful in identifying similar research areas [Garfield and Merton, 1979]. A citation index is a comprehensive result based on journal articles, keywords, publication dates, and abstracts, according to which the impact of a citation in a specific field can be determined [Robinson-García et al., 2015]. Such a method has applied in many other areas. For example, Kajikawa et al. used a citation-based method to study the structural changes in sustainable biomass and bioenergy [Kajikawa and Takeda, 2008] and Liu et al. used keyword co-word networks to identify the intellectual emerging trends in the research on innovation systems [Liu et al., 2015]. The successful application of the citation indexing method prompts the authors to apply it to make a literature analysis of coal mine technology. As illustrated in Figure 1.1, a generalized data analysis system composed of four interrelated links (objective determination, data collection, data preparation, and data analysis) is developed conceptually. Keyword co-word analysis has been widely used to examine and understand knowledge development dynamics [Tian et al., 2008]. A quantitative and visual knowledge map can be formed by combining keyword-based bibliometric and network analyses [Choi et al., 2011, Kim et al.,

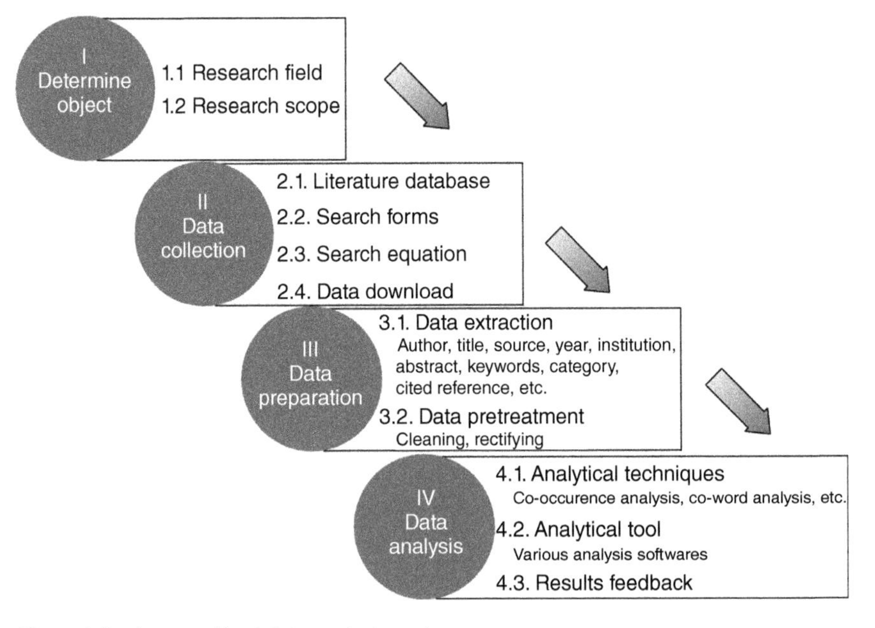

Figure 1.1 A generalized data analysis system.

Table 1.1 Selection criteria.

Search query:	TI = (coal mining) OR TS = (coal mining AND (method* OR technology* OR approach* OR systems* OR develop* OR trend*))
Language:	English
Publication type:	All document type
Time span:	1990–2017 *
Coverage:	Science citation indexes

Note that the WoS has no keywords for articles published prior to 1991. Our analysis of the changes in research frontiers and topics was therefore confined to post-1990 ISs research.

2016]. Therefore, a keyword analysis was carried out using scientific publication keyword co-word networks, for the sake of visualizing the global coal mining technological development dynamics and identify future trends.

The data acquisition module in the proposed analysis system inquired the Web of Science core collection database in January 2018 to identify the most relevant information. An advanced search was used, as shown in Table 1.1. When the search completed, articles, essays, book reviews, reviews, and editorial material were selected. After filtering, 3807 related articles were downloaded to form a text file, and then the CiteSpace analysis tool was used to identify all records and cited references to visualize the dynamics, patterns, and emerging trends of coal mining technology [Chen, 2016].

1.2.1.2 Knowledge Diagram

This chapter aims to reveal the development trend and research frontier of coal mining technology, thus a keyword co-word analysis was adopted. Keyword co-occurrence mapping is based on the keyword co-occurrence analysis method that explores theme variations across search fields by measuring the occurrence frequency of item pairs [Liu et al., 2015]. After standardization of similar or different words with the same meaning, CiteSpace was used to generate the keyword co-word network. As demonstrated in Figure 1.2, the simplified slice network generated by the minimum spanning tree (MST) algorithm consists of 790 nodes and 1021 links. The coal mining technology keywords knowledge diagram is then generated in the time zone view to focus on the evolution of knowledge over time, clearly showing how research was being updated and influences of mutual research [Chen et al., 2015]. As shown in Figure 1.2, the main keywords and associated frequencies are displayed along the time axis. Each node represents a different keyword, the size of each node denotes the co-occurrence frequency of the corresponding keyword, and each line indicates the co-occurrence relationship between the keywords.

As can be seen, coal miner (53) was a popular research focus around 1990, fully mechanized caving (13) became an important research topic from 1996, ecological and sustainable keywords have been prime research areas since 2004, and underground coal gasification (UCG) (94) has been focused on since 2006. After 2007,

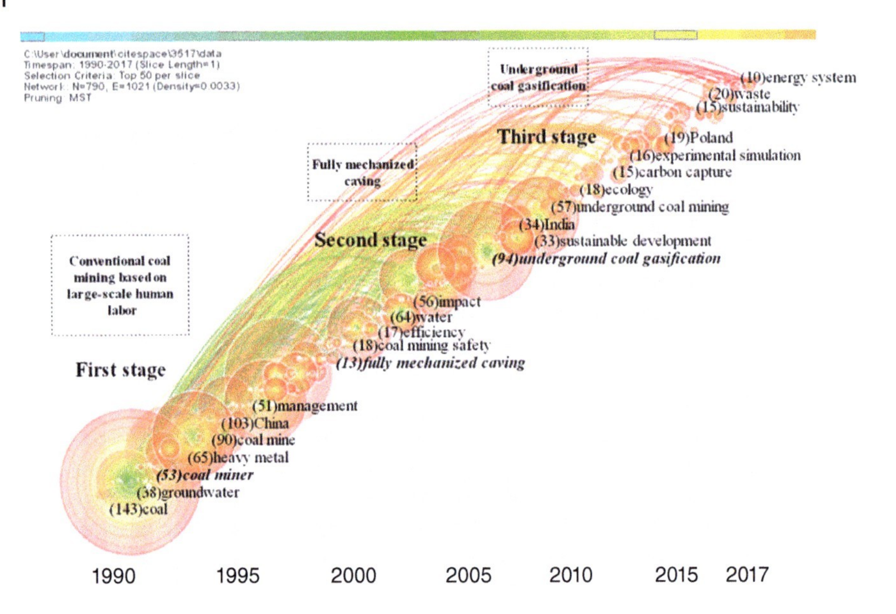

Figure 1.2 Timezone view for coal mining technology keywords diagram.

there was an obvious increase in renewable energy research with the appearance of keywords such as geothermal, solar energy utilization, and hydrogen production. Coal mining methods have also changed significantly over time, with opencast coal mining (11), long wall mining (14), and underground coal mining (57) appearing in sequence alongside some geographical focus such as China (103), United States (51), India (34), and Poland (19).

As can be seen from Figure 1.2, with the development history of coal mining technology, three main stages can be summarized. In the first stage, which lasted until early 2000, traditional coal mining based on large-scale human labor was the focus of research. After that, the second stage lasted for nearly 15 years took place, in which the fully mechanized caving technology was the most popular research topic, and in the third stage, from then to now, the UCG was the most important research area. Similar developing trends can be found in many other documents, not only in scientific literature, but also in reports or development plans. For example, a number of international conferences have discussed this issue and agreed that, a kind of clean coal technology, i.e. UCG, will be studied in many countries, such as Australia, New Zealand, India, Pakistan, Canada, Italy, the United States, and China [WCI, 2007, van der Riet, 2008]. In addition, the World Bank has reviewed clean coal mining technology from world experience and made implications for India and suggested that enough attention should be paid to UCG [The World Bank, 2008]. For China, one of the main coal producers, the government also encouraged to develop UCG in the future in the report of the 13th five-year plan for the development of coal industry [NEA NDRC, 2016]. Therefore, no matter from the results of literature mining, or the focus of relevant institutions, governments, and conferences, a common conclusion can be drawn that UCG will receive great attention in the future.

1.2.2 Three Periods of Coal Mining Technology

Technological innovation is essentially an iterative process with its push on industrial development and is triggered by the new market or service opportunities toward technological invention [Garcia and Calantone, 2002]. Technological innovation is in a cyclic process, where a new innovation is introduced for the first time and an improved innovation is reintroduced [Cheng et al., 2015]. This process is affected by multiple physical properties. With the increasing of those, a certain outcome point under the physical laws is obtained, namely S-curve [Cheng et al., 2015, Adner and Kapoor, 2016].

Additionally, from the systematic view, many other economic, social, institutional, and political factors can also be considered, and it may cause different technologies coexisting. Based on the accumulation of these factors, a technological paradigm can be naturally formed [Dosi, 1982]. Technology paradigm can determine the starting point and limit of new technology innovation cycles. Under the guidance of the specific paradigm, technological activities form the technological trajectory [Dosi, 1982]. As a result, technology evolution is a long-term series of technological paradigms, and each complete technological paradigm has an S-shaped curve (Figure 1.3). Based on the characteristics of different period, it could be roughly divided into three stages: competition, diffusion, and shift [Dosi, 1982, Christensen, 1992, 2013].

For the purpose of adequate appreciation of the long-term development of coal mining technology, it is necessary to explain the polymerization evolution of coal mining technology. Integrating literature analysis with S-curve theory of technology paradigm, coal mining technology paradigm can be divided into three stages:

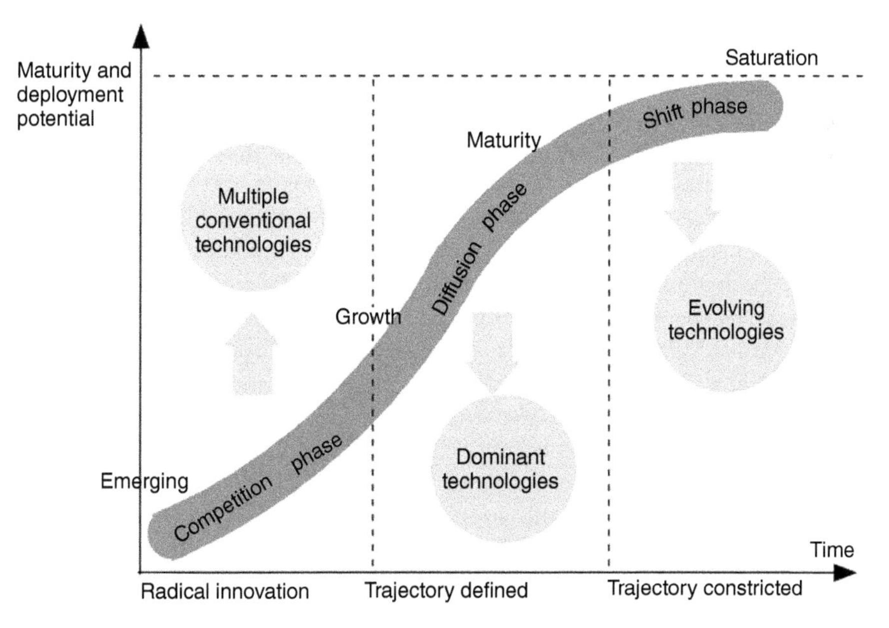

Figure 1.3 Three stages of the technological paradigm.

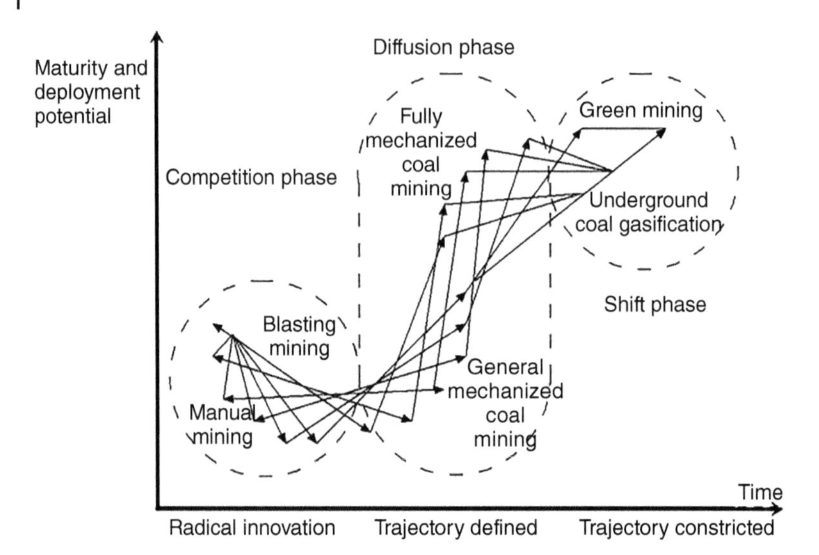

Figure 1.4 Technological paradigm for coal mining technology.

competition, diffusion, and shift, with each stage being characterized by different technologies, as described in Figure 1.4.

1.2.2.1 Competition Phase

Direct coal mining or manual mining is one of the oldest, most common, and versatile coal mining forms currently [Astakhov et al., 1990]. At first, coal mining was small scale, and coal was either buried on the surface or near the surface; consequently, drift mining and bell pit mining mainly depending on manual labor and the application of simple original equipment were the typical mining methods [Musson and Robinson, 1969]. Later, in order to satisfy the growing need brought about by the rapid development of international trade in the nineteenth century, the number of coal-fed steam engines for railways and ships were expanding. When coal became the major fuel supporting the industry, the demand for coal grew sharply and small-scale technologies were inadequate, consequently, coal mining developed from simple surface mining to surface and deep well mining. Even more important is that coal mining was implementing the machines, which greatly improve the production conditions. Taking an example, blasting technology was utilized for drilling and mining in horizontal tunneling lanes, which had a significant influence on the mining efficiency [Flinn et al., 1986]. Coal production reached 5 million tonnes in 1850. But manual labor remained the main mining technology as the machines were costly and unaffordable. Accident also occurred to coal mining and the miners were always in danger due to the worse working conditions and safety management.

1.2.2.2 Diffusion Phase

As mechanical and electrical technologies developed, safety production and high-efficient production were the requirements of the coal mining. The manual labor and the application of simple original equipment were gradually replaced by mechanized mining technologies. The coal mining machinery was specially designed to cut coal, the underground transportation adopted the conveyor, the ventilation adopted the fan, and the pumping equipment ensures the good gangue discharge. All of them improved the production efficiency and the overall safety production level of the coal mining [Stefanko, 1983].

Large complex systems were involved in modern coal mining, especially for underground coal mining with multiple production links, such as coal cutting, tunneling, transportation, ventilation and drainage, and surface production. But each mechanized working procedure was conducted as a separate production link and lacked comprehensive coordination. In order to integrate these procedures, fully mechanized mining technology occupied the modern coal mining, which lead to labor intensity reduction and safety improvement [Jinhua, 2006]. The United Kingdom was the first one to conduct fully mechanized coal mine equipped with self-advancing hydraulic supports in 1953, driving the development of various mechanized work faces and complete equipment packages in Germany, Japan, and China [Tian et al., 2006].

The stability of production became a new factor that restricted productivity after fully mechanized mining technology improved production efficiency significantly. In order to realize efficient production and reduce system risk, the stable automatic control production technology is adopted for the complete mining process. It had the ability to decrease the production staff needed down the shaft, further optimize the production system, increase mine safety, and save energy [Ralston et al., 2014]. At this time, information management had been employed to gather real-time production and supply information for fault prediction and disaster warnings [Guo et al., 2016].

1.2.2.3 Shift Phase

The efficient development and utilization of coal resources have played an important role in the world economic development; however, it meantime causes some issues such as environmental pollution and ecological damage and further limit sustainable development. With these problems reaching a critical level in the past few decades, the coal industry must move toward a new mode of sustainable development.

Latest studies have found that UCG technology provides a possible economic choice for extracting energy from coal resources while removing many environmental problems caused by deep mining [Son et al., 2016]. UCG technology converts in situ coal into a usable syngas to generate electricity or to produce liquid hydrocarbon fuels, natural gas surrogates, and valuable chemical products [Yang et al., 2016]. As is shown in Figure 1.5, UCG technology is realized by air and/or oxygen and steam injection into linked injection wells. The coal is then fired and a sequence of controlled chemical transformations takes place in the gasification channel, which is usually divided into three regions: oxidization, reduction and dry distillation, and pyrolysis [Samdani et al., 2016].

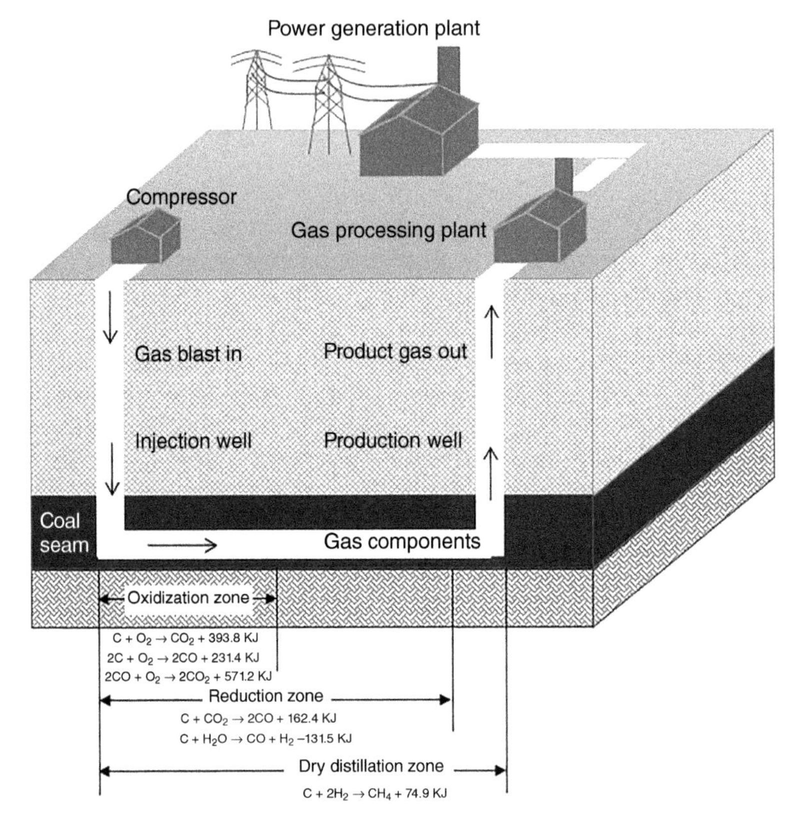

Figure 1.5 The UCG process.

In view of the process, UCG technology has many advantages over traditional coal mining and gasification technology [Su et al., 2016, Pei et al., 2016a]. First, laborer work underground is not needed and the general safety is improved. Second, when all the coal has been gasified, there is no need for surface gasifiers, consequently the surface footprints of the UCG plant are significantly reduced and related dust emissions and coal transport, treatment and storage costs are avoided. Third, UCG technology is capability to take advantage of coal seams which are so deep or so thin that they cannot be economically mined by conventional underground methods. According to the assessment, the United States, Australia, India, and China separately has over 5 million petajoules [PJ], 2.3 million PJ, 1.9 million PJ, and 2.2 million PJ of recoverable UCG syngas. With the application of UCG technology, there is a huge increase in the global recoverable coal reserves [PricewaterhouseCoopers, 2011, Su et al., 2016]. Based on the above advantages, UCG technology is recognized as a promising clean coal technology to help coal mining much securer, cleaner, and more economical. In addition to UCG technology, other potential technologies have also been come up to handle security and environmental issues. Qian presented a green coal mining technology system for China's coal industry which recommended taking measures to minimize the effects on the environment and develop an environmental-friendly recycling coal economy from the early stage of coal mining operations [Minggao, 2010].

1.3 Discussion

A paradigm is recognized as a comprehensive model that a specific scientific community must follow in a certain kind of scientific activities, which includes common world outlook, basic theory, paradigm, method, means, standard, etc. It was discovered through literature mining that the development of coal mining technology is on a trajectory that conforms to the S-curve of traditional technology paradigm, thus clarifying the paradigm of coal mining technology. Although there is few innovative coal mining technology paradigms from the literature analysis, latest researches pay attention to sustainable development, ecological sensitivity, energy system, renewable energy, and energy recovery and point out the direction of future technological innovation.

There is a need for a paradigm shift, which is a world view and behavior shared by a group of researchers engaged in a science. The term paradigm shift first appeared in The Structure of Scientific Revolution, the representative work of Thomas Kuhn. Paradigm shift, the fifth and final step in the Kuhn Cycle (Figure 1.6), is to break out of the original constraints and restrictions and to open up new possibilities by grafting on a higher level view such as a System Improvement Process [Thwink.org, 2014a, Ashkenazi and Lotker, 2014]. With coal expected to be a key driver of development, it will remain an important role in urbanization and industrialization [Thwink.org, 2014b]. However, exploiting and utilizing coal give rise to serious environmental issues. As coal occupies an important position, it cannot be abandoned. Therefore, the coal mine industry should keep a pace of sustainable development. UCG technology could assist in moving toward a path of coal industry sustainable development. UCG technology is a promising clean coal technology to help coal mining much securer, cleaner, and more economical with the ability to recover currently unmineable coal resources [Thwink.org, 2014a].

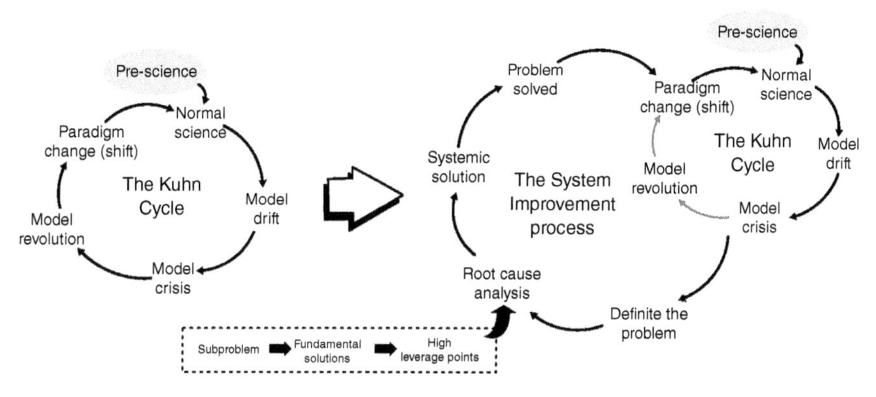

Figure 1.6 The Kuhn Cycle and the System Improvement Process.

Although this technology is still not satisfactory, it could be greener and more efficient when a paradigm shift happens.

Three main obstacles to achieve the UCG-based paradigm shift exist. First, realistic factors which are vital for complete UCG systems fall short of the requirement. Currently, the energy recovery rate is just approximately 30– 60%, at relatively low rate compared with open pit mining [Su et al., 2016]. Besides, the EROI index, which defines the relationship between the human-made capital energy required to produce energy and the amount of energy, is still unsatisfactory compared with other energy sources such as crude oil. Second, related environmental problems are caused as UCG technology involves geomechanical, hydrogeological, thermal, and geochemical process [Yang et al., 2007, Pei et al., 2016b]. Groundwater pollution is the most highlighted issue. A large number of pollutants, such as phenol, polycyclic aromatic hydrocarbons, benzene, carbon dioxide, ammonia, and sulfide, will be produced in the process of coal gasification. These pollutants diffuse and penetrate into the surrounding strata in the coal seam, which will pollute the surrounding groundwater. Moreover, underground caves are created in UCG technology and further rock and other materials cannot be supported; as a consequence, subsidence happens. Third, the desired syngas quality and composition are quite difficult to achieve [Imran et al., 2014, Pei et al., 2016b]. Syngas qualities have relations with the properties of the coal bearing strata, and the gasification conditions such as the size of the gasification cavities, the spacing, the moisture content, pressure, and temperature. Although hard technology can validly handle those obstacles [Pei et al., 2016b, Su et al., 2016], it is expensive and beyond capability of the developing countries which heavily rely on coal [Liu et al., 2007, Imran et al., 2014]. Therefore, integrating current technologies is more achievable option at present.

In order to realize sustainable and efficient coal mining and take full advantage of renewable energy, a more reasonable approach is necessary. Due to important role of UCG in future mining projects, an environmentally sustainable multiple energy comprehensive mining method based on ecological coal mining is proposed, graphically represented in Figure 1.7. Below are some highlights of the features for this approach.

As UCG technology causes carbon emissions, carbon capture and storage (CCS) technology is applied to capture these carbon emissions, and consequently the environment is improved [Asif and Muneer, 2007]. Combined UCG-CCS technology offers an approach to get better energy recovery from coal and avoid the hazardous environmental influences [Karki et al., 2010]. Specifically, after the UCG technology, there are large gaps deep underground which enable carbon emissions to be captured and stored [Khadse et al., 2007, Roddy and Younger, 2010, Tollefson and Van Noorden, 2012, Eftekhari et al., 2017]. When UCG is adopted at artificially high permeable areas such as depths underground greater than 700–800 m, carbon storage is considered to be attractive. Besides, Combined UCG-CCS technology is related to other carbon-intensive industries through the CO_2 pipeline grid, and power plants can utilize the UCG syngas with pre- and/or post-combustion capture.

For intra-regional self-supporting electricity supply for coal mining perspective, although wind and solar energy are widely applied to provide heating and cooling or

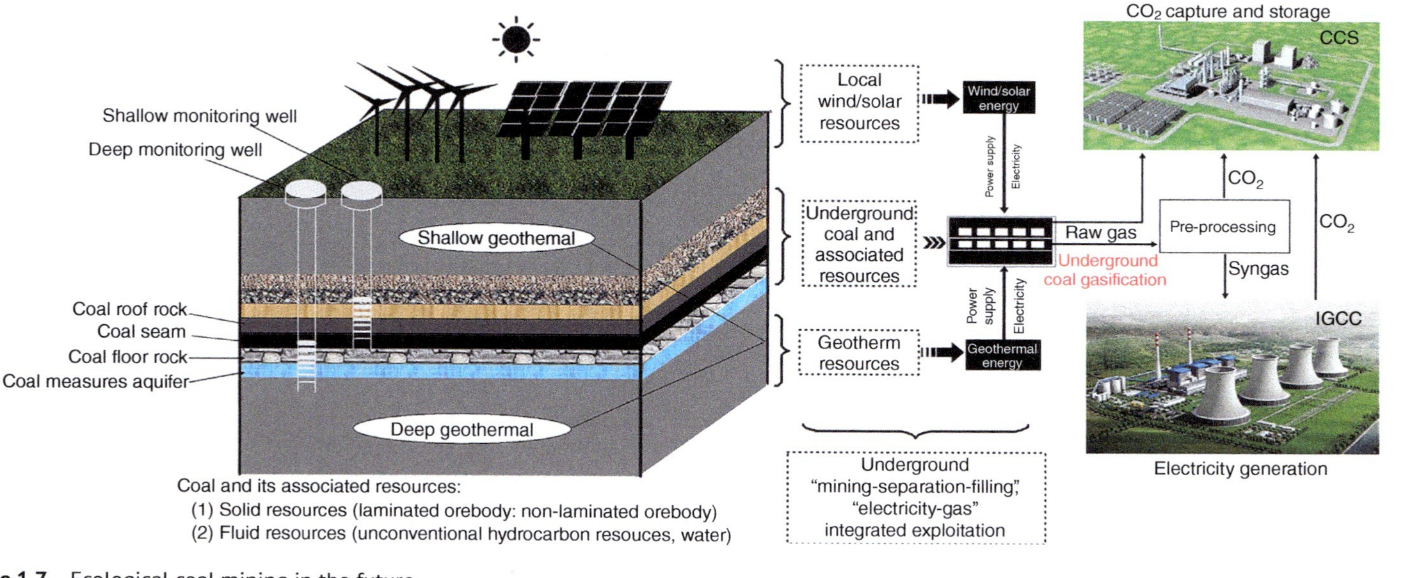

Figure 1.7 Ecological coal mining in the future.

to generate electricity, they are hardly employed in coal mining industry [Preene and Younger, 2014]. However, the integrated UCG system makes full use of the geology and geology conditions of the mine as well as the existing renewable energy sources, improving energy efficiency. Precisely, the integrated UCG system requires less or even no external power as it employs available intra-regional renewable energies to generate power [Wang et al., 2016, Hall et al., 2011, Ramos et al., 2015].

Moreover, the raw gas from UCG has too low energy content compared with generation standards, and requirement for pre-processing procedure is proposed for syngas separation and purification. This system is integrated with integrated gasification combined cycle (IGCC), which is able to save syngas transportation costs and reduce gasification and purification steps [Kintisch, 2007]. Besides, this system can add water monitor wells to monitor the underground water quality in case excessive concentration of groundwater pollutants occur.

To apply this integrated exploitation approach, it is vital to establish the regional polycentric energy systems which unite the scales and actors from different levels and areas [Kursun et al., 2015]. And this integrated exploitation method varies according to the particular geography and climate. In general, the exploitation approach eliminates the extreme environmental issues, ensures production stability, reduces energy consumption and ecological destruction, and realizes no coal on the ground [Lauber and Jacobsson, 2016].

References

Adner, R. and Kapoor, R. (2016). Innovation ecosystems and the pace of substitution: re-examining technology S-curves. *Strategic Management Journal* 37 (4): 625–648.

Ashkenazi, D. and Lotker, Z. (2014). The quasicrystals discovery as a resonance of the non-euclidean geometry revolution: historical and philosophical perspective. *Philosophia* 42 (1): 25–40.

Asif, M. and Muneer, T. (2007). Energy supply, its demand and security issues for developed and emerging economies. *Renewable and Sustainable Energy Reviews* 11 (7): 1388–1413.

Astakhov, A., Grübler, A., and Mookhin, A. (1990). Technology diffusion in the coal-mining industry of the USSR: an interim assessment. *Technological Forecasting and Social Change* 38 (3): 223–256.

Atay, M.T., Simonetti, B., and Coskun, S.B. (2014). Hybrid technology of hard coal mining from seams located at great depths. *Archives of Mining Sciences* 59 (3): 575–590.

Bise, C.J. (2013). *Modern American Coal Mining: Methods and Applications*. SME.

British Petroleum (2017). *BP Statistical Review of World Energy*. British Petroleum.

Chen, Y., Chen, C., Liu, Z. et al. (2015). The methodology function of cite space mapping knowledge domains. *Studies in Science of Science* 33 (2): 243–253.

Cheng, J.T.S., Jiang, I.-M., and Liu, Y.-H. (2015). Technological innovation, product life cycle and market power: a real options approach. *International Journal of Information Technology & Decision Making* 14 (01): 93–113.

Choi, J., Yi, S., and Lee, K. (2011). Analysis of keyword networks in MIS research and implications for predicting knowledge evolution. *Information & Management* 48 (8): 371–381.

Christensen, C.M. (1992). Exploring the limits of the technology S-curve. Part one: Component technologies. *Production and Operations Management* 1 (4): 334–357.

Christensen, C. (2013). *The Innovator's Dilemma: When New Technologies Cause Great Firms to Fail*. Harvard Business Review Press.

Court, V. and Fizaine, F. (2017). Long-term estimates of the energy-return-on-investment (EROI) of coal, oil, and gas global productions. *Ecological Economics* 138: 145–159.

De Bruijn, B. and Martin, J. (2002). Getting to the core of knowledge: mining biomedical literature. *International Journal of Medical Informatics* 67 (1): 7–18.

Dosi, G. (1982). Technological paradigms and technological trajectories: a suggested interpretation of the determinants and directions of technical change. *Research Policy* 11 (3): 147–162.

Eftekhari, A.A., Wolf, K.H., Rogut, J., and Bruining, H. (2017). Energy and exergy analysis of alternating injection of oxygen and steam in the low emission underground gasification of deep thin coal. *Applied Energy* 208: 62–71.

Flinn, M.W., Ashworth, W., and Pegg, M. (1986). *History of the British Coal Industry: 1946–1982: The Nationalized Industry*, vol. 5. Oxford University Press.

Garcia, R. and Calantone, R. (2002). A critical look at technological innovation typology and innovativeness terminology: a literature review. *Journal of Product Innovation Management* 19 (2): 110–132.

Garfield, E. and Merton, R.K. (1979). *Citation Indexing: Its Theory and Application in Science, Technology, and Humanities*, vol. 8. New York: Wiley.

Griffith, E.D. and Clarke, A.W. (1979). World coal production. *Scientific American* 240 (1): 38–47.

Guo, X., Wang, R., and Wu, Z. (2016). Research and application of WebGIS in coal mine information management system. *International Conference on Advanced Design and Manufacturing Engineering*.

Hall, C.A.S., Lambert, J.G., and Balogh, S.B. (2014). EROI of different fuels and the implications for society. *Energy Policy* 64: 141–152.

Hall, A., Scott, J.A., and Shang, H. (2011). Geothermal energy recovery from underground mines. *Renewable and Sustainable Energy Reviews* 15 (2): 916–924.

Imran, M., Kumar, D., Kumar, N. et al. (2014). Environmental concerns of underground coal gasification. *Renewable and Sustainable Energy Reviews* 31 (31): 600–610.

Ivanova, I.A. and Leydesdorff, L. (2015). Knowledge-generating efficiency in innovation systems: the acceleration of technological paradigm changes with increasing complexity. *Technological Forecasting and Social Change* 96: 254–265.

Jinhua, W. (2006). Present status and development tendency of fully mechanized coal mining technology and equipment with high cutting height in China. *Coal Science & Technology* 42: 1–4.

Kajikawa, Y. and Takeda, Y. (2008). Structure of research on biomass and bio-fuels: a citation-based approach. *Technological Forecasting and Social Change* 75 (9): 1349–1359.

Kapusta, K. and Stanczyk, K. (2011). Pollution of water during underground coal gasification of hard coal and lignite. *Fuel* 90 (5): 1927–1934.

Karki, N.R., Jha, D.K., and Verma, A.K. (2010). Rural energy security utilizing renewable energy sources: challenges and opportunities. *IEEE Conference*, India, pp. 551–556.

Khadse, A., Qayyumi, M., Mahajani, S., and Aghalayam, P. (2007). Underground coal gasification: a new clean coal utilization technique for India. *Energy* 32 (11): 2061–2071.

Kim, W., Khan, G.F., Wood, J., and Mahmood, M.T. (2016). Employee engagement for sustainable organizations: keyword analysis using social network analysis and burst detection approach. *Sustainability* 8 (7): 631.

Kintisch, E. (2007). Carbon emissions. Report backs more projects to sequester CO_2 from coal. *Science* 315 (5818): 1481.

Kursun, B., Bakshi, B.R., Mahata, M., and Martin, J.F. (2015). Life cycle and energy based design of energy systems in developing countries: centralized and localized options. *Ecological Modelling* 305: 40–53.

Kuusi, O. and Meyer, M. (2007). Anticipating technological breakthroughs: using bibliographic coupling to explore the nanotubes paradigm. *Scientometrics* 70 (3): 759–777.

Lauber, V. and Jacobsson, S. (2016). The politics and economics of constructing, contesting and restricting socio-political space for renewables-the German renewable energy act. *Environmental Innovation and Societal Transitions* 18: 147–163.

Li, J., Chen, C. (2016). *CiteSpace*. Press of Capital University of Economics and Business.

Liu, S.Q., Gang, J., Mei, M., and Dong, D. (2007). Groundwater pollution from underground coal gasification. *International Journal of Mining Science and Technology* 17 (4): 467–472.

Liu, Z., Yin, Y., Liu, W., and Dunford, M. (2015). Visualizing the intellectual structure and evolution of innovation systems research: a bibliometric analysis. *Scientometrics* 103 (1): 135–158.

Minggao, Q. (2010). On sustainable coal mining in China. *Journal of China Coal Society* 35 (4): 529–534.

Musson, A.E. and Robinson, E. (1969). *Science and Technology in the Industrial Revolution*. Manchester University Press.

NEA NDRC (2016). The 13th five-year plan for the development of coal industry.

Pei, P., Korom, S.F., Ling, K., and Nasah, J. (2016a). Cost comparison of syngas production from natural gas conversion and underground coal gasification. *Mitigation & Adaptation Strategies for Global Change* 21 (4): 629–643.

Pei, P., Nasah, J., Solc, J. et al. (2016b). Investigation of the feasibility of underground coal gasification in North Dakota, United States. *Energy Conversion and Management* 113: 95–103.

Preene, M. and Younger, P.L. (2014). Can you take the heat? The geothermal energy in mining. *Mining Technology* 123 (2): 107–118.

PricewaterhouseCoopers (2011). Industry review and an assessment of the potential of UCG and UCG value added products. www.lincenergy.com/data/media-news-articles/relatedreport02.pdf (accessed 25 May 2021).

Ralston, J., Reid, D., Hargrave, C., and Hainsworth, D. (2014). Sensing for advancing mining automation capability: a review of underground automation technology development. *International Journal of Mining Science and Technology* 24 (3): 305–310.

Ramos, E.P., Breede, K., and Falcone, G. (2015). Geothermal heat recovery from abandoned mines: a systematic review of projects implemented worldwide and a methodology for screening new projects. *Environmental Earth Sciences* 73 (11): 6783–6795.

Rashid, A., Asif, F.M.A., Krajnik, P., and Nicolescu, C.M. (2013). Resource conservative manufacturing: an essential change in business and technology paradigm for sustainable manufacturing. *Journal of Cleaner Production* 57 (20): 166–177.

Robinson-García, N., Jiménez-Contreras, E., and Torres-Salinas, D. (2015). Analyzing data citation practices using the data citation index. *Journal of the Association for Information Science and Technology* 67 (12): 2964–2975.

Roddy, D.J. and Younger, P.L. (2010). Underground coal gasification with CCS: a pathway to decarbonising industry. *Energy & Environmental Science* 3 (4): 400–407.

Saghafi, A. (2012). A Tier 3 method to estimate fugitive gas emissions from surface coal mining. *International Journal of Coal Geology* 100 (10): 14–25.

Samdani, G., Aghalayam, P., Ganesh, A. et al. (2016). A process model for underground coal gasification: part two growth of outflow channel. *Fuel* 181: 587–599.

Scott, B., Ranjith, P.G., Choi, S.K., and Khandelwal, M. (2010). A review on existing opencast coal mining methods within Australia. *Journal of Mining Science* 46 (3): 280–297.

Son, N.L.H., Anh, N.H., and Dong, H.N. (2016). Review of underground coal gasification technologies. *International Conference on Green Technology and Sustainable Development (GTSD)*, IEEE, pp. 69–73.

Stefanko, R. (1983). Coal mining technology: theory and practice.

Su, F.-q., Hamanaka, A., Itakura, K.-i. et al. (2016). Evaluation of coal combustion zone and gas energy recovery for underground coal gasification (UCG) process. *Energy and Fuels* 31 (1): 154–169.

The World Bank (2008). Clean coal power generation technology review: worldwide experience and implications for India.

Thwink.org (2014a). Kuhn Cycle. http://www.thwink.org/sustain (accessed 25 May 2021).

Thwink.org (2014b). A model in crisis: can civilization bypass the model revolution step or not? http://www.thwink.org/sustain/articles/018_ModelInCrisis/index.htm (accessed 25 May 2021).

Tian, Z.L., Zhang, C.W., and Li, J.H. (2006). Present status and tendency of fully mechanized coal mining equipment. *Coal Technology* 25: 1–2.

Tian, Y., Wen, C., and Hong, S. (2008). Global scientific production on GIS research by bibliometric analysis from 1997 to 2006. *Journal of Informetrics* 2 (1): 65–74.

Tollefson, J. and Van Noorden R. (2012). Slow progress to cleaner coal. *Nature* 484 (7393): 151–152.

van der Riet, M. (2008). Underground coal gagfication. *Proceedings of the SAIEE Generation Conference*. Midrand, South Africa: Eskom College.

Wang, W., Wang, Y., Song, W., and Shi, G. (2016). Evaluation of infrared heat loss of dust-polluted surface atmosphere for solar energy utilization in mine area. *International Journal of Hydrogen Energy* 41 (35): 15892–15898.

WCA (2017a). Environmental protection. http://www.worldcoal.org/coal/uses-coal/coal-electricity (accessed 25 May 2021).

WCA (2017b). Coal & electricity. http://www.worldcoal.org/coal/uses-coal/coal-electricity (accessed 25 May 2021).

WCI (2007). Coal Meeting the Climate Challenge. *Technical report*. World Coal Institute.

Yang, D., Koukouzas, N., Green, M., and Sheng, Y. (2016). Recent development on underground coal gasification and subsequent CO_2 storage. *Journal of the Energy Institute* 89 (4): 469–484.

Yang, L., Liu, S., Yu, L., and Jie, L. (2007). Experimental study of shaftless underground gasification in thin high-angle coal seams. *Energy and Fuels* 21 (4): 2390–2397.

Ye, R., Xiang, C., Lin, J. et al. (2013). Coal as an abundant source of graphene quantum dots. *Nature Communications* 4 (1): 94–105.

2

Developing Trending Toward Ecological Coal Utilization

Coal resource is widely used for power generation all over the world, exceeding 40% of worldwide electricity being produced from coal in 2015. The most important and common uses of coal are in industrial field, such as electricity generation, steel manufacturing, cement production, and as a liquid fuel. Attributes such as abundant yield, price affordability, and easy to be transported stored and used have been the main reasons that coal has been popularly used since before industrial revolution in nineteenth century [IEA, 2016]. The coal use for power generation has been continually growing since that time and it has been predicted that the reliance on coal will continue to boom now and well into the future [Clark and Jacks, 2007]. However, power stations usually generate electricity by combusting thermal coal, which would cause a large amount of carbon dioxide and other greenhouse gas emissions, with around 60% of CO_2 emissions from known fossil fuel reserves being directly attributable to coal [BP Global, 2009].

2.1 Background Introduction

In developing countries, taking China and India as an example, over 60% of the domestic power that provides energy supply for billions of people and also promotes local, regional, and even national economic growth is still being produced by coal power [OECD/IEA, 2015]. Based on International Energy Agency (IEA) analysis of official 2013 data, it was found that coal resource accounted for 29% of the total global primary energy supply and the emissions in emerging economies grew by 4%, largely because of explosive growth of coal consumption. With the worsening of environmental problems, there is a great necessary to drastically reduce CO_2 emissions, which poses the development and application of cleaner energy technologies into a primary position [Dovi et al., 2009, Negreanu and Mocanu, 2012, Minutillo and Perna, 2014, Lazaroiu et al., 2017]. The IEA addresses that greater efforts by governments and industry should be made to reduce coal generation related emissions and more efficient coal-based technologies needed to be developed to ensure that coal can be a cleaner source of energy in the decades to come [IEA, 2016]. The greatest challenge with these a large number of clean technologies is the integration of

Innovative Approaches towards Ecological Coal Mining and Utilization, First Edition.
Jiuping Xu, Heping Xie, and Chengwei Lv.

them into the energy system, as integrated energy systems based on coal ecological utilization are seen as the most possible future development trend of coal industry.

With the continuous development of coal industry and the great advances in scientific and technological, coal utilization technologies have also been constantly evolving. Over the past decade, significant achievements in coal industry have had a great effect on energy systems, with some developed countries such as Denmark now generating most of their power from renewable sources. However, generally speaking, the transition process from traditional methods to cleaner energy has been slow, primarily due to the lack of government actions. Therefore, if environmental related issues need to be resolved as soon as possible, governments should promulgate powerful regulations and policies which can force all the societies and industries to make use of clean energy [Jacobsson and Lauber, 2006]. Therefore, the coal industry is not only affected by market pull and technology push, but also by relevant policies and regulations. Taking a broad view of around the world, environmental policies published by enlightened and progressive governments have been gradually changing the world energy patterns, leading to the increasing demands for coal utilization technological developments to ensure more efficient and cleaner production.

Many significant researches on coal industry development and its related utilization technologies have been springing up in recent years. In 2000, Ebara presented a general framework of the foundational technologies for each coal type and at the same time predicted that these technologies would be put into use in Japan within the next 15 years [Ebara, 2000]. At that time, oxy-fuel coal combustion technology was being extensively discussed both in industry and academia which was seen as a feasible environmentally friendly means [Buhre et al., 2005, Hong et al., 2009]. In recent years, mixing coal with other renewable energy sources, especially biomass, has become a popular method to improve coal utilization efficiency [Lazaroiu et al., 2009, Pisa and Lazaroiu, 2012, Pisa et al., 2014]. By reviewing of some international researches, Phdungsilp found that cities were and will continue to be the main driving factor of energy use and the associated carbon emissions, and proposed a city-based integrated approach to energy and carbon emissions [Phdungsilp, 2010]. More recently, the concept of underground coal gasification (UCG) technology and subsurface energy systems has been put forward to eliminate problems of ash disposal and provide a more economical and efficient coal utilization method [Prabu and Jayanti, 2012, Kolditz et al., 2015]. To our knowledge, the existing research have paid main attentions to a specific technology or a particular field of coal use, however, very few of them have reviewed the overall and complete development process of coal utilization. In general, there is still a lack of universality and cohesion in the exploration of coal utilization technology development. Therefore, studying and revealing the current coal utilization development trajectory and the future tendencies through an analysis of the relevant research publications with the connections between their publish years and the associated keywords.

To explore sustainable development of coal industry, it is essential to understand the evolution history of it, existing status and future trends. Kuhn first defined paradigm theory in his study which laid the foundation for paradigm research. Later, in order to resolve economic problems with solutions based on principles of natural science, Dosi proposed the concepts of technological paradigm and technological trajectories, which have become classic theories in the field of literature analysis on innovation and technological change [Dosi, 1982]. Then, the idea of ecological paradigm was introduced by Bronfenbrenner in the 1970s, as a kind of reaction to the research in terms of restricted child psychological behaviors, family policies and educational practices [Bronfenbrenner, 1974, 1976, 1977]. Besides, Schwartz found that a dilemma associated with the relationship between people, nature and the social environment was the core of current environmental problems, leading to a value choice for various stakeholders [Schwartz, 1994]. Some researchers have emphasized to relieve this dilemma by protecting the ecological environment and living in harmony with the surroundings, which has been highlighted as a new ecological paradigm by environmental sociologists as it emphasized environmental factors, constrained human behaviors, and also promoted ecological values at the same time [Stern and Dietz, 1994]. Therefore, an ecological paradigm approach for coal utilization will be proposed to study its development paradigm and future trend when the concept of ecological paradigm is regarded as an appreciated method of solving the current environmental problems all around the world.

In this chapter, based mainly on literature mining methods, an optimized data analysis system (ODAS) is built to summarize the development of coal energy and elucidate the trends. By combining the paradigm concept with the coal utilization technological diffusion process, a coal utilization paradigm (CUP) is proposed, which can predict the future of coal utilization. The CUP can be summarized in three stages; the first two stages are defined as technological paradigms, while the third stage is an ecological paradigm. When the CUP is established, it could aid in reducing carbon emissions through the use of combined energy sources as well as giving guidance to the development of integrated energy systems.

2.2 Coal Utilization Evolution

Because of the explosive increase in the number of coal utilization research, it is difficult to screen out the most useful literature to seek for research focus and development path. Literature mining, therefore, is helpful in determining the most associated scientific research, especially in some particular areas such as energy systems and energy sources application [Scherf et al., 2005]. As a result, with the help of literature analysis tools, it is possible to understand the coal utilization development processes and then find out its technological paradigm which could make contributions to the prediction of possible future directions in this field. Based mainly on literature mining methods, an ODAS was designed to summarize the development of coal energy and predict future trends with coal utilization related literature by exploring the relationship between the published years and the article

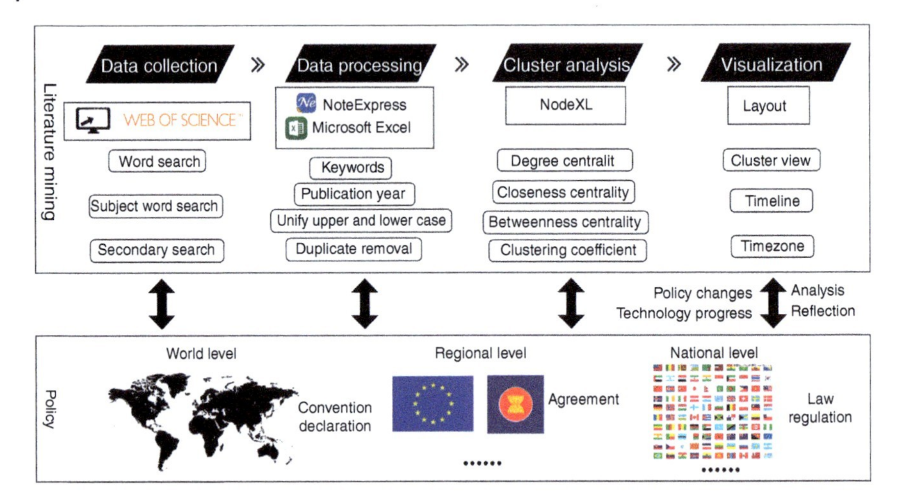

Figure 2.1 The structure of ODAS.

keywords. In the ODAS, there are five main sections and an additional section, the combination of which can help us better realize searching, refining, matching and analyzing of the relevant literature, as shown in Figure 2.1.

To be specific, the first section involved in the system named Data Collection, using the Web of Science (WoS) database which is regarded as one of the most reliable citation index as it covers almost all of the leading scholarly research and provides researchers, governments, and other faculties with convenient and effective access to the world leading citation databases. For example, Ridley et al. used the WoS to achieve a network analysis on biofuels in terms of the environmental and economic uncertainties [Ridley et al., 2012]. Then, Dataset Handling Section was designed where the free and open software sources NoteExpress and Microsoft Excel can be used to handle with obtained literature data which were vital research results achieved by scholars from various fields [Bin and San-Dang, 2010, Li et al., 2013]. In the Cluster Analysis Section, NodeXL is chosen to measure degree centrality, closeness centrality, betweenness centrality and the clustering coefficients of keywords. Finally, the Visualization Results Section and Data Update Section were designed to help realize result visualization of research analysis and also provide potential literature data updation in the future. Further, the Policy Analysis Section was a unique existence which provided greater depth analysis on coal utilization relevant policies and regulations than other literature mining system so as to better explore and match the paradigm development, being the main advantage of the ODAS.

In the process of literature mining, as the WoS has a huge amount of knowledge data, it is hard to select the most useful and needed papers if some appropriate screening criterions are not specified. In order to avoid overlook or duplication of critical and necessary documents, the methodology for this research has to obey the following two rules.

Rule 1: Searching for required keywords using the form of $A_i + B_j$, where A_i were keywords related to coal use itself such as coal utilization, coal consumption, and coal utilization technology, and B_i were index terms that included energy systems, coal energy and integrated energy systems, and so on. By this approach, it ensures that the mining process of literature data would be accurate and efficient.

Rule 2: To ensure high relevance of research topic, some unnecessary research should be filtered out from the titles and abstracts review, at the same time, only journal articles, review and book review with high quality will remain in the final screening results.

After carrying out the above steps, the initial records of 2089 articles were obtained. After carefully eliminating repeated and irrelevant articles using NoteExpress, 497 related articles were finally screened out. Then, importing these literature records into NoteExpress and letting the duplicates been removed, after which 2365 keyword items were extracted. After filtering the primary data, the analysis results of the keyword focus were laid out in years from 1992 to 2016 on the horizontal axis, as shown in Figure 2.2, from which a general trend of annual increase can be seen clearly. Through the ODAS review, it was concluded that the research in coal utilization met the principles of the technological paradigm, and its total development process could be summarized in three stages with the trajectory following an S-shaped curve, each of which was characterized by the corresponding technologies at the certain time [Ayres, 1988]. Figure 2.3 shows the S-shaped trajectory development process of coal utilization technological paradigm combining with the policy influences.

Based on the results of literature mining and paradigm theory, a novel concept of the CUP is proposed to present the technological evolution in terms of coal utilization, which to the best of our knowledge is a relatively new area of research so far. The CUP clearly shows the different technologies of coal utilization modes

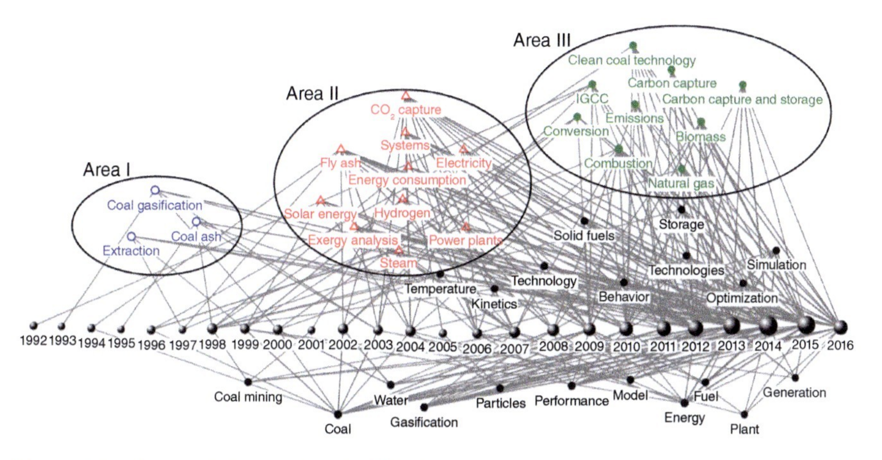

Figure 2.2 Keywords focus of coal utilization.

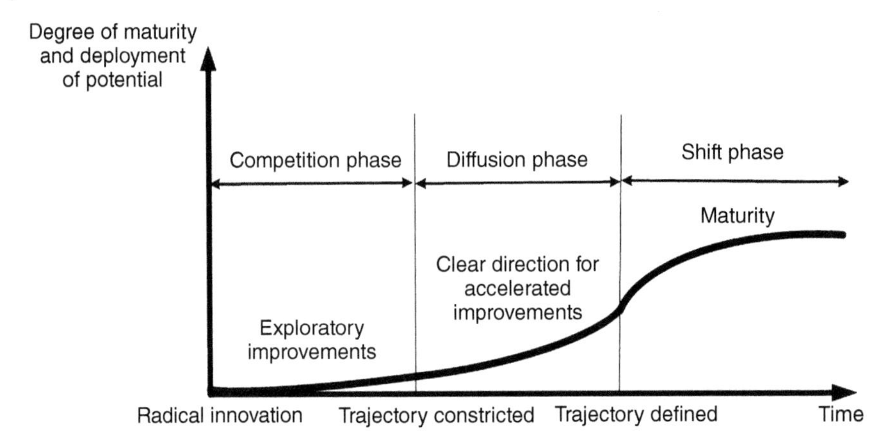

Figure 2.3 Three stages of the technological paradigm.

that can be divided into three stages: initial technological competition, fierce innovative diffusion, and disruptive integrated shift, meeting the paradigm rules. Generally speaking, the first stage emerged due to competition oriented where it mainly focused on early technological development and involved numerous possible trajectories and potential barriers associated with traditional way of coal utilization. In the second stage, the technology push and demand pull become one of the most important factors to determine the innovation diffusion of coal utilization technologies, respectively, corresponding to technological developments and market changes. Then, the shift phase showed the gradual maturity and stability of technology and also indicated the potential development trends of coal use technologies. However, the significant role of policy in guiding coal utilization technological innovation has often been ignored in the previous studies. Therefore, by combining the S-shaped curve with the proposed ODAS, a full CUP featured by the various technological developments and related police was presented, as shown in Figure 2.4.

2.2.1 Initial Technological Competition

The first stage of CUP was initial competition phase, involving many original coal utilization technologies. The oldest, most common, and most versatile utilization forms of coal utilization were direct-use relevant methods primarily focusing on physical treatment of coal, including coal dressing, coal washing, coal briquette, and other common operations. Then, at this stage, because of relatively immature understanding of energy use, researches about coal utilization for power generation concentrated mainly on simple coal combustion. The common use methods of coal resource were simple and rugged which have resulted in the environmental disruption issues such as coal ash diffusion, harmful gases emission, toxic waste water discharge, or other solid waste abandonment, and so on.

By reviewing the related literature, it can be seen that the exploration of coal utilization technologies has begun at an earlier time. For example, the coal briquette

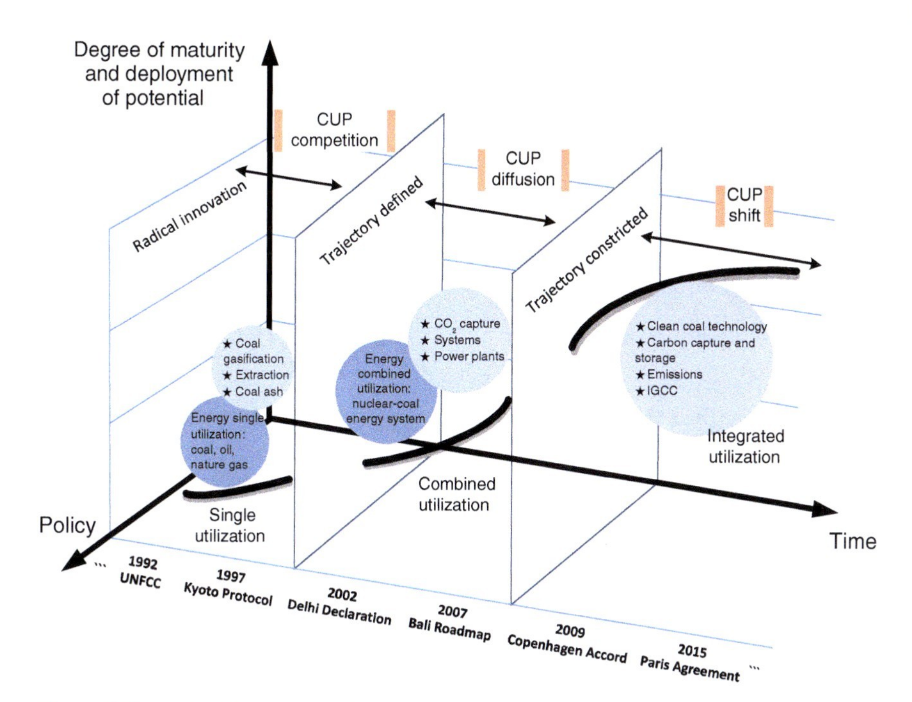

Figure 2.4 Coal utilization paradigm considering both technology and policy.

technology can be tracked back to the early nineteenth century in Europe. In the early 1980s, coal water slurry (CWS) technology began to be widely applied in some developed countries owing to the breakdown of oil crisis in 1970s, and then researches and development activities about CWS technology have been continuing over the next few decades [Aktaa and Woodburn, 2000]. During the twentieth century, in accordance with the demand of the increasing and various industrial applications (syngas, alternative gas, power generation, etc.), a number of different gasification methods of coal were developed, where the Winkler gasifier, the atmospheric K–T furnace, and the Lurgi pressurized gasifier were some of the early applications. As mentioned earlier, many earlier coal utilization technologies emerged in the initial competition phase, although they were at an early stage of exploration and development, to be more advanced to gradually replace the previous technologies was becoming a trajectory.

Over the last few decades, with more and more development of coal use technologies, environmental-related issues have arisen, such as overextension of resources caused by unsustainable production and problems about poor land management, resulting in the attentions to environmental protection, the efficient energy use and question of environmental benefits of traditional coal systems. Therefore, the first global environmental agreement, the Declaration of the United Nations Conference on the Human Environment, was enacted in 1972 to call attention to the world's emerging environmental problems. On the tenth anniversary of the United Nations Conference on the Human Environment (DUNCHE), the Nairobi Declaration was signed, aiming at global, regional, and national joint efforts for the environment protection and improvement. As a result, the single utilization of coal energy, which was the mainstream method in the competition stage, was found to have significant limitations, leading to increasing efforts toward new technologies, innovation methods, and more comprehensive systems. Therefore, technological innovation as the core of the direction of development is expected to make contribution to a new framework of energy system in the near future so as to change the utilization way coal resource.

2.2.2 Fierce Innovative Diffusion

Influenced primarily by market demands and related policy implementation, the second stage identified the prevailing dominant path of coal utilization, called as the comprehensive coal utilization phase. After the sufficient competition in the last phase, a dominant development trajectory driven by the interactions between the increasing energy demands and fierce industrial competition begun to rise, which can be likened to the contagion process of epidemic disease [Griliches, 1957].

During this stage, as a technical extension of the initial phase, coal was gradually combined with other energy resources to be used jointly so as to develop a relatively simple energy system that could capture CO_2as much as possible. It can be seen that in area II, although there were still some problems related to fly ash, there have been a major breakthrough in combination utilization of coal with other low emission renewable energy sources such as solar and hydrogen. However, the visualization results of keyword-year matching map indicated that the primary resource for power generation in this stage was still coal. Generally speaking, some indexes including energy utilization efficiency, primary energy consumption, and primary energy saving have been usually selected as evaluation indicators to measure the energy utilization in energy systems [Li et al., 2006, Mago and Chamra, 2009, Wang et al., 2011, Maraver et al., 2013]. Following these rules, coal utilization technologies have been constantly improved, making coal systems gradually evolve from the initial competition phase to this diffusion stage. Therefore, based on the analysis for keyword-year trend matching and paradigmatic evolution theory, it was not hard to find that the diffusion stage formulated an energy system that no longer contained a single resource, where a form of coal combining with other two or three energy sources was developing and maturing. As the attention began to focus on efficient and sustainable use of fossil resources so as to reduce the overall carbon footprint as much as possible, a nuclear-coal hybrid energy system was taken as a potential solution to be proposed [Chen et al., 2015], the framework of which is shown in Figure 2.5. In addition, solar energy was integrated into traditional coal-fired generation systems through solar-aided coal-fired power generation technologies, which has been proved to be an efficient approach to meet more and more strict emission reduction targets [Hu et al., 2010]. The schematic diagram for the solar-aided coal-fired power generation system is shown in Figure 2.6.

Technological development of coal industry in the diffusion stage was paid attention to the energy efficiency promotion and global environmental protection. Except for the development of the technology itself, a series of environmental

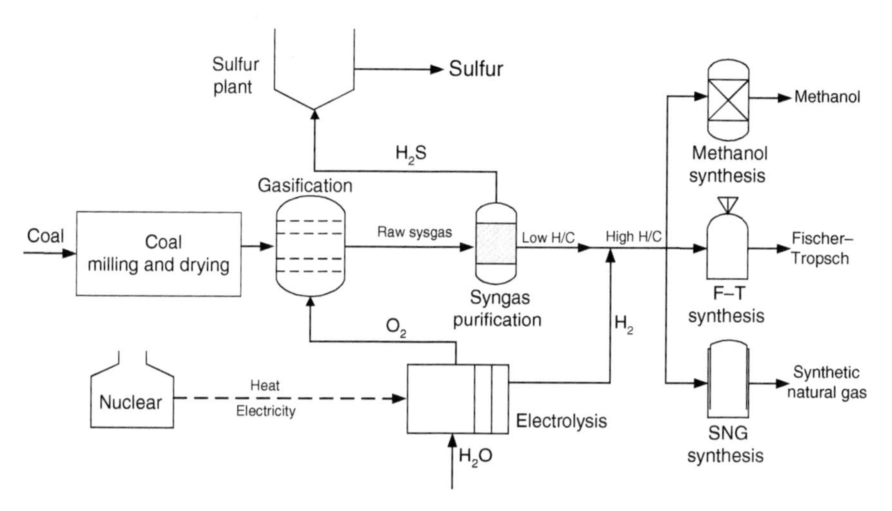

Figure 2.5 The framework of a nuclear-coal hybrid energy system.

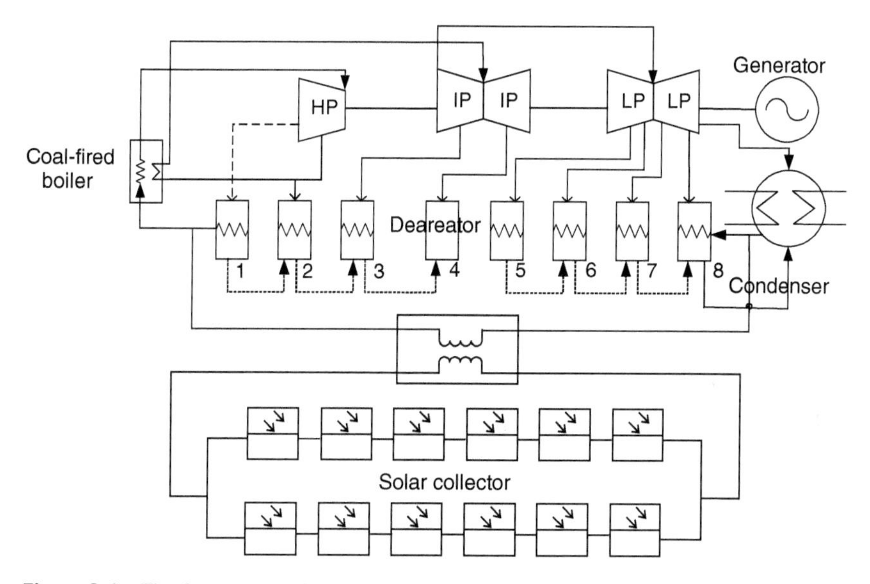

Figure 2.6 The framework of solar aided coal-fired power generation system.

policies were successively promulgated, promoting the world energy patterns to move in cleaner and more efficient direction. For example, the United Nations Framework Convention on Climate Change (UNFCCC) made stabilization of atmospheric concentrations of greenhouse gas as a global response to climate change. With such expectations for changes in energy use, research about coal utilization technology in the early twenty-first century studied advanced technologies to reduce emissions and tried to explore hybrid energy systems. In summary, the advantages and significant role of integrated energy systems that combine coal resource and other energy types began to gradually emerge in the CUP diffusion stage. However, limited by some practical problems such as technological barriers and rigorous requirements of environmental protection, paradigm curve needed to further develop to the next stage.

2.3 Coal Utilization Development Trends

Greatly influenced by technological progress as well as environmental policies, the technological paradigm progressively moved to the ecological paradigm. It was well known that we have explained the main driving forces for the development of coal utilization technologies were market-pull and technology-push in both initial competition stage and diffusion stage. However, the pursuit of highly efficient and low emission energy resources combined with coal became more evident when the paradigm moved to the third stage. These developments were regarded as ecologically oriented where the integrated coal utilization technologies were expected to ultimately break through the technical barriers and then enter a completely new stage of development.

2.3.1 Disruptive Integrated Shift

The paradigm inevitably entered into the shift phase owning to technological limitations in previous stages [Dalgleish and Foster, 1996] and the emergence of some disruptive technologies at this stage [Christensen, 2013]. In general, technological limitations come from the increasing market demands, and disruptive technologies are always prompted by full competition in a specific domain. Therefore, these two factors became the main reasons for the paradigm shift leading to revolutionary changes in industrial structures of coal utilization. At present, a large amount of advanced coal utilization technologies have been burgeoning, and coal industry were still maintaining its competitiveness in providing energy supply even if these technologies were not yet mature.

In area III, some keywords such as clean coal technology, carbon capture and storage (CCS), emissions, and integrated gasification combined cycle (IGCC) were found. There are many effective ways to generate clean coal power, with the IGCC being one of the more prominent technologies. The emergence of circulating fluidized bed combustion (CFBC), pressurized fluidized bed combustion (PFBC), and supercritical and ultra-supercritical power generation technologies, all of which are highly efficient and clean burning, have greatly improved coal utilization, and effectively reduced environmental damage. The flow charts for the IGCC and CFBC are shown in Figures 2.7 and 2.8.

As shown in area III, keywords such as clean coal technology, emissions, CCS, and IGCC got a lot of discussion. Among the many effective ways that work for generating clean energy power, IGCC deserved to be one of the most prominent technologies. In addition, the technology of CFBC were also highly efficient and clean burning of coal resource, which made contribution to effectively reduce

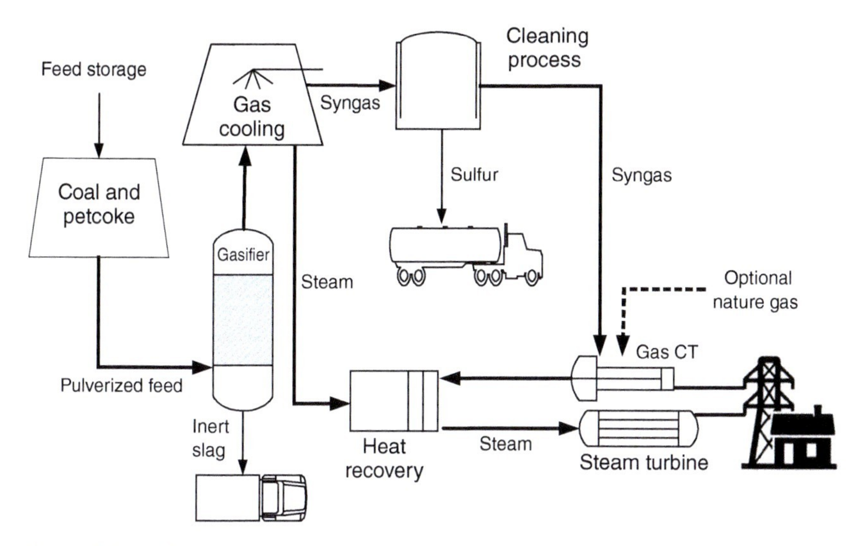

Figure 2.7 IGCC technological process.

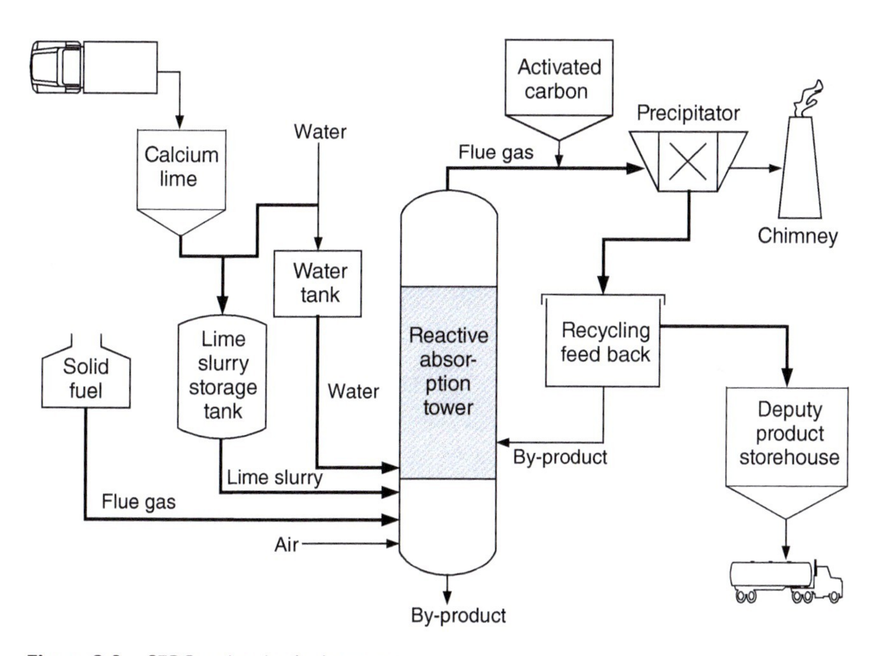

Figure 2.8 CFBC technological process.

environmental damage to the surroundings. Figures 2.7 and 2.8, respectively, show the flow charts for the IGCC and CFBC. In this stage, integrated energy systems with practicable energy storage parts were widely considered to stress environmental sustainability and also make adequate preparations for predictable growth in world energy demand. Because one of the advantages of these kind of integrated energy systems was its significant role in reducing CO_2 emissions [Dinca et al., 2009].

We defined this shift phase as eco-technological paradigm because coal utilization technologies and methods in this stage have the characteristics of ecological aims to reduce greenhouse gas emissions and other forms of pollution caused by coal use. While some important developments in cleaning technologies of coal utilization have been developed, it was only the beginning of the shift phase where further advanced technologies to achieve near zero emission were looking forward to be explored. Besides, in some researches, mathematic models have been used to forecast the possible development trends of coal and other energy utilization [Lazaroiu, 2007, Frentiu et al., 2009, Pisa et al., 2016].

It cannot be denied that coal utilization technology has been largely affected by growing environmental awareness as well as corresponding policies. Some international organizations enacted many environmental policies such as the Joint Declaration on climate change, the Cancun Agreement, and the Paris Agreement, which have won the broad consensus of the international community. Therefore, seeking for advanced energy conservation methods and environmentally friendly coal utilization approaches was becoming a common direction for the efforts of various communities. For example, in the future, a kind of hybrid energy system needs to be more secure, reliable, and flexible on both demand side and supply side, as well as to be low-carbon [Xu et al., 2016]. Therefore, a novel integrated system including UCG parts, syngas fueled solid oxide fuel cells parts, IGCC sections, and electrolyzer systems for hydrogen production have been proposed for practical applications [Bicer and Dincer, 2015].

In summary, a large number of integrated energy systems played a critical role in the CUP shift stage. As discussed earlier, although there are still many difficulties to be overcome in this stage, some alternative methods need to be developed because it is important to work toward long-term sustainability.

2.3.2 No-Coal-on-Ground Integrated Energy System

Based on the discussion above, it is obvious that the disruptive coal use technologies in the shift phase were leading to a new form of coal utilization that should be ecologically focused, highly efficient and nearly zero emissioned. Under this situation, coal resource would no longer to be simply utilized as the global concerns have widely highlighted the need for advanced technologies and integrated energy systems [Pisa et al., 2009]. By literature review and co-occurrence analysis, the technological revolution in the field of coal utilization has mainly presented in three types of innovation: co-production of coal-to-liquids and underground coal gasification, spray combustion explosion to power generation by coal and gas, and low carbon technologies to be near-zero emissions. Although some energy systems that were not too complex have emerged, the integrated energy systems that can combine various kinds of energy resources with different advanced technologies have become main current trends in the shift phase. What this kind of integrated energy system would be, how does it operate, and what the possible ecological benefits would have are all questions we are concerned about which will be answered in the following in detail.

The first question is that what a coal-based integrated energy system would look like. Generally speaking, the system needs to combine multiple energy resources such as wind, solar, heat, or gas on the basis of coal utilization to make them effective connection and operation. In this system, ecologically coal mining, gasification, other process measures, and corresponding energy transformation will be all operated underground in the future that can realize power transmission for ground energy demands and also energy storage if there were sufficient supply. Therefore, according to the above statement, a possible integrated energy system that can achieve the scenery of no-coal-on-the-ground would be similar to that shown in Figure 2.9.

Then, after knowing that the general type of the integrated energy system, something about how does this system operate is a key question needs to be resolved. Overall, the feasibility of the system would determine its possible operation mode. In such a system, the operating process can be roughly described as follows. First, the ground energies such as wind, solar, and others would generate power to supply electricity needed for the underground operations including coal mining, washing, gasification, and liquefaction. Thereinto, the co-production of gasification and coal-to-liquid can be realized by gasifying the coal using the related underground gasification technologies, after which some processes such as purification, F–T synthesis, and hydrogenation modification are going to happen to obtain final products that are coal chemicals, cleaning oil, and so on. Then, a series of underground operations on coal would also achieve power generation, and at the same time, the coal-bed methane and geothermal energy produced along with the coal mining

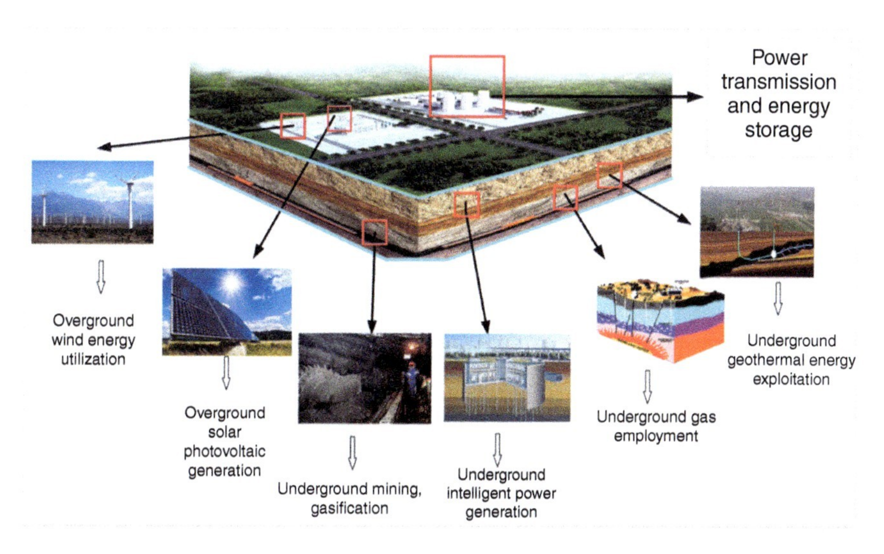

Figure 2.9 An imaginary no-coal-on-the-ground integrated energy system.

processes can be fully made use of. Finally, the energy generated by the system will be transmitted to the end users no matter they are above or below ground, and the energy storage parts of the system would help to store energy that may be left over.

Because this system no longer has the transportation and coal washing process of traditional mining on the surface, the costs for fuel oil and coal conversion can be significantly reduced which can make a contribution to the local ecological environment protection. Up to now, technology innovations in coal utilization have mainly concentrated on related clean technologies, especially clean coal combustion and clean coal conversion and their combination with other energy technologies. In the future, the integrated energy system based on coal resource will be seen as a critical solution for pollution reduction, environmental protection, and even climate change mitigation, which would be a new motivation for fast development of novel energy industries. As it can be seen in the research of Tokimatsu et al., zero emission of an energy system can be possible in this century [Tokimatsu et al., 2016]. Owing to the integrated energy system with advanced coal utilization technologies and comprehensive utilization of other energy resources are highly efficient and low emissions to a near-zero level, it strongly predicted coal resource will be continually used in this way to achieve ecological benefits required by world communities in the future.

2.4 Discussion

When the technological paradigm moved into the shift phase, the concept of eco-technological paradigm was proposed to describe the current state and the possible future development direction of coal utilization technology. However, the paradigm curve will continue to extend in this shift phase because some disruptive technologies have already appeared, but a big development bottleneck has not emerged. On the basis of existing technology, it is predicted that coal utilization will become increasingly ecologically focused and possibly be combined with other energy resources as integrated energy systems become more developed. In response to the flexible environmental changes, integrated energy systems with characteristics of low carbon emission and high energy efficiency are needed for sustainable development of the whole society where innovations in terms of coal use based comprehensive energy utilization technologies are becoming a center-stage issue. The analytical framework for the integrated energy system diffusion is shown in Figure 2.10.

Although some scholars and practical applications in industry have realized small-scale hybrid energy systems in practice, there was still a long way to go before full-featured integrated systems are widely available all around the world. A comprehensive discussion about the scenario of no-coal-on-the-ground integrated energy system for its forecast patterns, functions, and impacts has been proposed, which indicated that the urgent need for the extension and break of paradigm shift in the near future will be more obvious with the technological breakthroughs and strong demands of society.

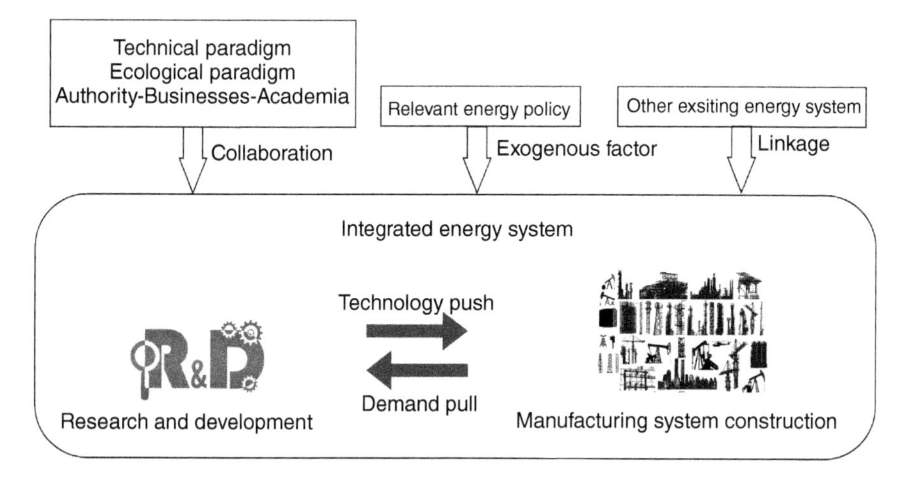

Figure 2.10 The analytical framework for the development of coal utilization towards integrated energy system.

In the future, taking account of new technologies, various resources, environmental impacts, relevant policies, and management characteristics all together to achieve the economic, ecological, and societal energy utilization for sustainable development will be common pursuit of the integrated energy system. Revolutionary technological changes described in the diffusion phase that can overturn current levels is what we need now. Similarly, government supports by implementing some relevant policies and regulations would play a great role in the development process of such systems, as proper policies can promote technological progress and maintain positive competitions of market demands. In addition, suitable, coherent, and intensive management methods are also necessary elements in the proposed systems because excellent management can ensure high efficiency of system operation. In brief, technological development will be linked to environmental policies, and innovative policy development will inspire technological progress in turn; therefore, an interaction between technological advance and policy supports is the inevitable direction of the trend for future coal utilization industry.

References

Aktaa, Z. and Woodburn, E.T. (2000). Effect of addition of surface active agent on the viscosity of a high concentration slurry of a low-rank British coal in water. *Fuel Processing Technology* 62 (1): 1–15.

Ayres, R.U. (1988). Barriers and breakthroughs: an expanding frontiers model of the technology-industry life cycle. *Technovation* 7 (2): 87–115.

Bicer, Y. and Dincer, I. (2015). Energy and exergy analyses of an integrated underground coal gasification with SOFC fuel cell system for multigeneration including hydrogen production. *International Journal of Hydrogen Energy* 40 (39): 13323–13337.

Bin, L. and San-Dang, G. (2010). Attentions on grey system theories by China scholars-based on literature metrology during 1982–2009. *Journal of Grey System* 22 (2): 137–146.

BP Global (2009). BP Sustainability Report 2015.

Bronfenbrenner, U. (1974). Developmental research, public policy, and the ecology of childhood. *Child Development* 45 (1): 1–5.

Bronfenbrenner, U. (1976). The experimental ecology of education. *Educational Researcher* 5 (9): 5–15.

Bronfenbrenner, U. (1977). Toward an experimental ecology of human development. *American Psychologist* 32 (7): 513.

Buhre, B.J.P., Elliott, L.K., Sheng, C.D. et al. (2005). Oxy-fuel combustion technology for coal-fired power generation. *Progress in Energy and Combustion Science* 31 (4): 283–307.

Chen, Q.Q., Tang, Z.Y., Lei, Y. et al. (2015). Feasibility analysis of nuclear-coal hybrid energy systems from the perspective of low-carbon development. *Applied Energy* 158: 619–630.

Christensen, C.M. (2013). *The Innovator's Dilemma: When New Technologies Cause Great Firms to Fail*. Harvard Business Review Press.

Clark, G. and Jacks, D. (2007). Coal and the industrial revolution, 1700–1869. *European Review of Economic History* 11 (1): 39–72.

Dalgleish, H.Y. and Foster, I.D.L. (1996). ^{137}Cs losses from a loamy surface water gleyed soil (Inceptisol); a laboratory simulation experiment. *Catena* 26 (3–4): 227–245.

Dinca, C., Badea, A.-A., Apostol, T., and Lazaroiu, G. (2009). GHG emissions evaluation from fossil fuel with CCS. *Environmental Engineering and Management Journal* 8 (1): 81–89.

Dosi, G. (1982). Technological paradigms and technological trajectories: a suggested interpretation of the determinants and directions of technical change. *Research Policy* 11 (3): 147–162.

Dovi, V.G., Friedler, F., Huisingh, D., and Klemes, J.J. (2009). Cleaner energy for sustainable future. *Journal of Cleaner Production* 17 (10): 889–895.

Ebara, N. (2000). R&D of coal utilization technology in Japan. *Fuel Processing Technology* 62 (2–3): 143–151.

Frentiu, T., Ponta, M., Mihaltan, A. et al. (2009). Qualitative assessment of heavy metals sources in pitcoal/biomass briquettes combustion using multivariate statistical analysis. *Extremes* 6: 10.

Griliches, Z. (1957). Hybrid corn: an exploration in the economics of technological change. *Econometrica, Journal of the Econometric Society* 25 (4): 501–522.

Hong, J., Chaudhry, G., Brisson, J.G. et al. (2009). Analysis of oxy-fuel combustion power cycle utilizing a pressurized coal combustor. *Energy* 34 (9): 1332–1340.

Hu, E., Yang, Y.P., Nishimura, A. et al. (2010). Solar thermal aided power generation. *Applied Energy* 87 (9): 2881–2885.

IEA (2016). International Energy Agency/Coal. Http://www.iea.org/topics/coal/ (accessed 26 May 2021).

Jacobsson, S. and Lauber, V. (2006). The politics and policy of energy system transformation and explaining the German diffusion of renewable energy technology. *Energy Policy* 34 (3): 256–276.

Kolditz, O., Xie, H., Hou, Z. et al. (2015). Subsurface energy systems in China: production, storage and conversion. *Environmental Earth Sciences* 73: 6727–6732.

Lazaroiu, G. (2007). Modeling and simulating combustion and generation of NO_x. *Fuel Processing Technology* 88 (8): 771–777.

Lazaroiu, G., Frentiu, T., Mihaescu, L. et al. (2009). The synergistic effect in coal/biomass blend briquettes combustion on elements behavior in bottom ash using ICP-OES. *Journal of Optoelectronics and Advanced Materials* 11: 713–721.

Lazaroiu, G., Pop, E., Negreanu, G. et al. (2017). Biomass combustion with hydrogen injection for energy applications. *Energy* 127: 351–357.

Li, H., Nalim, R., and Haldi, P.-A. (2006). Thermal-economic optimization of a distributed multi-generation energy system: a case study of Beijing. *Applied Thermal Engineering* 26 (7): 709–719.

Li, C., Xiong, K., and Wu, G. (2013). Process of biodiversity research of karst areas in China. *Acta Ecologica Sinica* 33 (4): 192–200.

Mago, P.J. and Chamra, L.M. (2009). Analysis and optimization of CCHP systems based on energy, economical, and environmental considerations. *Energy and Buildings* 41 (10): 1099–1106.

Maraver, D., Sin, A., Royo, J., and Sebastian, F. (2013). Assessment of CCHP systems based on biomass combustion for small-scale applications through a review of the technology and analysis of energy efficiency parameters. *Applied Energy* 102: 1303–1313.

Minutillo, M. and Perna, A. (2014). Renewable energy storage system via coal hydrogasification with co-production of electricity and synthetic natural gas. *International Journal of Hydrogen Energy* 39 (11): 5793–5803.

Negreanu, G. and Mocanu, C.R. (2012). Biomass briquettes from pitcoal-wood: boiler test facility combustion case study. *Journal of Environmental Protection and Ecology* 13 (2A): 1070–1081.

OECD/IEA (2015). World Energy Outlook 2015.

Phdungsilp, A. (2010). Integrated energy and carbon modeling with a decision support system: policy scenarios for low-carbon city development in Bangkok. *Energy Policy* 38 (9): 4808–4817.

Pisa, I. and Lazaroiu, G. (2012). Influence of co-combustion of coal/biomass on the corrosion. *Fuel Processing Technology* 104: 356–364.

Pisa, I., Lazaroiu, G., and Prisecaru, T. (2014). Influence of hydrogen enriched gas injection upon polluting emissions from pulverized coal combustion. *International Journal of Hydrogen Energy* 39 (31): 17702–17709.

Pisa, I., Lazaroiu, G., Mihaescu, L. et al. (2016). Mathematical model and experimental tests of hydrogen diffusion in the porous system of biomass. *International Journal of Green Energy* 13 (8): 774–780.

Pisa, I., Radulescu, C., Lazaroiu, G. et al. (2009). The evaluation of corrosive effects in co-firing process of biomass and coal. *Environmental Engineering and Management Journal* 8 (6): 1485–1490.

Prabu, V. and Jayanti, S. (2012). Integration of underground coal gasification with a solid oxide fuel cell system for clean coal utilization. *International Journal of Hydrogen Energy* 37 (2): 1677–1688.

Ridley, C.E., Clark, C.M., LeDuc, S.D. et al. (2012). Biofuels: network analysis of the literature reveals key environmental and economic unknowns. *Environmental Science and Technology* 46 (3): 1309–1315.

Scherf, M., Epple, A., and Werner, T. (2005). The next generation of literature analysis: integration of genomic analysis into text mining. *Briefings in Bioinformatics* 6 (3): 287–297.

Schwartz, S.H. (1994). Are there universal aspects in the structure and contents of human values. *Journal of Social Issues* 50 (4): 19–45.

Stern, P.C. and Dietz, T. (1994). The value basis of environmental concern. *Journal of Social Issues* 50 (3): 65–84.

Tokimatsu, K., Konishi, S., Ishihara, K. et al. (2016). Role of innovative technologies under the global zero emissions scenarios. *Applied Energy* 162: 1483–1493.

Wang, J., Jing, Y., Zhang, C., and Zhai, Z.J. (2011). Performance comparison of combined cooling heating and power system in different operation modes. *Applied Energy* 88 (12): 4621–4631.

Xu, J., Li, L., and Zheng, B. (2016). Wind energy generation technological paradigm diffusion. *Renewable and Sustainable Energy Reviews* 59: 436–449.

3

Multiple Coal Seam Coproduction-Oriented Equilibrium Approach Toward Coal–Water Conflict

As the most abundant and widely used energy resource, coal shared 27.6% of the world's primary energy consumption amount and, with the consideration of the rapid economic development around the world, the demand on coal is predicted to continuously increase in the following years, which will result in the increasing production of coal. Under this situation, the coal–water conflict that was mainly caused by the underground coal mining will be aggravated. For detail, with the consideration of safety, significant amounts of groundwater are discharged during underground coal mining and, at the same time, many pollutants are involved in the drainage that cause further damage to the local water environment. Under the pressure of the huge demand amount and encouraged by the high economic profit, most of the large-scale coal production regions in the world, especially in the developing countries and regions like China and India, are now conducting extensively exploitation with little attention to the protection of the local water environment. This situation has already resulted in serious water environmental problems, for example, groundwater level depression and regional water quality deterioration. Many research have been conducted to solve this problem, however, the situation is still not satisfactory, and some innovative approaches are needed. The equilibrium strategy that was first proposed in the modern economic and trading research has been known as one the most useful tools in solving the conflicts and has already been employed in many other fields, such as the regional water resources allocation and facility location problem. Encouraged by these excellent works, in this chapter, the equilibrium strategy will be used to solve the coal–water conflict. To make the equilibrium strategy to be quantitatively researched, a bilevel programming model will be built, and with the consideration of the realistic situation, the coefficients in the proposed model will be treated as the fuzzy form for the purpose of describing the uncertainty in the coal mining activities. Then, the model will be defuzzified by the fuzzy expected value theory and the fuzzy possibility measure, and a solution-based approach on the Karush–Kuhn–Tucker (KKT) condition is designed to search for the solutions. Finally, a real-world case study in Yulin coal field, China, is presented to work as the foundation of the discussion and policy recommendations.

Innovative Approaches towards Ecological Coal Mining and Utilization, First Edition.
Jiuping Xu, Heping Xie, and Chengwei Lv.

3.1 Background Review

For the purpose of solving the coal–water conflict and improving the water environmental quality in the large-scale coal fields, an equilibrium strategy-based innovative approach will be researched in this chapter, and to make it more convenient to be understood, topics on some basic backgrounds and key problems, which include the multiple coal seam production system, the mining quota allocation framework, and the uncertainty environment are introduced in this section.

3.1.1 Multiple Coal Seam Production System

As it is known, for most of the large-scale coal production regions in the world, there always exists multiple coal seams, and with the development of the coal mining technology, especially for the safety engineering technology, coal mining in these coal seams at the same time is now becoming the main mode of coal production [Ward, 2001]. Such production mode promotes the yield of coal in these regions greatly and makes the machinery and manpower of each colliery more efficient, and under the pressure of the increasing of coal demand, the multiple coal seam production will still be the main production mode in the following years. For modern collieries, their production plans always related to each sub-coal seam. Generally speaking, as geological conditions are not the same, coal quality from different coal seams is not all the same and the unit price of coal in the market is mainly based on the quality of coal, which results in the difference in the unit revenue of coal for each colliery who conduct multiple coal seam production [Ward, 2001].

On the other hand, affected by the differences in the geological conditions, the damage that caused by coal mining activities on the local groundwater environment is also different for the water quality and quantity perspective [Falcon, 1989]. Generally speaking, the unit coal drainage coefficient at different coal seams is different, which is mainly caused by the different water content and permeability coefficient, and this difference directly resulted in different damage degree to the local groundwater quantity and the groundwater-level depression degree. As for the water quality, because the drainage from different coal seams is generated from different stratum, the contaminated components and the content are not all the same, especially for the acidic and heavy metals. This results in the different influence on the local water quality and brings different degrees of deterioration to the local water environment after being drained [Falcon, 1989].

3.1.2 Mining Quota Allocation Scheme

For most of the large-scale coal production regions in the world, there always exist two kinds of the decision makers whose decision will affect the coal production plans in the certain region directly, the local authority and the collieries. Acting on behalf of the public, the local authority always has a relative higher decision priority and the public considers are their interests and key points when making decisions. Both the economic and environment perspectives should be taken into consideration when the authorities are making their decisions, and on the other hand, all the collieries consist of the other critical roles in the large-scale coal production region. They are all independent of each other, and as enterprises, they individually pursue the largest possible economic profits and paid little attention to environmental protection. To be summarized, when considering the developing plan of the large-scale coal production region, both the local authority and the collieries play critical roles, however, according to their different decision preferences, which were oriented by the differences in their representations of the region, their purposes will be not all the same. The local authority focus on the development of the regional economic under the environmental carrying capacity constraint, and the collieries would like pursuing the possible highest profit with the limitations from the authority.

Under this situation, some conflicts always take place among these two kinds of decision makers. For the purpose of conducting the most suitable coal production plan for the large-scale coal production region, some innovative approach should be researched. Mining quota allocation, which means allocating a certain mining quota by the local authority to sub-collieries for the purpose of controlling and influencing their production plans, has already been applied in many large-scale coal fields. Under the coal mining quota allocation scheme, the local authority first decides on the initial mining quota for each sub-colliery according to historical data and then the collieries in the region develop their own production plans according to the decision from the local authority. Furthermore, the decisions of collieries will be sent back to the authority and after collecting all the plans, the authority will make some further adjustments on the initial mining quota scheme and the adjusted mining quota scheme will be transmitted back to the sub-colliery for further improving their own production plans. Such decision procedure will be conducted for many times and finally a mining quota allocation scheme that fulfill both the requirements of the authority and sub-collieries will be reached. In this chapter, a bilevel programming model is proposed to describe such decision procedure in which the authority has the higher decision priority and will act as the decision maker of the upper level model, and the sub-collieries in the area are the subordinates and work as the decision makers of the lower level model.

3.1.3 Uncertain Condition

Uncertainties are present along with the coal mining activities, and for the purpose of conducting more suitable and realistic innovative approach for solving the conflict between local water environmental protection and coal mining, uncertainties should be highlighted. For example, when evaluating the influence on groundwater level caused by coal mining in a certain coal seam, the unit coal mining drainage coefficient is the most widely used index. Affected by the complex geological condition, the changing of the rainfall and the water content as well as the differences in the production process, it is difficult to determine this coefficient into a certain constant. In another word, this coefficient may change from case to case or even changed by times. However, such changes are following certain rules, and the value of such uncertain coefficient is always described in linguistic terms, such as the range of the value is between a and d, with the most possible value being between c and d (where $a \leq b \leq c \leq d$). Such condition is similar to the fuzzy theory that was proposed by Zadeh, which has already proven to be effective in coping with this kind of uncertain condition [Zadeh, 1965]. For this reason, in this article, fuzzy variable is employed to characterize the uncertainty encountered in using the equilibrium approach to solve the conflicts between coal and water. Many other coefficients such as the concentration of the pollutants in the coal mining drainage will also be measured in the fuzzy form.

3.2 Modeling

The mathematical expression of using equilibrium approach to solve the conflict between coal and water is given in this section. As what was discussed earlier, the bilevel programming model will be employed as the fundamental model for this problem, and the local authority will be the upper level decision maker and the sub-collieries are the lower level decision makers. The key point that the authority focuses when making such decisions is to conduct economic development under some certain constraints for both the environmental protection perspective and the social consideration perspective. Thus, the main objective of the authority is pursing the economic benefit and, for environmental protection perspective, the total amount of coal mining drainage limitation in the region is the main constraint that the local authority faced, and at the same time, the pollutants concentration in the mining drainage is another constraint. Furthermore, when the authority allocates mining quotas to each sub-colliery, the total allocation amount should not exceed the total proved coal reserve, and it also has the responsibility to ensure the basic market demand. As the lower level decision makers, the sub-collieries are independent of each other and individually pursue the possible highest economic profit. However, they cannot make their production plans arbitrary and some limitations have to be fulfilled. The coal mining quota amount allocated by the authority is the main limitation that collieries faced. The environmental protection restriction about the concentration of main pollutants in the coal mining drainage is another limitation that collieries have to be met, and, at the same time, the production capacity should also be taken into consideration when they are making their own production plans. The detailed modeling procedure will be described later (Figure 3.1).

3.2.1 Motivation for Employing Uncertain Variables

The need to address uncertainty when developing production and environmental protection plans is widely recognized, as uncertainties exist in a variety of system components. It is the responsibility of the regional authority to decide how to allocate mining quota to competing users. At the same time, the competing collieries also need to develop reasonable production plans under the consideration of groundwater protection to be allocated as high a quota as possible. To achieve an optimal decision, all programming parameters need to be determined. While

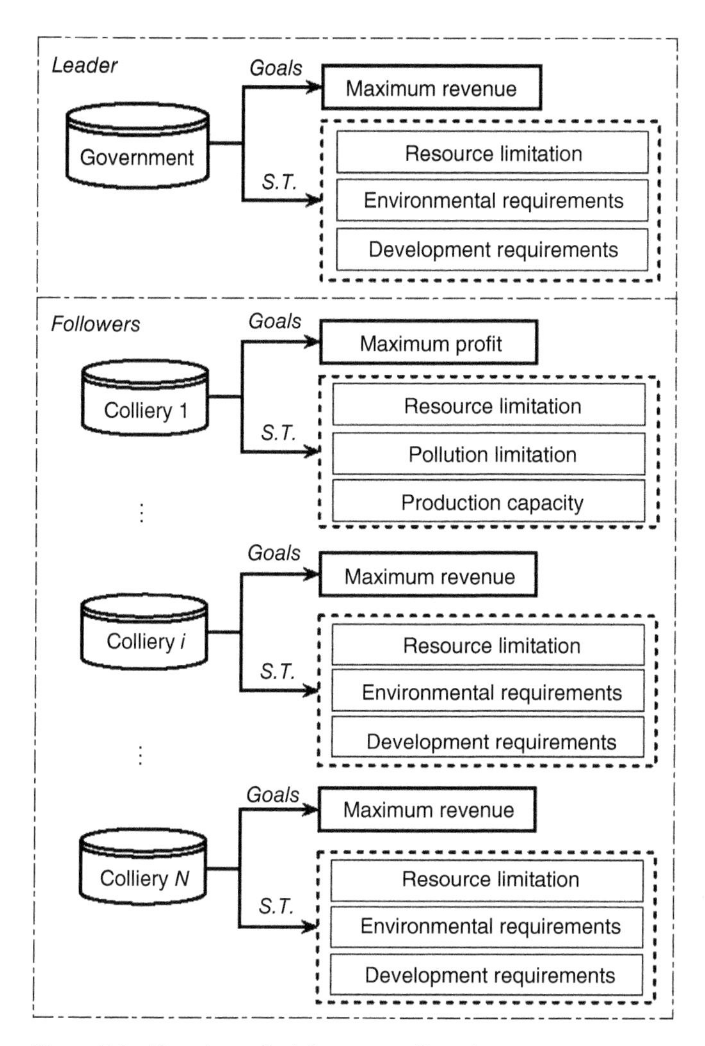

Figure 3.1 Flowchart of mining quota allocation.

some of these parameters, such as the effluent standards and the price of the coal, can be easily determined using statistical methods and historical data, some other parameters, such as the per tonne coal drainage coefficient and the concentration of main pollutants in the mining water, are more difficult to determine because of the unique natural environment and coal field operating methods. Therefore, to determine these parameters, information from engineers and workers is needed. The observed data are based on the engineer's and worker's experience, different people have different conclusions. The collected data from interviews with the engineers and workers are described in linguistic terms, such as the range of the value is between l_1 and l_4, with the most likely value being between l_2 and l_3, where $l_1 \leq l_2 \leq l_3 \leq l_4$. For this reason, fuzzy theory is suitable for dealing with such situations. In this chapter, fuzzy variables are employed to characterize the uncertainty in using the equilibrium strategy solving the coal–water conflict, as they are natural environment operating methods and human behavioral influences.

3.2.2 Typical Fuzzy Variables in the Proposed Method

Due to the inherent uncertainty and the system complexity, there are some typical fuzzy variables in the proposed method.

(1) *Per tonnes coal mining drainage coefficient*: The per tonnes coal mining drainage coefficient is a typical fuzzy variable, which represents the amount of displacement when producing one unit of coal. Displacement fluctuates because of many factors, such as rainfall, operating methods, and geological structure, so historical data are not consistent. Therefore, to gain this information, interviews with engineers and workers are necessary. Through these interviews, this value is estimated to be in a region with the minimum value represented by r_1 and the maximum value represented by r_4 and the most possible value being between r_2 and r_3, $(r_1 \leq r_2 \leq r_3 \leq r_4)$. Based on the discussion earlier, in this chapter, let $\widetilde{E_{ij}}$ be the per tonnes coal drainage coefficient when colliery i produces in the coal stream j, which we describe as a trapezoidal fuzzy number $\widetilde{E_{ij}} = (r_{ij1}, r_{ij2}, r_{ij3}, r_{ij4})$, where $r_{[\bullet]}$ are the parameters in the membership function of the trapezoidal fuzzy number $\widetilde{E_{ij}}$. Further, the regional authority needs to determine the per tonnes coal mining drainage coefficient of colliery i. Let $\widetilde{E_i}$ represents the per tonnes coal drainage coefficient for exploitation at colliery i. It is obvious this is related to $\widetilde{E_{ij}}$, and the exploitation in coal seam j of colliery i can be represented by X_{ij}. Through fuzzy theory, we know that this is still a trapezoidal fuzzy number and can be calculated using the following formula: $\widetilde{E_i} = \frac{\sum_{j=1}^{m} \widetilde{E_{ij}} X_{ij}}{\sum_{j=1}^{m} X_{ij}}$. According to the trapezoidal fuzzy number arithmetic proposed by Xu and Zhou, we can describe the per tonnes coal drainage coefficient for exploitation at when colliery i as;

$$\widetilde{E_i} = \left(\frac{\sum_{j=1}^{m} r_{ij1} X_{ij}}{\sum_{j=1}^{m} X_{ij}}, \frac{\sum_{j=1}^{m} r_{ij2} X_{ij}}{\sum_{j=1}^{m} X_{ij}}, \frac{\sum_{j=1}^{m} r_{ij3} X_{ij}}{\sum_{j=1}^{m} X_{ij}}, \frac{\sum_{j=1}^{m} r_{ij4} X_{ij}}{\sum_{j=1}^{m} X_{ij}} \right) \tag{3.1}$$

(2) *The concentration of main pollutants in the mine water*: As there are many contaminants in mine water, there are many criteria for evaluating mine water quality. Among these, suspended solids (SS) and the chemical oxygen demand (COD) are recognized as the most important indices. For this reason, in this chapter, we consider SS and COD as the main objectives. However, it is difficult to determine an exact value because of the inherent uncertainty, which is influenced by the geological structure, mining methods, and many other factors. Therefore, data regarding the concentration of main pollutants in the mine water can only be derived from a combination of information from interviews with the engineers and historical data, from which it is found that the concentration of main pollutants in the mine water is also a typical trapezoidal fuzzy number. Let $\widetilde{\theta_{ij}^1}$ be the concentration of *SS* in the mining water from coal stream j at colliery

i; and $\widetilde{\theta_{ij}^2}$ be the concentration of *COD* in the mining water from coal stream j at colliery i, then we get $\widetilde{\theta_{ij}^1} = (v_{ij1}, v_{ij2}, v_{ij3}, v_{ij4})$, where $v_{[\bullet]}$ are the parameters in membership function of the trapezoidal fuzzy number $\widetilde{\theta_{ij}^1}$ and $\widetilde{\theta_{ij}^2} = (w_{ij1}, w_{ij2}, w_{ij3}, w_{ij4})$, where $w_{[\bullet]}$ are the parameters in the membership function of the trapezoidal fuzzy number $\widetilde{\theta_{ij}^2}$.

3.2.3 Assumptions and Notations

3.2.3.1 Assumptions

Before developing the programming model, some assumptions are given:

(1) This chapter only focuses on the single production cycle decision problem.
(2) The tax rate that each colliery paid is all the same.
(3) Before processing, the coal mining drainage from different coal seams will be mixed.

3.2.3.2 Notations

The following symbols are used in this chapter.

3.2.4 Lower Level Decision-Making Model

Each sub-colliery in a certain large-scale coal field is the decision maker of the lower level decision-making model. With the consideration of the essence of the enterprise, in the lower level decision making model, the largest possible economic profit will be set as the objective function. Each sub-colliery will pursue their possible highest benefit under the constraints from the authority as well as some other limitations. For detail, when the sub-colliery is making their own production plans, the constrains that should be fulfilled are production capacity, mining quota limitations, environmental protection constraints, and, the same time, the nonnegative constraints that also need to be satisfied.

3.2.4.1 Objective Function

As the market-based enterprise, pursuing of the possible highest economic profit, is the most priority of each sub-colliery when they are making their own decisions. Generally speaking, the economic profit for a colliery can be described as the sales revenue minus tax and costs roughly. Tax is charged based on a fixed percentage of the sales revenue, so in this chapter, let TP represents the tax percentage and SP_{ij} be the per tonne price of coal produced from coal seam j at colliery i. Therefore, the total tax that colliery i should pay is $TP \sum_{j=1}^{m} SP_{ij}PA_{ij}$, where PA_{ij} is the quantity that colliery i exploits from coal stream j. Colliery costs are mainly made up of production costs and sewage treatment costs. Let T_{ij} be the total cost per tonne of coal when colliery i produces from coal stream j, so the production cost of colliery i can be described as $\sum_{j=1}^{m} T_{ij}PA_{ij}$.

Indices

i = Index of the sub-collieries, $i = 1, 2, \ldots, n$, and

j = Index of coal seams, $j = 1, 2, \ldots, m$;

Parameters

R^u = The recoverable amount of coal in the region;

D^L = The lowest amount of coal that the region should supply;

SP_{ij} = The price of coal from coal stream j at colliery i;

C_i = The average price of coal produced at colliery i

T_{ij} = The total cost per tonne of coal when colliery i produces in coal stream j;

U_i = The maximum amount of groundwater tolerated for exploitation at colliery i;

O_{ij}^U = The maximum mining capacity of colliery i for coal stream j;

φ^1 = The government stipulated emission standard for SS;

φ^2 = The government stipulated emission standard for COD;

TP = The stipulated government tax rate;

ξ = Penalty coefficient

H = The maximum carrying capacity that the groundwater can undermine in the region;

N_i^1 = The maximum processing capacity of COD at colliery i;

N_i^2 = The maximum processing capacity of SS at colliery i;

P_i = Sewage treatment unit cost at colliery i

Uncertain parameters

$\widetilde{E_{ij}}$ = Per tonnes coal drainage coefficient when colliery i produces from coal stream j;

$\widetilde{E_i}$ = Per tonnes coal drainage coefficient for exploitation at colliery i;

$\widetilde{\theta_{ij}^1}$ = The concentration of SS in the mining water from coal stream j at colliery i;

$\widetilde{\theta_{ij}^2}$ = The concentration of COD in the mining water from coal stream j at colliery i;

Decision variables

MQ_i = Mining quota that the government allocates to each sub-colliery i to exploit;

PA_{ij} = Production amount that colliery i conducts from coal stream j;

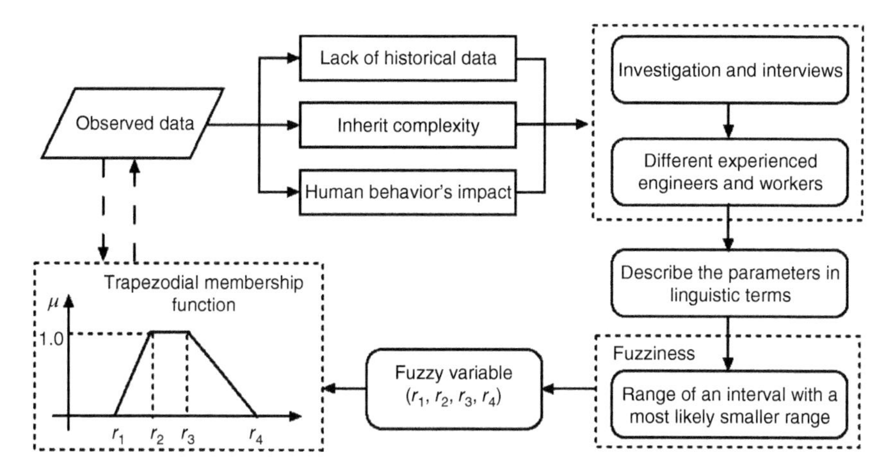

Figure 3.2 Flowchart for the construction of the fuzzy membership functions.

The total sewage treatment cost at colliery i is determined by the per unit sewage treatment cost (P_i), the per tonnes coal drainage coefficient ($\widetilde{E_{ij}}$), and the quantity that colliery i exploits (PA_{ij}), which can be written as $P_i \sum_{j=1}^{m} \widetilde{E_{ij}} PA_{ij}$. As the parameters $\widetilde{E_{ij}}$ are more difficult to determine because of many factors such as rainfall, operating methods, and geological structure, to determine this information, interviews with engineers and workers are necessary. Through these interviews, this value is estimated to be in a region with the minimum value represented byr_1, the maximum value represented by r_4, and the most possible value being between r_2 and r_3, $(r_1 \leq r_2 \leq r_3 \leq r_4)$. For this reason, fuzzy theory is suitable for dealing with such uncertain situations(Xu and Zhou, 2011], and it be described as a trapezoidal fuzzy number $\widetilde{E_{ij}} = (r_{ij1}, r_{ij2}, r_{ij3}, r_{ij4})$, where $r_{[\bullet]}$ are the parameters in the membership function of the trapezoidal fuzzy number $\widetilde{E_{ij}}$. Figure 3.2 shows a flowchart for the construction of the fuzzy membership functions. Based on the above, the objective function for the collieries is as follows:

$$\max G_i = \sum_{j=1}^{m} \left[(1 - TP)\, SP_{ij} - T_{ij}\right] PA_{ij} - P_i \sum_{j=1}^{m} \widetilde{E_{ij}} PA_{ij} \tag{3.2}$$

3.2.4.2 Constraints

Some constraints must be fulfilled that include the mining quota restrictions, mining ability restrictions, and the mining drainage quality restrictions when the collieries making their own decisions on the production plans.

(1) *Mining quota restrictions*: Because the decision-making process in the proposed equilibrium strategy-based innovative bilevel programming model is sequential, on the subordinate side, the collieries make their own decisions about the quota

the government allocates. The government also adjusts the mining quota for each colliery to achieve its own optimal objectives. Therefore, the total mining quantity at each colliery should not exceed the quota. Let MQ_i be the quota that the government allocates to colliery i and PA_{ij} represent the mining amount at coal seam j of colliery i, so the mining quota constraint for each colliery is as follows:

$$\sum_{j=1}^{m} PA_{ij} \leq MQ_i \tag{3.3}$$

(2) *Mining ability restrictions*: For each colliery, their mining ability at each coal seam is always predefined and cannot be quickly improved. In another word, for each colliery, their mining ability at each coal seam is limited which should be taken into consideration when they make the decisions on the production plans. Therefore, when colliery i makes a decision about the quantity that should be mined from coal seam j, this quantity must not exceed its mining ability O_{ij}^U. Obviously, the amount that the colliery i exploits from coal stream j, which is represented by PA_{ij}, is nonnegative, so the mining ability constraint is as follows:

$$0 \leq PA_{ij} \leq O_{ij}^U \tag{3.4}$$

(3) *Mining drainage quality restrictions*: While mining drainage contents, many contaminants will cause serious damage to the local water environment and results in the deterioration of water quality if it is drained without any treatments. Among all the contaminants in the coal mining drainage, SS and COD are recognized as the most important indices [Banks et al., 1997, Johnson, 2003]. Thus, in this chapter, they are included as the main quality restrictions. Similar to the parameters $\widetilde{E}_{ij}$, the concentration of main pollutants in the mine water is also a typical trapezoidal fuzzy number. That is, let $\widetilde{\theta_{ij}^1} = (v_{ij1}, v_{ij2}, v_{ij3}, v_{ij4})$, $\widetilde{\theta_{ij}^2} = (w_{ij1}, w_{ij2}, w_{ij3}, w_{ij4})$ be the concentration of SS, and COD in the mining water from coal stream j at colliery i, where $v_{[\bullet]}$, $w_{[\bullet]}$ are the parameters in the membership function of the above trapezoidal fuzzy numbers, respectively. In China, the mine water discharge standards state that the COD, represented by φ^1, in the treated mine water should not exceed 150 mg/l, and the concentration of SS, represented by φ^2, should not exceed 100 mg/l for water quality level II or 100 and 70 mg/l for water quality level I [State of Environmental Protection Agency, 2005]. If the mine water discharged from the colliery does not meet these standards, the government deprives them of their production rights. Since the sewage treatment capacity at each colliery is limited, the pollutant concentrations should be no more than the sewage treatment capacity constraint. Let N_i^1 be the COD treatment capacity at colliery i and N_i^2 be the SS treatment capacity at colliery i, then the environmental protection constraint is as follows;

$$\begin{aligned} &\sum_{j=1}^{m} \widetilde{E}_{ij} PA_{ij} \widetilde{\theta_{ij}^1} \Big/ \sum_{j=1}^{m} \widetilde{E}_{ij} PA_{ij} \leq \varphi^1 + N_i^1 \\ &\sum_{j=1}^{m} \widetilde{E}_{ij} PA_{ij} \widetilde{\theta_{ij}^2} \Big/ \sum_{j=1}^{m} \widetilde{E}_{ij} PA_{ij} \leq \varphi^2 + N_i^2 \end{aligned} \tag{3.5}$$

3.2.5 Upper Level Decision Making Model

The local government of the large-scale coal field acts as the authority and has the higher decision-making priority, so is in the upper level position in the proposed equilibrium strategy-based innovative bilevel programming model. At the upper level decision maker, the local government has the obligation to promote the financial revenue; however, this promotion should ensure the environmental self-repair capacity. Therefore, in this chapter, to build the mathematical model, we set the financial revenue as the objective function and environmental protection and some other objective conditions as the constraints.

3.2.5.1 Objective

Generally speaking, the financial revenue of a regional authority consists of local tax revenue, central tax returns, and transfer payments. As for the coal mining industry, their contribution to local financial revenue mainly comes from the tax and with consideration that the collieries pay tax based on their sales revenue, in this chapter, let TP be the tax rate, SP_i be the average price of coal for colliery i, and MQ_i be the mining quota for colliery i given by the authority, so the regional financial revenue of the coal industry is $TP\sum_{i=1}^{n} SP_iMQ_i$. Therefore, the regional financial revenue for the coal industry is as shown:

$$\max F_1 = TP\sum_{i=1}^{n} SP_iMQ_i \tag{3.6}$$

3.2.5.2 Constraints

Even though the local government acts as the decision maker of upper level programming model in the proposed innovative bilevel model, some constraints should also be highlighted when they make their own decisions. In this chapter, the constraints from the environmental protection, the realistic constraint, and the market operation constraint perspectives will be focused.

(1) *Groundwater protection restrictions*: Works as the upper level decision maker and acts on behalf of the local public, the regional government has the responsibility in the local environmental protection, especially for the large-scale coal field. One of the most serious environmental issues that the local government facing is extensive coal mining drainage and the groundwater level depression. Therefore, in this chapter, the local groundwater protection will be set as the main constraint for the local government when they are making the mining quota allocation decisions. However, it is very difficult to estimate the exact value of the constraint due to the inherent uncertainty and complexity. The regional authority, therefore, needs to use linguistic terms to describe their intentions, such as we would like to try our best to control the amount of extruded groundwater within a certain degree. That is to say, the government cannot be sure if a certain degree is entirely reasonable or can be easily achieved. Therefore, the government sets the environmental protection as a prescribed possibilistic level. These linguistic terms are described as fuzzy chance constraint, which

was originally proposed by Liu and Iwamura Liu and Iwamura [1998]. In this chapter, let H be the maximum carrying capacity that the groundwater can tolerate in the region and γ be the possibilistic level, so then the environmental protection constraint is:

$$Pos\left\{\left[\sum_{i=1}^{n} \widetilde{E_i} MQ_i \leq H\right]\right\} \geq \gamma \tag{3.7}$$

where Pos is the possibility measure proposed by Dubois and Prade Dubois and Prade [1983]. This function indicates that the government requires the amount of the excluded groundwater to be controlled as less than H under the given possibilistic level γ.

(2) *Total recoverable amount restriction*: When the local authority makes their own decision on the mining quota allocation scheme, the total allocation amount should be less than the recoverable amount in the region, which means that the quota quantity for all collieries in the region must not exceed the amount of coal in the region. In this chapter, let R^U be the recoverable amount of coal in the region that the total quota allocated to the collieries should not exceed.

$$\sum_{i=1}^{n} MQ_i \leq R^U \tag{3.8}$$

(3) *Satisfy market demand*: Works as the foundation energy resource, coal is critical for the stability and development of the region; therefore, the government has the responsibility to ensure that the coal production quantity from all collieries should meet the basic market demand, in this chapter that was represented by D^L and this constraint can be described as follows:

$$\sum_{i=1}^{n}\sum_{j=1}^{m} PA_{ij} \geq D^L \tag{3.9}$$

3.2.6 Global Optimization Model

In the proposed equilibrium strategy-oriented bilevel programming model, to cope with the coal–water conflict in the large- scale coal field, works on behalf of the public, the local authority has a relative higher decision priority, which is the upper level decision maker. A reasonable mining quota allocation scheme to maximize total social benefit will be made by the local authority, and this scheme is done under constraints on coal storage quantity, basic market demand, and environmental protection. At the lower level decision model, each sub-colliery decides on their own production plans for the exploitation of each coal seam based on their own economic benefit objectives under the mining quota constraints, the environmental protection constraints, and production capacity. In this problem, the regional authority and the sub-collieries conflicted both in their respective objectives and the constraints. The regional authority attempts to maximize total social benefit through a consideration of the environmental pollution from mining quota allocations, while each sub-colliery only seeks to gain the largest economic benefits. To deal with this conflict and by integrating Eqs. (3.2)–(3.9), the global model can be formulated as in Eq. (3.10).

$$\max F = TP \sum_{i=1}^{n} SP_i MQ_i$$
$$\text{s.t.} \begin{cases} \sum_{i=1}^{n} MQ_i \leq R^U \\ \sum_{i=1}^{n}\sum_{j=1}^{m} PA_{ij} \geq D^L \\ Pos\left\{\left[\sum_{i=1}^{n} \widetilde{E_i} MQ_i \leq H\right]\right\} \geq \gamma \\ \forall i = 1, 2, \ldots, n \\ \forall j = 1, 2, \ldots, m \\ \max G_i = \sum_{j=1}^{m} \left[(1 - TP)\, SP_{ij} - T_{ij}\right] PA_{ij} - P_i \sum_{j=1}^{m} \widetilde{E_{ij}} PA_{ij} \\ \text{s.t.} \begin{cases} \sum_{j=1}^{m} PA_{ij} \leq MQ_i \\ 0 \leq PA_{ij} \leq O_{ij}^U \\ \sum_{j=1}^{m} \widetilde{E_{ij}} \widetilde{\theta_{ij}^1} PA_{ij} \Big/ \sum_{j=1}^{m} \widetilde{E_{ij}} PA_{ij} \leq \varphi^1 + N_i^1 \\ \sum_{j=1}^{m} \widetilde{E_{ij}} \widetilde{\theta_{ij}^1} PA_{ij} \Big/ \sum_{j=1}^{m} \widetilde{E_{ij}} PA_{ij} \leq \varphi^2 + N_i^2 \\ \forall i = 1, 2, \ldots, n \\ \forall j = 1, 2, \ldots, m \end{cases} \end{cases} \quad (3.10)$$

The mathematical essence of the proposed model in Eq. (3.10) is the bilevel programming model with one leader and multiple followers under the fuzzy environment. A subjective parameter γ that expresses the government's efforts to protect the groundwater resources is employed. It is hard to solve such model directly using the existing solution method. Therefore, in the following section, a model transformation procedure will be designed and a relative algorithm will be proposed to solve it.

3.3 Solution Approach

In this section, a novel model transformation method that integrated the fuzzy expected valve measure, the fuzzy chance constraint method, and the KKT approach will be proposed to work as the foundation of the solution approach of the proposed model as shown in Eq. (3.10).

3.3.1 Parameters Defuzzification

In the proposed model, there are fuzzy parameters in both the objective functions and the constraints, and these are all trapezoidal fuzzy numbers, which means that the objectives or the constraints are classes of alternatives whose boundaries are not sharply defined. Fuzzy goals and fuzzy constraints can be defined precisely as fuzzy sets in the spaces of alternatives. Thus, this chapter introduces the fuzzy expected value measure *Me* which embeds an optimistic–pessimistic parameter to determine the combined attitudes of the decision makers [Xu and Zhou, 2011]. Figure 3.2 shows the construction of the fuzzy membership functions.

Through the fuzzy measure *Me*, the fuzzy variable, $\widetilde{E_{ij}}$, which represents the per tonne coal drainage coefficient when colliery i exploits in the coal seam j, is also described as a trapezoidal fuzzy number $\widetilde{E_{ij}} = (r_{ij1}, r_{ij2}, r_{ij3}, r_{ij4})$, and can be transformed into a crisp form as follows:

$$\widetilde{E_{ij}} \rightarrow E[\widetilde{E_{ij}}] = \frac{1-\lambda}{2}(r_{ij1} + r_{ij2}) + \frac{\lambda}{2}(r_{ij3} + r_{ij4}) \tag{3.11}$$

where λ is the optimistic–pessimistic index to determine the combined attitudes of the decision makers. From Eq. (3.11), we can then transform the fuzzy parameter $\widetilde{E_i}$ into its crisp form as follows:

$$\begin{aligned}\widetilde{E_i} \rightarrow E[\widetilde{E_i}] = &\frac{1-\lambda}{2}\left(\frac{\sum_{j=1}^{m} r_{ij1}PA_{ij} + \sum_{j=1}^{m} r_{ij2}PA_{ij}}{\sum_{j=1}^{m} PA_{ij}}\right)\\ &+ \frac{\lambda}{2}\left(\frac{\sum_{j=1}^{m} r_{ij3}PA_{ij} + \sum_{j=1}^{m} r_{ij4}PA_{ij}}{\sum_{j=1}^{m} PA_{ij}}\right)\end{aligned} \tag{3.12}$$

We can then transform the fuzzy parameter $\widetilde{\theta_{ij}^1}$ and $\widetilde{\theta_{ij}^2}$ into their crisp forms as follows:

$$\widetilde{\theta_{ij}^1} \rightarrow E[\widetilde{\theta_{ij}^1}] = \frac{1-\lambda}{2}(v_{ij1} + v_{ij2}) + \frac{\lambda}{2}(v_{ij3} + v_{ij4}) \tag{3.13}$$

$$\widetilde{\theta_{ij}^2} \rightarrow E[\widetilde{\theta_{ij}^1}] = \frac{1-\lambda}{2}(w_{ij1} + w_{ij2}) + \frac{\lambda}{2}(w_{ij3} + w_{ij4}) \tag{3.14}$$

Using fuzzy expected valve measure, *Me*, all fuzzy parameters in this chapter can be transformed into their crisp forms. However, there is an uncertain constraint, as shown in Eq. (3.7), which makes the proposed model that cannot be solve directly. To cope with this issue, this uncertain constraint should be transformed into its equivalent crisp form. In this chapter, the fuzzy chance constraint method, which was researched in [Xu and Zhou, 2011], will be used to determine the crisp form of Eq. (3.7) as shown follows;

$$(1-\gamma)\sum_{i=1}^{n} r_{i1}MQ_i + \gamma\sum_{i=1}^{n} r_{i2}MQ_i - H \leq 0 \tag{3.15}$$

Through the transformation progress for Eqs. (3.11)–(3.15), the proposed bilevel model with fuzzy parameters and fuzzy constraint (as shown in Eq. (3.10)) is transformed into its equivalent crisp form as follows:

$$\max F = TP\sum_{i=1}^{n} SP_i MQ_i$$

$$\text{s.t.}\begin{cases} \sum_{i=1}^{n} MQ_i \leq R^U \\ \sum_{i=1}^{n}\sum_{j=1}^{m} PA_{ij} \geq D^L \\ (1-\gamma)\sum_{i=1}^{n} r_{i1}MQ_i + \gamma\sum_{i=1}^{n} r_{i2}MQ_i - H \leq 0 \forall i = 1,2,\dots,n \\ \forall j = 1,2,\dots,m \\ \max G_i = \sum_{j=1}^{m} \left[(1-TP)\,SP_{ij} - T_{ij}\right] PA_{ij} - P_i\sum_{j=1}^{m} APA_{ij} \\ \text{s.t.}\begin{cases} \sum_{j=1}^{m} PA_{ij} \leq MQ_i \\ 0 \leq PA_{ij} \leq O_{ij}^U \\ \sum_{j=1}^{m} ABPA_{ij} \Big/ \sum_{j=1}^{m} APA_{ij} \leq \varphi^1 + N_i^1 \\ \sum_{j=1}^{m} ACPA_{ij} \Big/ \sum_{j=1}^{m} APA_{ij} \leq \varphi^2 + N_i^2 \\ \forall i = 1,2,\dots,n \\ \forall j = 1,2,\dots,m \end{cases} \end{cases} \tag{3.16}$$

$$\text{where}\begin{cases} A = \frac{1-\lambda}{2}(r_{ij1} + r_{ij2}) + \frac{\lambda}{2}(r_{ij3} + r_{ij4}) \\ B = \frac{1-\lambda}{2}(v_{ij1} + v_{ij2}) + \frac{\lambda}{2}(v_{ij3} + v_{ij4}) \\ C = \frac{1-\lambda}{2}(w_{ij1} + w_{ij2}) + \frac{\lambda}{2}(w_{ij3} + w_{ij4}) \end{cases} \tag{3.17}$$

It is easy to see that this crisp bilevel programming model must have the same solution as the formal model. Further, it is clear that this new model is easier to solve than the formal one, so we are now able to find the optimal solution to the proposed model by solving this new crisp bilevel model.

3.3.2 KKT Condition Transformation

Even though the proposed model as shown in Eq. (3.10) has already been simplified by using the fuzzy expected value measure, and the fuzzy chance constraint method, which was designed in Section 3.2.6, is still quite difficult to solve because of an NP hard problem [Ben-Ayed and Blair, 1990]. Many previous works have discussed bilevel programming solution approaches [Colson et al., 2005], of which the most popular is the KKT approach. The KKT approach has been proven to be a valuable analysis tool with a wide range of successful applications for bilevel programming [Lu et al., 2006]. The fundamental strategy for the KKT approach is that it replaces the follower problem with its KKT conditions and appends the resultant system to the leader problem [Hanson, 1981, Lu et al., 2007]. Based on a previous work by Shi et al., we transformed the bilevel model into a single-level model as shown in Eq. (3.18) [Shi et al., 2005].

Though the detail transformation procedure, the proposed model in Eq. (3.10) is now transformed into its equivalent form with crisp parameters and constraints. The mathematical essence of it is now becoming the single-level program with a unique objective function and certain parameters. Thus, based on the existing solution algorithms for such programming, the optimal solution can be determined [Lu et al., 2006].

3.4 Case Study

In this section, the Yulin Coal Field in Yulin Area, Shaanxi Province, China, is used as an example to demonstrate the practical applicability and efficiency of the proposed model.

3.4.1 Presentation of Case Problem

Yulin is located in the north of Shaanxi Province. This area includes four major collieries: Jinniu (Jn) colliery, Qishan (Qs) colliery, Bailu (Bl) colliery, and Shibadun (Sbd) colliery. Each colliery has three main coal seams. The locations of the mines

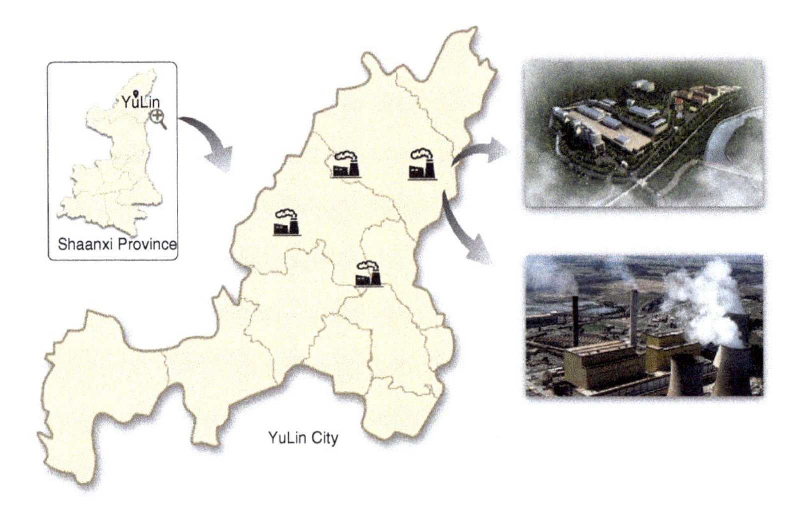

Figure 3.3 Presentation of the case region.

are shown in Figure 3.3.

$$\max F = TP \sum_{i=1}^{n} SP_i MQ_i$$

$$\text{s.t.} \begin{cases} \sum_{i=1}^{n} MQ_i \le R^U \\ \sum_{i=1}^{n} \sum_{j=1}^{m} PA_{ij} \ge D^L \\ (1-\gamma) \sum_{i=1}^{n} r_{i1} MQ_i + \gamma \sum_{i=1}^{n} r_{i2} MQ_i - H \le 0 \\ \sum_{j=1}^{m} PA_{ij} \le MQ_i \\ X_{ij} \le O_{ij}^U \\ \sum_{j=1}^{m} ABPA_{ij} \Big/ \sum_{j=1}^{m} APA_{ij} \le \varphi^1 + N_i^1 \\ \sum_{j=1}^{m} ACPA_{ij} \Big/ \sum_{j=1}^{m} APA_{ij} \le \varphi^2 + N_i^2 \\ u_1 + u_2 + u_3 \sum_{j=1}^{m} [AB - (\varphi^1 + N_i^1)A] + u_4 \sum_{j=1}^{m} [AC - (\varphi^2 + N_i^2)A] - u_5 \\ \quad = \sum_{j=1}^{m} [(1-S)C_{ij} - T_{ij}] - P_i \sum_{j=1}^{m} A - \xi \sum_{j=1}^{m} A \\ u_1 g_1(PA_{ij}, MQ_i) + u_2 g_2(PA_{ij}, MQ_i) + u_3 g_3(PA_{ij}, MQ_i) \\ \quad + u_4 g_4(PA_{ij}, MQ_i) + u_5 g_5(PA_{ij}, MQ_i) = 0 \\ g_1(X_{ij}, MQ_i) = MQ_i - \sum_{j=1}^{m} PA_{ij} \ge 0 \\ g_2(X_{ij}, MQ_i) = O_{ij}^U - PA_{ij} \ge 0 \\ g_3(X_{ij}, MQ_i) = (\varphi^1 - N_i^1) \sum_{j=1}^{m} APA_{ij} - \sum_{j=1}^{m} ABPA_{ij} \ge 0 \\ g_4(X_{ij}, MQ_i) = (\varphi^2 - N_i^2) \sum_{j=1}^{m} APA_{ij} - \sum_{j=1}^{m} ACPA_{ij} \ge 0 \\ g_5(X_{ij}, MQ_i) = PA_{ij} \ge 0 \\ \forall i = 1, 2, \dots, n \\ \forall j = 1, 2, \dots, m \end{cases} \tag{3.18}$$

$$\text{where} \begin{cases} A = \frac{1-\lambda}{2}(r_{ij1} + r_{ij2}) + \frac{\lambda}{2}(r_{ij3} + r_{ij4}) \\ B = \frac{1-\lambda}{2}(v_{ij1} + v_{ij2}) + \frac{\lambda}{2}(v_{ij3} + v_{ij4}) \\ C = \frac{1-\lambda}{2}(w_{ij1} + w_{ij2}) + \frac{\lambda}{2}(w_{ij3} + w_{ij4}) \end{cases} \tag{3.19}$$

In previous years, the exploitation of coal-oriented mineral resources has led to the development of transportation, electricity, communication, construction, and other industries. These changes have made Yulin one of the most dynamic regions in China. However, the region has a dry and semiarid climate with relatively poor water resources and a fragile ecological environment. The exploitation of coal resources could cause the leakage of groundwater into the mine from the aquifer located directly above the upper portion of the coal bearing strata in Yulin, which would aggravate the water shortage faced by the collieries. More seriously, if this mine water is not used well, a large amount of potentially useful mine water will be discharged in vain. The result is a waste of precious water resources, pollution of the environment, and even destruction of the ecological balance. With the development of the coal mines, significant quantities of mine waste have been generated, and this contaminated wastewater was mostly untreated, further contaminating the surrounding water environment. Because of the water from the mine drainage, the groundwater supplement is insufficient, which affects plant growth and adds to environmental degradation. All in all, the coal–water conflict is now becoming a major obstacle to social development at Yulin. Therefore, in this chapter, Yulin coal field is researched as case region to demonstrate the practicability of the proposed equilibrium strategy- oriented innovative approach toward the coal–water conflict.

3.4.2 Data Collection

Detailed data for the research region were obtained from the Statistical Yearbook of Chinese coal industry [The Statistical Yearbook of Chinese Coal Industry, 2013], the Statistical Yearbook of Chinese energy [The Statistical Yearbook of Chinese Energy, 2013], and field research. Certain parameters taken from the Statistical Yearbooks and data published by the companies are shown in Table. 3.1.

Table 3.1 Crisp parameters in the proposed model.

	SP_{ij} (yuan)	O_{ij}^{U} (tonnes)	T_{ij} (yuan)
Jn colliery 3♯	795	320	198
Jn colliery 15-1♯	805	400	203
Jn colliery 15-2♯	810	225	215
Qs colliery 3♯	735	145	190
Qs colliery 15-1♯	760	140	199
Qs colliery 15-2♯	775	205	205
Bl colliery 3♯	740	170	220
Bl colliery 15-1♯	760	285	223
Bl colliery 15-2♯	775	345	237
Sbd colliery 3♯	790	65	200
Sbd colliery 15-1♯	795	85	215
Sbd colliery 15-2♯	815	115	223

As the fuzzy parameter $\widetilde{E_i}$ is calculated using the fuzzy parameter $\widetilde{E_{ij}}$, so from the data shown in Table 3.2, we can determine the fuzzy parameter $\widetilde{E_i}$, details for which are shown in Table 3.3.

Besides the detailed data for each coal seam at each colliery, some further parameters based on the whole coal mining industry were still needed, which were sourced from the Chinese Coal Industry Development Plan in the Twelfth Five-Year Plan National Energy Administration [2010] and the enterprise development plan, as shown in Table 3.4. The parameter U_i represents the mining drainage threshold amount for each colliery authorized by the authority. However, different authorities may have different attitudes toward environmental protection, so it is difficult to determine an exact threshold value U_i. Even though the attitudes of the various authorities are different, their intentions to protect the environment are consistent. Therefore, in this chapter, without loss of generality, we set the threshold value based on historical data. That is to say, we set the threshold value at 90% of the actual mining drainage amount from the last production cycle of each colliery, so the value of U_i was determined, as shown in Table 3.5.

3.4.3 Results for Different Scenarios

In this section, based on the different water quality control level chosen by the authority, we present the corresponding optimization results using the proposed Stackelberge-Nash equilibrium - multiple coal seam mining programming (SN-MCMP) model.

3.4.3.1 Scenario 1: Water Quality Standards I

When the authority stipulates the water quality standards at level I, the sewage threshold φ^1 and φ^2 are set at 70 and 100 mg/l, respectively. Under this scenario, the solution to the proposed model with a confidence of $\gamma = 1$ is presented in Table 3.6, where $\gamma = 1$ indicates the regional authority's environmental protection constraint that must be satisfied. This situation indicates that the regional authority has such an absolutely conservative environmental management strategy (Table 3.7).

3.4.3.2 Scenario 2: Water Quality Standards II

When the authority sets the water quality standards at level I, the sewage threshold φ^1 and φ^2 are set at 100 and 150 mg/l, respectively. Under this scenario, the solution to the proposed model under a confidence level of $\gamma = 1$ is presented in Table 3.8, which also shows the optimization results for each sub-colliery.

The findings of the sensitivity analysis under the two scenarios are given in Tables 3.7 and 3.9, which shows the results of three candidates under confidence levels $\gamma = 0.9$, $\gamma = 0.8$, and $\gamma = 0.7$, respectively, where γ expresses the authority's environmental protection attitude, and $\gamma = 0.9$ shows that there is a 0.9 confidence that the total mining drainage in the region is less than H. Compared with the set maximum allowed amount of mining drainage at certain value, to control it less than H to some degree seemed to be more reasonable due to the inherent uncertainty of such decision problem. With a decrease in the value γ, the attitude of

Table 3.2 Fuzzy parameters in the proposed model.

	$\widetilde{\theta_{ij}^1}$ (mg/l)	$\widetilde{\theta_{ij}^2}$ (mg/l)	$\widetilde{E_{ij}}$ (m^3/tonnes)
Jn 3♯	(134.59, 137.75, 140.50, 142.35)	(144.78, 146.25, 149.45, 150.65)	(0.95, 0.99, 1.02, 1.05)
Jn 15-1♯	(144.70, 148.25, 149.35, 153.10)	(155.85, 157.90, 160.15, 162.30)	(0.99, 1.02, 1.07, 1.16)
Jn 15-2♯	(164.85, 167.95, 170.20, 172.15)	(177.85, 181.35, 182.95, 185.00)	(1.08, 1.15, 1.17, 1.21)
Qs 3♯	(115.35, 116.75, 119.20, 122.05)	(145.25, 147.73, 148.85, 150.20)	(0.95, 0.98, 1.01, 1.04)
Qs 15-1♯	(127.75, 130.05, 132.45, 134.10)	(152.15, 153.35, 157.45, 158.55)	(1.02, 1.05, 1.08, 1.19)
Qs 15-2♯	(140.15, 142.35, 145.55, 147.05)	(167.95, 170.15, 173.80, 177.05)	(1.12, 1.14, 1.17, 1.21)
Bl 3♯	(115.95, 117.05, 118.65, 120.95)	(140.15, 143.05, 146.50, 148.90)	(0.99, 1.02, 1.03, 1.07)
Bl 15-1♯	(123.95, 126.58, 127.98, 130.05)	(152.15, 154.30, 155.30, 157.65)	(0.99, 1.05, 1.11, 1.16)
Bl 15-2♯	(134.80, 137.15, 140.45, 143.85)	(164.15, 167.55, 171.05, 172.15)	(1.13, 1.15, 1.24, 1.28)
Sbd 3♯	(124.95, 126.85, 128.75, 131.05)	(135.00, 135.20, 135.60, 135.80)	(0.96, 1.00, 1.03, 1.04)
Sbd 15-1♯	(140.35, 142.95, 144.15, 146.95)	(141.05, 141.35, 142.65, 143.90)	(0.99, 1.04, 1.10, 1.18)
Sbd 15-2♯	(154.95, 157.85, 160.05, 162.40)	(153.80, 154.90, 155.35, 156.50)	(1.12, 1.15, 1.17, 1.21)

Table 3.3 The data of $\widetilde{E}_i$.

	$\widetilde{E_i}$ (m^3/tonnes)
Jn mine	(1.07, 1.11, 1.16, 1.18)
Qs mine	(0.98, 1.01, 1.12, 1.16)
Bl mine	(1.00, 1.08, 1.25, 1.28)
Sbd mine	(1.03, 1.05, 1.12, 1.17)

Table 3.4 Other input crisp parameters of the proposed model.

S	H (10^4 tonnes)	R^U (10^4 tonnes)	D^L (10^4 tonnes)
0.17	2500	2875	2210

Table 3.5 Crisp parameters of each collieries in the research region.

	SP_i (yuan)	N_i^1 (mg/l)	N_i^2 (mg/l)	P_i (yuan)
Jn colliery	805	70	50	2.25
Qs colliery	755	60	60	2.00
Bl colliery	765	60	60	2.15
Sbd colliery	790	150	120	2.00

Table 3.6 Result of SN-MCMP under confidence level $\gamma = 1$, water quality I.

Total benefit: $F(PA_{ij}, MQ_i)$ (10^4 yuan)	Colliery	Total mining quota: MQ_i (10^4 tonnes)	Coal seam 1: PA_{i1} (10^4 tonnes)	Coal seam 2: PA_{i2} (10^4 tonnes)	Coal seam 3: PA_{i3} (10^4 tonnes)	Benefit: $G_i(PA_{ij}, MQ_i)$ (10^4 yuan)
301 859.61	Jn colliery	605.52	200.00	400.00	6.52	271 260.00
	Qs colliery	388.69	120.00	140.00	128.69	167 300.00
	Bl colliery	678.05	160.00	300.00	218.05	280 850.00
	Sbd colliery	177.07	60.00	70.00	47.07	78 957.00
	Total	2265.71				

Table 3.7 Results of SN-MCMP under different confidence levels, water quality I.

γ	**Total benefit: $F(PA_{ij}, MQ_i)$ (10^4 yuan)**	**Colliery**	**Total mining quota: MQ_i (10^4 yuan)**	**Coal seam 1: PA_{i1} (10^4 tonnes)**	**Coal seam 2: PA_{i2} (10^4 tonnes)**	**Coal seam 3: PA_{i3} (10^4 tonnes)**	**Benefit: $G_i(PA_{ij}, MQ_i)$ (10^4 yuan)**
0.9	320 122.57	Jn	626.08	200.00	400.00	26.08	279 990.00
		Qs	405.00	120.00	140.00	145.00	174 350.00
		Bl	708.54	160.00	300.00	248.54	293 670.00
		Sbd	213.24	60.00	70.00	83.24	95 363.00
		Total	2402.42				
0.8	330 995.23	Jn	643.47	200.00	400.00	43.47	287 750.00
		Qs	417.93	120.00	140.00	157.93	180 020.00
		Bl	726.72	160.00	300.00	266.72	301 320.00
		Sbd	231.26	60.00	70.00	101.26	103 540.00
		Total	2483.92				
0.7	337 424.92	Jn	658.31	200.00	400.00	58.31	294 370.00
		Qs	430.86	120.00	140.00	170.86	185 640.00
		Bl	730.60	160.00	300.00	270.60	302 950.00
		Sbd	247.91	60.00	70.00	117.91	111 090.00
		Total	2532.22				

Table 3.8 Result of SN-MCMP under confidence level $\gamma = 1$, water quality level II.

Total benefit: $F(PA_{ij}, MQ_i)$ (10^4 yuan)	**Colliery**	**Total mining quota: MQ_i (10^4 tonnes)**	**Coal seam 1: PA_{i1} (10^4 tonnes)**	**Coal seam 2: PA_{i2} (10^4 tonnes)**	**Coal seam 3: PA_{i3} (10^4 tonnes)**	**Benefit: $G_i(PA_{ij}, MQ_i)$ (10^4 yuan)**
311 218.87	Jn colliery	626.08	200.00	400.00	26.08	279 990.00
	Qs colliery	392.00	120.00	140.00	132.00	168 733.20
	Bl colliery	690.00	160.00	300.00	230.00	285 961.00
	Sbd colliery	195.00	60.00	70.00	65.00	87 087.50
	Total	2335.25				

Table 3.9 Results of SN-MCMP under different confidence levels, water quality II.

γ	Total benefit: $F(PA_{ij}, MQ_i)$ (10^4 yuan)	Colliery	Total mining quota: MQ_i (10^4 yuan)	Coal seam 1: PA_{i1} (10^4 tonnes)	Coal seam 2: PA_{i2} (10^4 tonnes)	Coal seam 3: PA_{i3} (10^4 tonnes)	Benefit: $G_i(PA_{ij}, MQ_i)$ (10^4 yuan)
0.9	333 523.70	Jn	652.17	200.00	400.00	52.17	291 640.00
		Qs	435.00	120.00	140.00	175.00	187 520.00
		Bl	726.72	160.00	300.00	266.72	301 320.00
		Sbd	222.25	60.00	70.00	92.25	99 449.00
		Total	2503.09				
0.8	351 742.50	Jn	678.26	200.00	400.00	78.26	303 280.00
		Qs	465.34	120.00	140.00	205.34	200 650.00
		Bl	754.00	160.00	300.00	294	312 800.00
		Sbd	249.27	60.00	70.00	119.27	111 710.00
		Total	2639.01				
0.7	358 236.82	Jn	686.95	200.00	400.00	86.95	307 160.00
		Qs	480.00	120.00	140.00	220.00	207 020.00
		Bl	772.18	160.00	300.00	312.18	320 440.00
		Sbd	250.00	60.00	70.00	120.00	112 040.00
		Total	2689.13				

the authority toward environmental protection is more relaxed, that is, the larger of the value γ, the more conservative the environmental protection attitude.

3.5 Discussion

The current study balanced the coal–water conflict and aimed to improve the water environmental quality in the large-scale coal fields. A studied area is Yulin in this chapter. The policy implications for Yulin are currently being discussed by researchers and policy makers. Meanwhile, in this chapter, some managerial insights based on the above quantitative analysis are given as follows.

3.5.1 Propositions and Analysis

First of all, based on the results solved by the proposed model, we find that the collieries still engage in full load mining in the more environmentally friendly coal seams to compete for a higher mining quota, even though the unit revenue is smaller. Hence, we conclude that mining quota competition mechanism enables environmental friendly exploitation, in the future, setting a mining quota allocation mechanism is suggested for authorities.

Second, in addition to the mining quota allocation mechanism, water quality control is found to be another efficient way to achieve at environmental friendly exploitation. Based on the quantified results, we find decreasing trends of total coal mining industry financial revenue and total mining drainage by 3%, when the water quality control is set at level I ($\gamma = 1$). A main reason derives from the proposed model. As we know, the optimal solution is found in the boundary or the vertex of the feasible solution space, which is largely influenced by the constraints. Hence, when the water quality standard changes, the environmental protection constraint changes correspondingly. Therefore, when such a constraint changes, the revenue and total mining quota also changes, which demonstrates that water quality control constraints also play an important role in guiding the collieries to engage in environmentally friendly exploitation. Similar results can also be found when the authority sets $\gamma = 0.9$, $\gamma = 0.8$, and $\gamma = 0.7$, respectively.

Third, collieries have different sensitivities toward changing water quality standards. By comparing the detailed results for each sub-colliery under the two scenarios, we can see when the authority intensifies the water quality control level from level II to level I ($\gamma = 1$), the mining quota at the five sub-collieries decrease with by 3.2%, 0.9%, 1.8%, 9.1%, and 3.7%, respectively, with the Sbd colliery being the most sensitive one. This shows that the collieries' sensitivities to changes in the water quality standard are not all the same. The reason for this difference is that when the water quality standards change by a certain degree, the smaller the unit concentration of pollutants, the larger the floatation mining quota the colliery can get using the proposed model. Note that in this case, Sbd colliery is the most sensitive one to both water quantity control and water quality control; however, in other cases, one colliery may not showing such intense sensitivity to the two criteria at the same time.

Fourth, relatively relaxed environmental protection constraints increase the revenue, but cause more damage to the environment. When the authority sets the environmental protection confidence at $\gamma = 0.9$, as in scenario 2, the total financial revenue of the regional coal mining industry achieves a 7.1% growth. However, at the same time, the mining drainage amount is 221.26×10^4 tonnes more than when $\gamma = 1$, which is a 9.2% increase. The increase in financial revenue (7.1%) is lower than the total mining drainage (9.2%). Similar conclusions can also be found when $\gamma = 0.8$ and $\gamma = 0.7$ in both scenarios 1 and 2, which shows that when the authority has a relatively relaxed environmental protection policy, both revenue and the total mining drainage increase and the growth proportion of the former is lower than the latter. The reason for this is that when the authority sets the mining quotas allocation mechanism using the proposed model, the collieries engage in full exploitation in the most environmentally friendly coal seams, and when the environmental protection constraints are relaxed, the mining quotas at each sub-colliery increase. The increasing part will be allocated to those coal seams with a higher drainage coefficient, which results in a higher economic benefit, but, due to the larger drainage coefficient, the total mining drainage will increases in a higher ratio.

Finally, due to the different geological structures and production processes, the sensitivity to different environmental protection constraints between any two collieries may be different, not only in this case but also in a wider range. Scenarios are defined in this chapter by relaxing the environmental protection confidence (i.e. from 1 to 0.9), the quantified results show that collieries have different sensitivities toward different environmental protection constraints. We analyze the reason is the difference in geological structures and production processes. Due to this reason, the unit drainage coefficients of each colliery are not the same. At the meantime, we find that the collieries with a smaller unit drainage coefficient show more sensitivities toward a change of environmental protection constraints. Indeed, by analyzing the proposed model, we can see that when γ changes, the smaller $\widetilde{E_i}$ is, the more MQ_i can be changed. This is why the collieries with smaller unit drainage coefficient are more sensitive to changes in environmental protection constraints.

3.5.2 Management Recommendations

Based on what has been discussed and analyzed earlier, in this section, we give some management recommendations.

First of all, in large-scale coal fields, mining quota allocation mechanisms based on environmental protection should be established. Without a mining quota allocation scheme, sub-collieries would extensively exploit the coal that would lead to serious environmental damage. Using the proposed model, a quota compete mechanism can be built. As discussed earlier, such a mechanism can guide sub-collieries to conduct environmentally friendly exploitation. Further, under environmental constraints but seeking as much quota as possible, sub-collieries may seek to improve their mining technology to decrease the unit mining drainage.

Second, the authority can select their own mining quota allocation mechanism using the model; that is to say, the authority can fully consider their actual situation and choose a desired environmental protection confidence γ and water quality standard. However, as discussed above, a relaxed environmental protection strategy brings higher marginal environmental costs and more serious environmental damage. Therefore, we recommend that for developed regions, the authority should have the strictest attitude when using the model (i.e. set $\gamma = 1$ and choose the highest water quality standard). For developing regions, the authority can gradually tighten up the environmental protection attitude. In other words, the authority in developing regions should have a dynamic attitude when using the model to solve the coal–water conflict. For example, to encourage economic development, the authority can set $\gamma = 0.8$ and choose a relatively lower water quality standard; however, after a period of time, they should change their attitude toward the environmental protection and upgrade the requirements (i.e. set $\gamma = 0.9$ and choose a higher water quality standard). In the next period, the requirements should be upgraded again, finally reaching the most strict environmental protection requirements.

References

Banks, D., Younger, P.L., Arnesen, R.T. et al. (1997). Mine-water chemistry: the good, the bad and the ugly. *Environmental Geology* 32 (3): 157–174.

Ben-Ayed, O. and Blair, C.E. (1990). Computational difficulties of bilevel linear programming. *Operations Research* 38 (3): 556–560.

Colson, B., Marcotte, P., and Savard, G. (2005). Bilevel programming: a survey. *4OR* 3 (2): 87–107.

Dubois, D. and Prade, H. (1983). Ranking fuzzy numbers in the setting of possibility theory. *Information Sciences* 30 (3): 183–224.

Falcon, R.M.S. (1989). Macro-and micro-factors affecting coal-seam quality and distribution in southern Africa with particular reference to the No. 2 seam, Witbank coalfield, South Africa. *International Journal of Coal Geology* 12 (1): 681–731.

Hanson, M.A. (1981). On sufficiency of the Kuhn–Tucker conditions. *Journal of Mathematical Analysis and Applications* 80 (2): 545–550.

Johnson, D.B. (2003). Chemical and microbiological characteristics of mineral spoils and drainage waters at abandoned coal and metal mines. *Water, Air, & Soil Pollution: Focus* 3 (1): 47–66.

Liu, B. and Iwamura, K. (1998). Chance constrained programming with fuzzy parameters. *Fuzzy Sets and Systems* 94 (2): 227–237.

Lu, J., Shi, C., and Zhang, G. (2006). On bilevel multi-follower decision making: general framework and solutions. *Information Sciences* 176 (11): 1607–1627.

Lu, J., Shi, C., Zhang, G., and Dillon, T. (2007). Model and extended Kuhn–Tucker approach for bilevel multi-follower decision making in a referential-uncooperative situation. *Journal of Global Optimization* 38 (4): 597–600.

National Energy Administration (2010). Chinese Coal Industry Development Plan in the Twelfth Five-Year Plan. *Technical report*. National Energy Administration.

Shi, C., Lu, J., and Zhang, G. (2005). An extended Kuhn–Tucker approach for linear bilevel programming. *Applied Mathematics and Computation* 162 (1): 51–63.

State of Environmental Protection Agency (2005). Coal Industry Emission Standards. *Technical report*. State of Environmental Protection Agency.

The Statistical Yearbook of Chinese Coal Industry (2013). National Bureau of Statistics of the Peoples Republic of China. *Technical report*. The Statistical Yearbook of Chinese Coal Industry.

The Statistical Yearbook of Chinese Energy (2013). National Bureau of Statistics of the Peoples Republic of China. *Technical report*. The Statistical Yearbook of Chinese Energy.

Ward, C.R. (2001). Analysis and significance of mineral 799 matter in coal seams. *International Journal of Coal Geology* 50 (1): 135–168.

Xu, J. and Zhou, X. (2011). *Fuzzy-like Multiple Objective Decision Making*, vol. 4. Heidelberg: Springer-Verlag.

Zadeh, L.A. (1965). Fuzzy sets. *Information and Control* 8 (3): 338–353.

4

Seasonal Changes-Oriented Dynamic Strategy Toward Coal–Water Conflict Resolutions

As the first source in energy structure, coal accounts for about 68.7% of the total domestic energy consumption. This energy structure will be maintained in the near future [Hu and Xiao, 2013, Shen et al., 2012, Li and Leung, 2012]. In recent years, due to the increasing energy demands pressure of energy demand and considerable profits of the coal industry, many coal fields in China have been exploited in large quantities without sufficient control, which has caused serious damage to the local ecological environment, especially the underground aquifer [Younger and Wolkersdorfer, 2004, Yuan et al., 2008, Feng et al., 2009]. However, water resource is important to the comprehensive development of the coal field, the destruction of the aquifer results in a large amount of groundwater discharged into the coal mine which leads to the so-called water coal conflict. Many studies have pointed out that the contradiction between water and coal has seriously hindered the sustainable development of large coal fields [Younger, 2001, Wolkersdorfer and Bowell, 2005, Glauser et al., 2005, Mudd, 2008, Bian et al., 2009, Sun et al., 2012, Silva et al., 2013]. In recent years, with the rapid development of the Chinese economy, coal consumption is also growing rapidly, and the coal–water conflict now becoming increasingly serious. Therefore, the sustainable development of large-scale coal fields is facing more and more threats [Chen and Xu, 2010, Yuan et al., 2014].

4.1 Background Expression

To deal with this issue, there has been a worldwide multilateral effort, especially policies and regulations. For example, the Chinese government has issued the Mineral Resources Law of the People's Republic of China and the Environmental Protection Law of the People's Republic of China. However, due to information asymmetry and other objective reasons, many policies and laws formulated from a macro-perspective are often inflexible in different fields and conditions. Therefore, to determine the precise constraints for individual collieries in different regions is always extremely difficult. Many scholars have studied to solve this problem. Baker developed a simulation model based on a created wetland to reduce and treat coal mine drainage and applied this proposed method in a real world case in the United States [Baker et al., 1991]. Younger and Wolkersdorfer designed a decision

Innovative Approaches towards Ecological Coal Mining and Utilization, First Edition.
Jiuping Xu, Heping Xie, and Chengwei Lv.

framework for long-term mine water management at the catchment scale based on an Environmental Information System [Younger and Wolkersdorfer, 2004]. Rapantova et al. analyzed the features and problems of numerical modeling applications in mining environment and then presented three modeling case studies [Rapantova et al., 2007]. Qiao et al. established a three-dimensional groundwater flow model and then applied it to several scenarios to explore the quantitative influence of mining activities on the water environment [Qiao et al., 2011]. More recently, Xu et al. proposed a multi-coal seam mining-based bilevel programming model to solve the coal–water conflict at a large-scale coal field with full consideration of economic development and environmental protection [Xu et al., 2015a]. These excellent researches had different perspectives to protect the groundwater resources in coal fields and had already reached some achievements. However, in the specific practical problem, the coal–water conflict has still not been fully solved and further development is necessary.

In fact, the key factor to solve the coal–water conflict is to formulate some rules and plans for the resource development while fully considering economic development and social security, so as to prevent extensive production without planning. That is, methods which allow for the equilibrium between coal exploitation and environmental protection are critical in solving such conflicts. Equilibrium strategies have been regarded as a powerful tool to solve such problems. In fact, equilibrium strategies have been applied to solve conflicts in many other fields and have already achieved some significant results. Based on equilibrium strategy, Labbe et al. proposed a bilevel model to optimize highway pricing with full consideration of the conflict between the highway authority and the users [Labbé et al., 1998]. Similarly, Kalashnikov et al. proposed a stochastic bilevel programming model to solve the problem of unbalanced natural gas arbitrage [Kalashnikov et al., 2010]. Angulo et al. used a kind of equilibrium optimization method to solve the contradiction among the social, economic, and environmental factors in the problem of transportation network expansion and promotes the development of traffic network problems [Angulo et al., 2014]. Further, Xu et al. recently proposed a tripartite equilibrium optimization method for carbon emission allowance allocations in the power-supply industry [Xu et al., 2015b]. These excellent previous research results have inspired the study of this chapter, which try to establish an optimization method based on equilibrium strategy to solve the coal–water conflict, so as to ensure the sustainable development of coal fields in China.

To solve the coal–water conflict by equilibrium optimization, several barriers need to be overcome. First, full consideration needs to be given to all the rights of all stakeholders, who are often in competition with each other. This chapter uses a bilevel programming model with one leader and multiple followers to describe the trade-offs between the stakeholders. In addition, the changing seasons have significant impact on coal mining activity, so this aspect must be considered within the equilibrium strategy in the search for solution of the coal–water conflict. Further, because almost all coal mining activities are uncertain and the environmental impact of coal mining is difficult to accurately measure, the planning method with crisp parameters is not suitable for such problems. Therefore, uncertainty theory must be integrated into the bilevel programming. On this basis, an equilibrium strategy-based bilevel programming method in an uncertain environment is proposed to solve the coal–water conflict and to explore the sustainable development of large-scale coal fields in China.

4.2 Methodology

This section introduces some basic background and description, which lays the foundation for the development of groundwater damage optimization method based on equilibrium strategy.

4.2.1 Key Problem Statement

Generally speaking, there are three categories of the stakeholders whose decisions can affect the development of large-scale coal fields: the authority, the collieries, and influential independent organizations (i.e. environmental protection organizations), within which there are both simultaneous conflicts and cooperation. The regional authority, who acts on behalf of the public, has relative priority in determining the development plan for the entire coal field. Under the dual pressure of energy demands and the environmental damage limitation, the authority has an obligation to formulate a scientific and reasonable coal mining plan to balance both of them to achieve the aim of sustainable development. Therefore, when makes decisions, the authority must consider both economic development and environmental protection. The main purpose is to ensure the sustainable development of local economy within the scope of environmental sustainability. Each coal mine is independent of each other and pursues the maximum profit. As an enterprise, the first goal is to obtain profit. Therefore, under the incentive of huge potential interests, all coal mines try to fully develop the mine, and almost no consideration is given to environmental protection. Although this kind of profit-seeking behavior helps the collieries to obtain the maximum possible benefits, it is not conducive to the long-term sustainable development of the whole coal field because of the serious environmental damage caused by it. The main ways for independent organization to influence coal field development is to supervise the behaviors of the authority and the collieries for the public. And they need to regulate two core supervisory areas, one is whether the development plan of the coal field is environmental friendly, the other is whether the plan is equitable for each colliery. The action of these three stakeholders is crucial for the sustainable development of the coal field. However, it is difficult to describe the relationship between them with mathematical model of quantitative research. This chapter focuses on the trade-offs and cooperation between the authority and the coal mining enterprises. In other words, the actions of the authorities and the coal mines have been affirmed by independent organizations.

The main methods for the authorities to guide the overall direction of coal field development have been the allocation of mining quotas to each colliery. In fact, the authority can establish a mining quota competition scheme based on an environmental protection to avoid extensive colliery exploitation. In addition, since coal mining activities are very sensitive to the seasonal changes, especially from an environmental point of view, the collieries can determine suitable production plan for each season to improve the overall environmental protection performance and compete for mining quota as much as possible. The mining plan of each mine is fed back to the authorities, who adjust the plan according to the performance of each mine. After a further adjustment, the mining quota allocation scheme is returned to collieries, and the coal mine changed its mining plan based on the new quota. Both the authority and all collieries will obtain an acceptable final scheme after repeated this process several times. This kind of decision-making is similar to a Stackelberg game, which can be abstracted as bilevel programming method with the authority and the collieries. The authority, as the leader, has the right to establish an environmental protection based on mining quota competition scheme, in which those collieries that have a better attitude toward environmental protection gain greater mining quota allocations. And the colliery, as the followers, must improve its mining plan to achieve more environmental friendly exploitation so as to be able to gain a greater share of the mining quota. Because this leader–follower behavior between the two decision-maker levels is a Stackelberg game, an equilibrium solution between the authority and the collieries can be determined. The structure of this bilevel programming is shown in Figure 4.1.

In addition, the uncertainty in the coefficient of drainage per tonnes of coal is also a concern when the local authorities make a preliminary decision on the mining quota of the coal mine. However, it is difficult for the regional authority to obtain specific data for each colliery on the per tonnes coal drainage coefficient, so the definition of these data is vague. Fuzzy theory established by Zadeh has been proved to be a powerful tool to solve such problems [Zadeh, 1965]. Therefore, this chapter uses fuzzy set theory and related calculation methods to determine the best method to deal with this fuzzy environment.

4.2.2 Modeling

A mathematical description for the above-mentioned issue is given in this section.

4.2.2.1 Assumption

(1) This is a single production period decision problem; therefore, at the beginning of the next production period, the decision progress will be reset.
(2) Mining quota transactions are not considered here.
(3) It is assumed that the behavior of the authority and the collieries have been affirmed by the independent organizations.

4.2.2.2 Notations

The following symbols are used in this chapter.

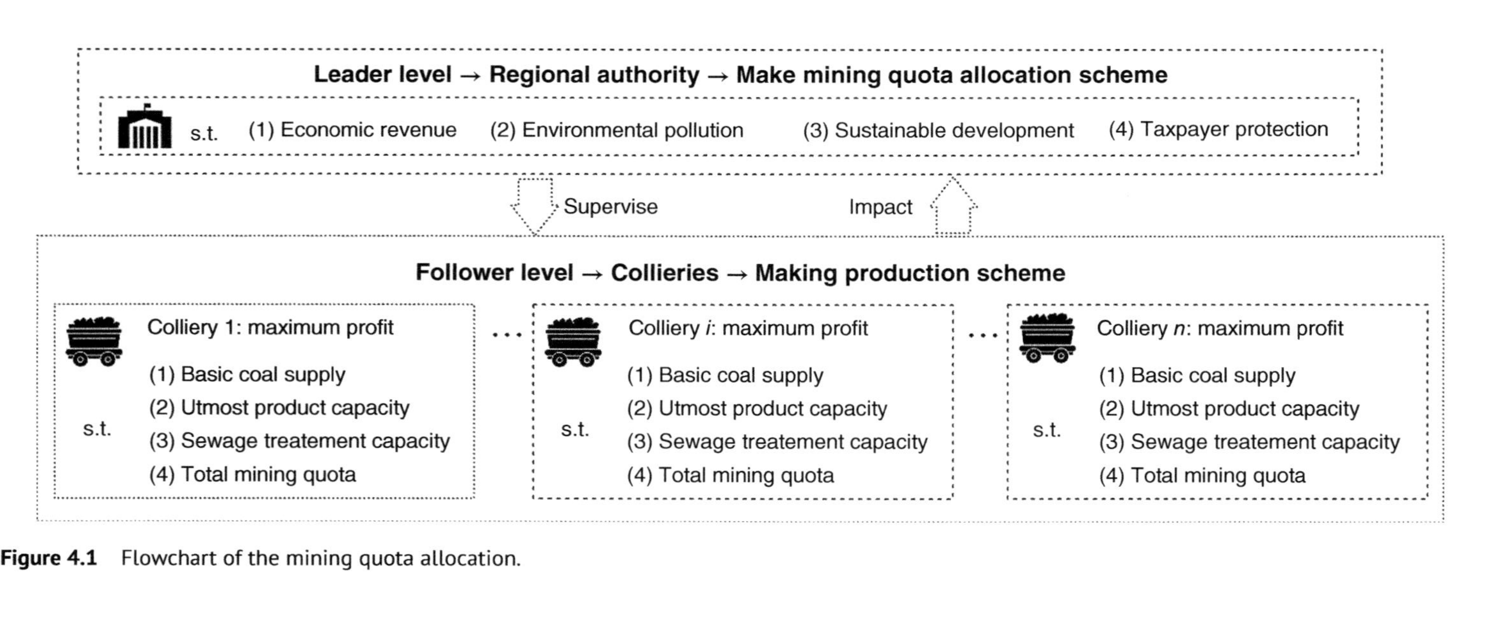

Figure 4.1 Flowchart of the mining quota allocation.

Indices

i = Collieries index, $i = 1, 2, \ldots, n$, and
j = Season index, $j = 1, 2, \ldots, m$;

Parameters

R^u = Recoverable amount of coal in the region;
Q_i = Production quantity to maintain the basic operations at colliery i;
D^L = Lowest amount of coal that the region should supply;
V_i = Satisfaction degree of colliery i toward the allocated mining quota
C_{ij} = Price of coal products at colliery i in the jth season;
C_i = Average price of coal produced at colliery i
T_{ij} = Total cost per tonne of coal when colliery i produces in season j;
D_{ij} = Demand coal production at colliery i in season j
O_{ij}^U = Maximum mining capacity at colliery i for season j;
S_1 = Colliery income tax rate;
S_2 = Water environment protection tax rate;
S_3 = Resource exploitation tax rate;
H = Maximum amount of groundwater allowed to drain in the coal field;
P_i = Sewage treatment unit costs at colliery i;
B_{ij} = Highest waste water processing ability at colliery i in season j;
β = Minimal satisfaction degree chosen by the regional authority;
γ = Environmental protection confidence level chosen by the regional authority;
λ = The optimistic–pessimistic parameter of the decision-maker, which reflects its attitude toward the trapezoidal fuzzy number, which in this study it is set at 0.5, indicating that the authority has a neutral attitude toward the fuzzy data;
$r[\bullet]$ = The parameters in the trapezoidal fuzzy number membership function.

Uncertain parameters

$\widetilde{E}_{ij}$ = Per tonne coal drainage coefficient when colliery i produces in season j;
$\widetilde{E}_i$ = Per tonne coal drainage coefficient for exploitation at colliery i;

Decision variables

Y_i = Quantity that the government allows colliery i to exploit;
X_{ij} = Quantity that colliery i exploits from season j.

4.2.2.3 Logical Representation for the Collieries

The collieries are independent decision-makers and individually pursue the largest possible profits under the mining quota constraints proposed by the government. The objective constraints that each colliery must meet are production capacity, basic supply security, mining quota limitations, and sewage treatment constraints. The nonnegative constraints also need to be satisfied.

4.2.2.3.1 Maximize Total Benefit

For the market-based collieries, the pursuit of the highest profit, which comes mainly from coal production and sales, is priority when the managers make decisions. The taxes imposed on the collieries can be roughly divided into three categories: income taxes, resource exploitation taxes, and environmental protection taxes. Colliery income tax is charged based on a fixed percentage of sales revenue, so in this chapter let S_1 represents the income tax rate and C_{ij} be the per tonne price of coal produced in season j at colliery i. Therefore, the total income tax that colliery i should pay is $S_1 \sum_{j=1}^{m} C_{ij}X_{ij}$, where X_{ij} is the quantity that colliery i exploits in season j. In China, the resource exploitation taxes imposed on collieries depends on the resource mining quota allocation from the local authority. Let S_3 be the resource exploitation tax and Y_i be the mining quota for colliery i, then the total resource exploitation tax that the colliery should pay is S_3Y_i. The environmental protection tax is based on the waste emission quantities at each colliery. In this chapter, groundwater environmental protection is the concern, so let S_2 be the environmental groundwater protection tax rate, so the total environmental groundwater protection tax that colliery i should pay can be expressed as $S_2 \sum_{j=1}^{4} \widetilde{E_{ij}}X_{ij}$.

The costs of each colliery are mainly made up of production costs and sewage treatment costs. Let T_{ij} be the total cost per tonne of coal when colliery i produces in season j, so the production cost of colliery i can be described as $\sum_{j=1}^{m} T_{ij}X_{ij}$. The Chinese government stipulates that coal mining sewage must be processed to a certain standard before drainage [Chen and Xu, 2010, You et al., 2010]. Therefore, the collieries must pay for this provision, which is another cost which affects the collieries total profit. The total sewage treatment cost at colliery i is determined by the per unit sewage treatment cost (P_i), the per tonne coal drainage coefficient ($\widetilde{E_{ij}}$), and the quantity of coal that colliery i exploits (X_{ij}), which can be written as $P_i \sum_{j=1}^{m} \widetilde{E_{ij}}X_{ij}$. As the parameters $\widetilde{E_{ij}}$ are more difficult to determine because of many factors such as rainfall, operating methods, and geological structure, to determine this information, interviews with engineers and workers are necessary [Younger and Wolkersdorfer, 2004]. Through these interviews, the value is estimated to be in a region with the minimum value represented by r^1 and the maximum value represented by r^4 and the most possible value being between r^2 and r^3, $(r^1 \leq r^2 \leq r^3 \leq r^4)$. For this reason, fuzzy theory has been selected as it can deal with such uncertain situations, and it can be described as a trapezoidal fuzzy number $\widetilde{E_{ij}} = (r_{ij}^1, r_{ij}^2, r_{ij}^3, r_{ij}^4)$, where $r_{[\bullet]}$ are the parameters in the membership function of the trapezoidal fuzzy number $\widetilde{E_{ij}}$ [Grzegorzewski, 2008, Xu and Zhou, 2011]. Figure 4.2 shows a flowchart for the construction of the fuzzy membershipfunctions.

As there are trapezoidal fuzzy parameters in the objective functions, the objectives are classes of alternatives where the boundaries are not sharply defined, making these difficult to calculate [Liu and Liu, 2002, Ganesan and Veeramani, 2006]. As such, this chapter introduces the fuzzy measure *Me*, which embeds an optimistic–pessimistic parameter to determine the combined attitudes of the decision-makers [Xu and Zhou, 2011]. Based on the method proposed by Xu and

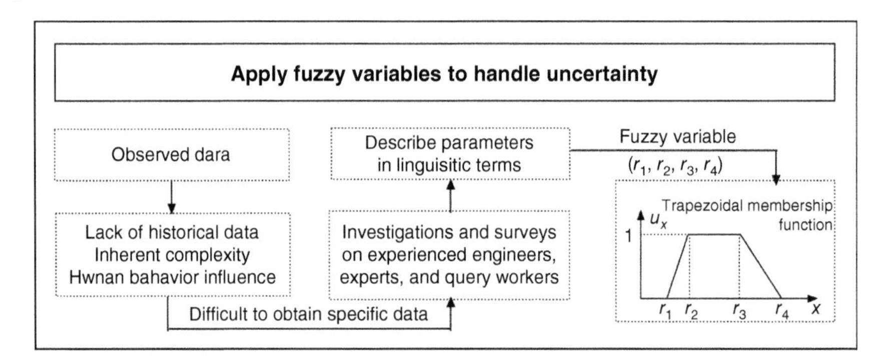

Figure 4.2 Flowchart for the construction of the fuzzy membership functions.

Zhou Xu and Zhou [2011], the fuzzy variable $\widetilde{E_{ij}}$ can be transformed into a crisp form as follows:

$$\widetilde{E_{ij}} \rightarrow E[\widetilde{E_{ij}}] = \frac{1-\lambda}{2}\left(r_{ij}^1 + r_{ij}^2\right) + \frac{\lambda}{2}\left(r_{ij}^3 + r_{ij}^4\right) \tag{4.1}$$

Therefore, the objective function can also be written as:

$$\max G_i = \sum_{j=1}^{m} [(1 - S_1)C_{ij} - T_{ij}]X_{ij} - P_i \sum_{j=1}^{m} E[\widetilde{E_{ij}}]X_{ij} - S_2 \sum_{j=1}^{m} E[\widetilde{E_{ij}}]X_{ij} - S_3 Y_i \tag{4.2}$$

4.2.2.3.2 Colliery Limitations

Since each colliery is influenced by the government and the market, there are some conditions that should be satisfied.

(1) *Mining quota restrictions*: As the mathematical logic of the model is actually a leader–follower game, on the subordinate side, the collieries must make their own decisions under the restrictions of the authority. Therefore, the total mining quantity from each season in each colliery should not exceed the quota that the authority allocated. Let Y_i be the quota that the authority allocates to colliery i, so the mining quota constraint for each colliery is as follows:

$$\sum_{j=1}^{m} X_{ij} \leq Y_{ij} \tag{4.3}$$

(2) *Mining quantity restrictions*: For each colliery, take colliery i as an example, its mining ability is limited which cannot be quickly improved. Therefore, when colliery i makes a decision about the quantity that should mined in season j, this quantity must not exceed its mining ability O_{ij}^U. On the other side, collieries need to have some social responsibility as their production is the basic energy for China. Therefore, the decision of colliery on the quantity that should be produced in seam j should meet the basic demand of the market (represented by D_{ij}), so the mining quantity constraint is as follows:

$$D_{ij} \leq X_{ij} \leq O_{ij}^U \tag{4.4}$$

(3) *Sewage treatment capacity*: Sustainable development in coal fields is affected by the high concentrations of pollutants in the mining drainage pollution [Jarvis and Younger, 2000, Wolkersdorfer and Bowell, 2005]. The Chinese government has enacted the Environmental Protection Law of the People's Republic of China which includes provisions on mining sewage pollution restrictions. This law states that collieries which do not meet these provisions will be punished and even be deprived of the right to produce. Therefore, mining drainage from colliery i in season j cannot exceed its sewage treatment capacity (represented by B_{ij}) in each season. Also use the method which was proposed by Xu and Zhou Xu and Zhou [2011], and this equation can be written as follows:

$$E[\widetilde{E_{ij}}]X_{ij} \leq B_{ij} \tag{4.5}$$

4.2.2.4 Logical Representation for the Authority

In China, the government acts as the authority and has the higher decision-making priority and, so is in the leadership position when developing the comprehensive regional development plan. At the leader level, to achieve sustainable development, the government has the obligation to ensure the financial revenue as well as the environmental self-repair capacity. Therefore, the financial revenue is set as the objective function and environmental protection and some other objective conditions are set as the constraints to build the mathematical model.

4.2.2.4.1 Expected Target for the Regional Authority

As a developing country, China prioritizes economic development. Therefore, the regional authority must have economic development as their primary objective. The collieries' contribution to regional financial revenue comes mainly from taxes, which can be roughly divided into three categories, as previously mentioned. By summing the revenue from these taxes and using the method proposed by Xu and Zhou to deal with the fuzzy parameters in this function, the regional financial revenue for the coal industry can be calculated as follows:

$$MaxF = S_1 \sum_{i=1}^{n} \sum_{j=1}^{m} C_{ij}X_{ij} + S_2 \sum_{i=1}^{n} \sum_{j=1}^{m} E[\widetilde{E_{ij}}]X_{ij} + S_3 \sum_{i=1}^{n} Y_i \tag{4.6}$$

4.2.2.4.2 Regional Authority Limitations

As the leader and priority decision-maker, the regional authority must consider all macro-elements when developing the mining quota allocation plan. Therefore, there are the following constraints.

(1) *Environmental protection restrictions*: To pursue sustainable development, groundwater protection in large-scale coal fields is critical. Therefore, the authority is duty-bound to control the groundwater damage caused by the coal mining activities. However, it is very difficult to estimate the exact value of the constraint due to the inherent uncertainty and complexity [Tiwary, 2001, Qiao et al., 2011]. Most of the times, the authority would like to describe this constraint in linguistic terms, such as we would like to try our best to control the

amount of mining drainage within a certain range. The fuzzy chance constraint, which was originally proposed by Liu and Iwamura, has been proven to be an efficient mathematical tool to deal with such situations [Rong and Lahdelma, 2008, He et al., 2008]. Let H be the maximum carrying capacity that the groundwater can tolerate in the region and γ be the probabilistic confidence level, so then the environmental protection constraint is

$$Pos\left\{\left[\sum_{i=1}^{n} \widetilde{E_i}\, Y_i \leq H\right]\right\} \geq \gamma \tag{4.7}$$

where *Pos* is the possibility measure proposed by Dubois and Prade [1983], and the parameter $\widetilde{E_i}$ denotes the per tonne coal mining drainage coefficient at colliery i. This function indicates that the government requires the amount of the mining drainage to be controlled at less than H under the given probabilistic confidence level γ.

Based on the work by Xu and Zhou Xu and Zhou [2011], Eq. (4.7) can be transformed into its crisp forms as:

$$(1-\gamma)\sum_{i=1}^{n} r_i^1 Y_i + \gamma \sum_{i=1}^{n} r_i^2 Y_i - H \leq 0 \tag{4.8}$$

(2) *Objective restriction*: Some conditions must be satisfied when the authority makes decisions. The total mining quota for all collieries in the region must not exceed the recoverable amount in the coal field, represented by R^U. On the other hand, the authority also has the responsibility to provide sufficient coal to ensure the stability and development of the region, so the coal production from all collieries should meet the basic market demand D^L. Based on the above, this restriction can be shown as follows;

$$D^L \leq \sum_{i=1}^{n} Y_i \leq R^U \tag{4.9}$$

(3) *Protect taxpayers*: Since collieries are taxpayers, the regional authority has the responsibility to protect their basic rights and interests. The authority protects the taxpayers in many different ways, with the basic operating rights being the most fundamental. In this chapter, the basic operating rights are explained as the authority should provide the collieries with a mining quota that guarantees the basic operations if the collieries meet the requirements. In addition, as the collieries are obliged to pay a resource exploitation tax, which is determined by the mining quota allocated by the authority, when deciding on the mining quota allocation scheme, the authority should not allocate a mining quota larger than the mining capacity of the colliery. Let Q_i be the basic mining quota at colliery i, and this restriction can be written as follows;

$$Q_i \leq Y_i \leq \sum_{j=1}^{m} O_{ij}^U \tag{4.10}$$

(4) *Equality guarantee*: Previous researches have identified equality as a critical factor in sustainable development [Hopwood et al., 2005, Robert et al., 2005, Tsoutsos et al., 2009]. However, even though there is still no uniform equality measure method, several different measures have been suggested in research [Marsh and Schilling, 1994, Golany and Tamir, 1995, Cho and Lee, 2014]. These methods have been proven to be efficient in measuring equality in allocation problems. Recently, Xu et al., proposed a kind of satisfaction degree-based equality measurement which proved to be efficient in such quota allocation problems [Xu et al., 2015a]. This method is used to measure the equality and to define the satisfaction degree of each colliery as follows;

$$V_i = \begin{cases} 0, & Y_i \leq Q_i \\ \frac{Y_i - Q_i}{\sum_{j=1}^{m} O_{ij}^U - Q_i}, & Q_i < Y_i \leq \sum\limits_{j=1}^{m} O_{ij}^U \\ 1, & Y_i > \sum\limits_{j=1}^{m} O_{ij}^U \end{cases} \tag{4.11}$$

It is assumed here that if the satisfaction degree of each colliery achieves a certain level, then the mining quota allocation scheme equality is guaranteed. Let β be the minimized satisfaction degree chosen by the regional authority. Then the equality guarantee constraint can be described as follows:

$$V_i \geq \beta \tag{4.12}$$

4.2.2.5 Global Optimization Model for the EP-MQC

In this mining quota allocation process, the regional authority first chooses a reasonable mining quota allocation based on historical data to maximize the total financial revenue. This is done with the constraints of coal storage quantity, environmental protection, equality guarantee, and also with the consideration of the taxpayers basic rights and interests. The collieries then decide on their own production plans among each season based on their own economic benefit objectives with the constraints of mining quota, environmental protection, and production capacity. On one side, the authority's initial allocation scheme impacts the decisions of each colliery, and on the other side, the collieries' decisions will further make the authority adjusts its initial scheme. As can be seen, this process requires interaction and no single body can independently decide on the mining quota allocation scheme. After several recursive interactive processes, a solution which satisfies all stakeholders is finally achieved. In this problem, the regional authority and the collieries are conflicted in their respective objectives as the authority wishes to develop the economy within the environmental carrying capacity and the collieries wish to fully exploit the coal resources to gain the highest benefit. To describe this interactive process and to resolve the conflict between the authority and collieries, Eqs. (4.1)–(4.12) are integrated, so the model can be formulated as a global model, as shown in Eq. (4.13).

$$
\begin{array}{l}
MaxF = S_1 \sum_{i=1}^{n} \sum_{j=1}^{4} C_i X_{ij} + S_2 \sum_{i=1}^{n} \sum_{j=1}^{4} E[\widetilde{E_{ij}}] X_{ij} + S_3 \sum_{i=1}^{n} Y_i \\
\text{s.t.} \left\{
\begin{array}{l}
D^L \leq \sum_{i=1}^{n} Y_i \leq R^U \\
Q_i \leq Y_i \leq \sum_{j=1}^{4} O_{ij}^U, \quad \forall i \in \Psi \\
(1-\gamma) \sum_{i=1}^{n} r_i^1 Y_i + \gamma \sum_{i=1}^{n} r_i^2 Y_i - H \leq 0 \\
V_i \geq \beta, \quad \forall i \in \Psi \\
MaxG_i = \sum_{j=1}^{m} [(1-S_1)C_i - T_{ij}] X_{ij} - P_i \sum_{j=1}^{m} E[\widetilde{E_{ij}}] X_{ij} - S_2 \sum_{j=1}^{m} E[\widetilde{E_{ij}}] X_{ij} - S_3 Y_i \\
\text{s.t.} \left\{
\begin{array}{ll}
\sum_{j=1}^{m} X_{ij} \leq Y_i, \quad \forall i \in \Psi & \\
D_{ij} \leq X_{ij} \leq O_{ij}^U, & \forall i \in \Psi, \forall j \in \Phi \\
E[\widetilde{E_{ij}}] X_{ij} \leq B_{ij}, & \forall i \in \Psi, \forall j \in \Phi
\end{array}
\right. \\
\Psi \in \{1, 2, \ldots, n\} \\
\Phi \in \{1, 2, \ldots, m\}
\end{array}
\right.
\end{array}
\tag{4.13}
$$

In the model, γ expresses the government's efforts to protect the groundwater resources. Based on fuzzy chance constraints theory, γ is determined by the authority within an interval from 0 to 1 (i.e. $0 \leq \gamma \leq 1$) [Liu and Iwamura, 1998]. The closer the γ level is to 1, the stricter the environmental protection attitude that the authority has. For the different coal fields, based on their own individual situations, the authority chooses a specific γ. When some unforeseen circumstances significantly impact the coal mining activities, like a sudden change in the climate, the authority can also adjust γ to maintain equanimity.

The parameter β represents the minimum degree of satisfaction for each colliery, which is also set by the authority. Based on the mathematical form for the satisfaction degree measure used in this chapter, it can be seen that $0 \leq \beta \leq 1$. The mining quota allocated to each colliery may result in differences in their satisfaction degrees; however, the authority has a duty to ensure that the minimum satisfaction degree is in a reasonable range to maintain equality in the allocation scheme.

4.2.3 Model Transformation

The proposed model is a bilevel linear programming model and the bilevel model has been proven to be an non-deterministic polynomial (NP)-hard problem even in its simplest form [Ben-Ayed and Blair, 1990]. The solution approaches of the bilevel model have been widely discussed and have resulted in some valuable achievements. Of these, the Karush–Kuhn–Tucker (KKT) approach has been the most popular and efficient in solving bilevel models with linear objective functions and constraints [Colson et al., 2005]. Generally speaking, the fundamental strategy for the KKT approach is that it replaces the follower's problem with its KKT conditions and appends the resultant system to the leader's problem [Shi et al., 2005]. The proposed bilevel programming model is linear, so, based on work by Shi et al., Shi et al. [2005], the model can be transformed into a single-level model as follows;

$$MaxF = S_1 \sum_{i=1}^{n} \sum_{j=1}^{4} C_i X_{ij} + S_2 \sum_{i=1}^{n} \sum_{j=1}^{4} E\left[\widetilde{E_{ij}}\right] + S_3 \sum_{i=1}^{n} Y_i$$

$$\text{s.t.} \begin{cases} \sum_{i=1}^{n} Y_i \leq R^U, \\ Y_i \geq Q_i, \quad \forall i \in \Psi \\ (1-\gamma) \sum_{i=1}^{n} r_i^1 Y_i + \gamma \sum_{i=1}^{n} r_i^2 Y_i - H \leq 0 \\ V_i \geq \beta, \quad \forall i \in \Psi \\ \sum_{j=1}^{4} X_{ij} \leq Y_i, \quad \forall i \in \Psi \\ D_{ij} \leq X_{ij} \leq O_{ij}^U, \quad \forall i \in \Psi, \forall j \in \Phi \\ E\left[\widetilde{E_{ij}}\right] X_{ij} \leq B_{ij}, \quad \forall i \in \Psi, \forall j \in \Phi \\ u_{ij}^1 - u_{ij}^2 + u_{ij}^3 + E\left[\widetilde{E_{ij}}\right] u_{ij}^4 = (1 - S_1)C_i - T_{ij} - P_i E\left[\widetilde{E_{ij}}\right] - S_2 E\left[\widetilde{E_{ij}}\right] \\ u_{ij}^1 g_1(X_{ij}, Y_i) + u_{ij}^2 g_2(X_{ij}, Y_i) + u_{ij}^3 g_3(X_{ij}, Y_i) + u_{ij}^4 g_4(X_{ij}, Y_i) = 0 \\ g_1(X_{ij}, Y_i) = Y_i - \sum_{j=1}^{4} X_{ij} \geq 0, \quad \forall i \in \Psi \\ g_2(X_{ij}, Y_i) = X_{ij} - D_{ij} \geq 0, \quad \forall i \in \Psi, \forall j \in \Phi \\ g_3(X_{ij}, Y_i) = O_{ij}^U - X_{ij} \geq 0, \quad \forall i \in \Psi, \forall j \in \Phi \\ g_4(X_{ij}, Y_i) = B_{ij} - E\left[\widetilde{E_{ij}}\right] X_{ij} \geq 0, \quad \forall i \in \Psi, \forall j \in \Phi \\ \Psi \in \{1, 2, \dots, n\} \\ \Phi \in \{1, 2, \dots, m\} \end{cases} \tag{4.14}$$

Previous research has proven that the linear bilevel programming model and its relative single-level model transformed using the KKT approach have the same solutions [Bialas and Karwan, 1984, Lu et al., 2006]. Therefore, the new model as shown in Eq. (4.14) can be used to calculate rather than using Eq. (4.13). Although the model shown in Eq. (4.14) is extremely complex, it is a single-level program with a unique objective function and certain parameters. Therefore, based on the existing solution algorithms for such programming and running it in the MATLAB 7.0, the solution of the model can be determined [Dantzig, 2016, Boyd and Vandenberghe, 2004].

4.3 Case Study

In this section, to demonstrate the practical applicability and efficiency of the above proposed model, the Yanzhou coal field in Yanzhou City, Shandong Province, China is used as an example.

4.3.1 Presentation of the Case Region

The Yanzhou coal field, located in southwest Shandong Province, is famous for its fine varieties of thermal coal with low content of sulfur, phosphorus, and ash but high calorific value. As one of the new developing large-scale coal fields, the total area of Yanzhou is more than 3400 km^2 and the proven coal reserves are 9.1 billion tonnes. There are five major collieries (i.e. $i = 1, 2, \dots, n$) in the area: Nantun (Nt) colliery, Xinglongzhen (Xlz) colliery, Baodian (Bd) colliery, Dongtan (Dt) colliery, and Jier (Je) colliery.

The coal field belongs to temperate, semi-humid monsoon area, and the rainfall changes greatly. Driven by the huge profits of mining coal mine, the coal mine has been working at full load. With the development of coal resources, more and more collapsed and damaged land have been found. As of 2015, the total mining subsidence area of Yanzhou coal field is 192.64 km^2, and the groundwater level in this area has dropped sharply, which has caused serious environmental problems [Bian et al., 2009]. The coal–water conflict in this area is the main obstacle to sustainable development. Therefore, taking Yanzhou coal field as an example, the practicability of this method is verified.

4.3.2 Data Collection

The input data and parameters for the proposed model can be divided roughly into two categories: the uncertain data and the crisp data.

The uncertain data (i.e. the mining drainage coefficient $\widetilde{E_{ij}}$ and $\widetilde{E_i}$) were derived using the following steps: (i) interviews with engineers and workers from each colliery in the Yanzhou coal field were conducted, in which they were asked to given a range for each uncertain parameter; (ii) then the initial collected data were analyzed, and some extreme values (i.e. much larger or smaller than most of the data) were eliminated; (iii) the minimum value of the remaining data set was set as the lower bound for each uncertain parameter (i.e. r_{ij}^1 and r_i^1), and the maximum value was set as the upper bound (i.e. r_{ij}^4 and r_i^4); and (iv) without a loss of generality, the values with the highest frequency were assumed to be the most possible range for each uncertain parameter (i.e. form r_{ij}^2 to r_{ij}^3 and form r_i^2 to r_i^3). Using this method, $\widetilde{E_{ij}}$ and $\widetilde{E_i}$ were determined as shown in Tables 4.1 and 4.2.

Table 4.1 Unit coal drainage coefficient when colliery i produces in season j: $\widetilde{E_{ij}}$ (m^3/tonnes).

	The 1st season	The 2nd season	The 3rd season	The 4th season
Nt colliery	(1.25, 1.27, 1.31, 1.33)	(1.26, 1.29, 1.35, 1.37)	(1.35, 1.38, 1.43, 1.46)	(1.26, 1.30, 1.33, 1.36)
Xlz colliery	(1.06, 1.09, 1.12, 1.15)	(1.09, 1.13, 1.15, 1.19)	(1.18, 1.22, 1.25, 1.27)	(1.13, 1.15, 1.17, 1.19)
Bd colliery	(1.00, 1.05, 1.14, 1.18)	(1.02, 1.06, 1.10, 1.18)	(1.08, 1.12, 1.20, 1.25)	(1.01, 1.05, 1.14, 1.19)
Dt colliery	(1.13, 1.15, 1.19, 1.22)	(1.15, 1.18, 1.23, 1.25)	(1.19, 1.23, 1.26, 1.31)	(1.16, 1.20, 1.22, 1.25)
Je colliery	(1.18, 1.21, 1.23, 1.26)	(1.19, 1.23, 1.26, 1.28)	(1.22, 1.26, 1.29, 1.31)	(1.20, 1.23, 1.26, 1.29)

Table 4.2 Unit coal drainage coefficient for exploitation at colliery i: $\widetilde{E_i}$ (m^3/tonnes).

	Average mining drainage
Nt colliery	(1.27, 1.29, 1.33, 1.35)
Xlz colliery	(1.08, 1.12, 1.16, 1.21)
Bd colliery	(1.05, 1.07, 1.10, 1.18)
Dt colliery	(1.17, 1.20, 1.23, 1.25)
Je colliery	(1.21, 1.24, 1.27, 1.30)

Table 4.3 The input parameters of the environmental protection based mining quota competition (EP-MQC) model.

S_1 (%)	S_2 (RMB/tonnes)	S_3 (RMB/tonnes)	H (10^7 tonnes)	R^U (10^9 tonnes)
17	8	1.5	3.804 59	2.85

The crisp data input into the proposed model were mainly obtained from the Statistical Yearbooks (i.e. Statistical Yearbook of Chinese coal industry, and the Statistical Yearbook of Chinese energy), the Chinese Coal Industry Development Plan in the Twelfth Five-Year Plan, and data published by the coal companies. Some parameters were obtained through a market survey, such as the price C_i and the basic coal production demand D_{ij}. All crisp parameters are shown in Tables 4.3–4.5.

It has to be noted that it is difficult to determine an exact value for the maximal amount of groundwater that is allowed to drain which is represented by H. Even though the attitudes of the various authorities are different, their intentions to protect the environment are consistent. In this chapter, without loss of generality, this threshold value H was set based on historical data. That is to say, H was determined from the actual mining drainage quantities from the last production cycle and the authority's environmental protection attitude toward the next production cycle. For example, it was assumed that the authority had the aim to reduce total mining drainage in the whole coal field by 10%, so the H was determined to be 90% of the actual mining drainage amount from the last production cycle. In this way, the value of H was determined and is shown in Table 4.3.

4.3.3 Results Under Different Situations

By inputting the data into the proposed model and running the proposed solution method on MATLAB 7.0, the results were determined. In Table 4.6, it can be seen that for a confidence level $\gamma = 1$ and a satisfaction degree of $\beta = 0.5$, the total financial revenue is 2.9497×10^9 RMB, with the total mining drainage in this coal field being 38.0459×10^7 m^3. Further, to demonstrate the effectiveness of the proposed model, sensitivity analyses were conducted by adjusting some of the subjective parameters, such as the groundwater protection confidence level

Table 4.4 Crisp parameters used in EP-MQC model.

	T_{ij} (RMB/tonnes)	D_{ij} (10^6 tonnes)	O_{ij}^U (10^6 tonnes)	B_{ij} (10^6 tonnes)
Nt 1st season	215	0.4	1.4	1.265
Nt 2nd season	218	0.5	1.3	1.265
Nt 3rd season	225	0.3	1.1	1.265
Nt 4th season	217	0.5	1.3	1.265
Xlz 1st season	197	1.5	2.1	2.415
Xlz 2nd season	199	1.4	2.1	2.415
Xlz 3rd season	205	1.2	1.9	2.415
Xlz 4th season	200	1.6	2.0	2.415
Bd 1st season	215	1.5	2.5	2.185
Bd 2nd season	218	1.3	2.4	2.185
Bd 3rd season	225	1.1	2.1	2.185
Bd 4th season	220	1.4	2.3	2.185
Dt 1st season	209	2.0	2.6	2.760
Dt 2nd season	211	1.7	2.5	2.760
Dt 3rd season	216	1.5	2.4	2.760
Dt 4th season	213	1.8	2.6	2.760
Je 1st season	200	0.5	1.5	1.275
Je 2nd season	202	0.5	1.4	1.275
Je 3rd season	207	0.4	1.3	1.275
Je 4th season	203	0.7	1.4	1.275

Table 4.5 The crisp parameters of the EP-MQC model.

	C_i (RMB/tonnes)	P_i (RMB/m^3)	Q_i (10^6 tonnes)
Nt colliery	525	8.0	2.3
Xlz colliery	460	7.0	6.0
Bd colliery	510	7.0	5.5
Dt colliery	435	7.0	7.0
Je colliery	485	7.0	2.2

γ and the allocation satisfaction degree β. The results under different confidence levels γ and satisfaction degree $\beta = 0.5$ are shown in Tables 4.7–4.10 show the calculation results when the basic satisfaction degree β is changed. It should be noted that in this chapter, the minimize satisfaction degree was set at 0.5 (i.e. $\beta = 0.5$). From Table 4.8, it can be seen that when the regional authority sets the strictest groundwater protection regulation (i.e. $\gamma = 1$), the highest satisfaction degree that all the collieries can reach at the same time is 0.6 (i.e. $\beta = 0.6$). When $\gamma = 0.9$, the highest satisfaction degree is 0.7 as shown in Table 4.9, and when

Table 4.6 Result of EP-MQC model under confidence level $\gamma = 1$, satisfaction degree $\beta = 0.5$.

Total benefit: $F(X_{ij}, Y_i)$ (10^9 RMB)	Colliery	Total mining quota: Y_i (10^6 tonnes)	1st season: X_{i1} (10^6 tonnes)	2nd season: X_{i2} (10^6 tonnes)	3rd season: X_{i3} (10^6 tonnes)	4th season: X_{i4} (10^6 tonnes)	Benefit: $G_i(X_{ij}, Y_i)$ (10^9 RMB)
2.9497	Nt	3.700	1.400	1.300	0.300	0.700	1.5365
	Xlz	7.050	2.100	2.100	1.200	1.650	2.5662
	Bd	9.205	2.500	2.400	2.005	2.300	3.7368
	Dt	8.550	2.600	2.500	1.500	1.950	2.9314
	Je	3.900	1.500	1.300	0.400	0.700	1.4983
	Total	32.405					

$\gamma = 0.8$ the highest satisfaction degree is 0.75. It can be calculated that when $\gamma = 0.7$, the highest satisfaction degree will be 0.9 and when γ is loosened to 0.6, all the collieries can conduct full load production. That is to say, when $\gamma \leq 0.6$, the groundwater protection constraint loses its function and the proposed model is without significance.

4.4 Discussion

In this section, the results from the proposed model are discussed.

4.4.1 Propositions and Analysis

First of all, we conclude that environmental protection-based mining quota competition mechanism leads to more sustainable exploitation. Table 4.6 shows that all the five collieries conduct full load production in the first and second season, that is, collieries try their best to exploit greater quantities in the more environmental friendly seasons. However, in the third season, they only exploit enough coal to meet the basic market requirements. If full load exploitation has already taken place in the season with the smallest per unit mining drainage and there is still mining quota remaining, this is then allocated to the season with the second smallest per unit mining drainage. Such a process continues until all mining quota is allocated. This is not occasional behavior, as trail can be observed through an analysis of the mathematical form and properties of the proposed model. To maximize total financial revenue under the acceptable mining drainage constraint (i.e. Eqs. (4.7) and (4.8)), the regional authority allocates more mining quota to the colliery with a smaller mining drainage coefficient which is represented by $\widetilde{E_i}$ and on the another hand, $\widetilde{E_i}$ is determined by X_{ij} and $\widetilde{E_{ij}}$ and it can be seen that a larger of X_{ij} with smaller $\widetilde{E_{ij}}$, the smaller that $\widetilde{E_i}$ is. In such a situation, to gain as greater share of the mining quota, all collieries will try their best to increase exploitation in the season with

Table 4.7 Results of EP-MQC model under different confidence levels γ, satisfaction degree $\beta = 0.5$.

γ	Total benefit: $F(X_{ij}, Y_i)$ (10^9 RMB)	Colliery	Total mining quota: Y_i (10^6 tonnes)	1st season: X_{i1} (10^6 tonnes)	2nd season: X_{i2} (10^6 tonnes)	3rd season: X_{i3} (10^6 tonnes)	4th season X_{i4} (10^6 tonnes)	Benefit: $G_i(X_{ij}, Y_i)$ (10^9 RMB)
0.9	3.1182	Nt	3.700	1.400	1.300	0.300	0.700	1.5365
		Xlz	8.100	2.100	2.100	1.900	2.000	2.9479
		Bd	9.300	2.500	2.400	2.100	2.300	3.7753
		Dt	9.350	2.600	2.500	1.650	2.600	3.2056
		Je	3.900	1.500	1.300	0.400	0.700	1.4983
		Total	34.35					
0.8	3.1995	Nt	3.700	1.400	1.300	0.300	0.700	1.5365
		Xlz	8.100	2.100	2.100	1.900	2.000	2.9479
		Bd	9.300	2.500	2.400	2.100	2.300	3.7753
		Dt	10.10	2.600	2.500	2.400	2.600	3.4624
		Je	4.100	1.500	1.400	0.400	0.800	1.5751
		Total	35.30					
0.7	3.3529	Nt	3.850	1.400	1.300	0.300	0.850	1.5988
		Xlz	8.100	2.100	2.100	1.900	2.000	2.9479
		Bd	9.300	2.500	2.400	2.100	2.300	3.7753
		Dt	10.10	2.600	2.500	2.400	2.600	3.4624
		Je	5.600	1.500	1.400	1.300	1.400	2.1509
		Total	36.95					
0.6	3.4769	Nt	5.100	1.400	1.300	1.100	1.300	2.1172
		Xlz	8.100	2.100	2.100	1.900	2.000	2.9479
		Bd	9.300	2.500	2.400	2.100	2.300	3.7753
		Dt	10.10	2.600	2.500	2.400	2.600	3.4624
		Je	5.600	1.500	1.400	1.300	1.400	2.1509
		Total	38.20					

the smaller mining drainage coefficient to minimize $\widetilde{E_i}$. In all, under such a mining quota competition mechanism, to seek as high a mining quota as possible, the most environmental friendly seasons are the first choice for the collieries.

In addition, collieries with smaller unit mining drainage have priority in competing for quotas when environmental protection constraints change. In general, when the authority relaxes the environmental protection confidence changes, all the five collieries show different sensitivities, to be specific, the Xlz and the Bd collieries are the most sensitive and have priority when competing for the increased mining quota which resulting from the relaxed environmental protection constraint.

Table 4.8 Results of EP-MQC model under different satisfaction degree β, confidence levels $\gamma = 1$.

β	Total benefit: $F(X_{ij}, Y_i)$ (10^9 RMB)	Colliery	Total mining quota: Y_i (10^6 tonnes)	1st season: X_{i1} (10^6 tonnes)	2nd season: X_{i2} (10^6 tonnes)	3rd season: X_{i3} (10^6 tonnes)	4th season X_{i4} (10^6 tonnes)	Benefit: $G_i(X_{ij}, Y_i)$ (10^9 RMB)
0.5	2.9497	Nt	3.700	1.400	1.300	0.300	0.700	1.5365
		Xlz	7.050	2.100	2.100	1.200	1.650	2.5662
		Bd	9.205	2.500	2.400	2.005	2.300	3.7368
		Dt	8.550	2.600	2.500	1.500	1.950	2.9314
		Je	3.900	1.500	1.300	0.400	0.700	1.4983
		Total	32.405					
0.55	2.9388	Nt	3.840	1.400	1.300	0.300	0.840	1.5946
		Xlz	7.155	2.100	2.100	1.200	1.755	2.6044
		Bd	8.554	2.500	2.400	1.354	2.300	3.4729
		Dt	8.705	2.600	2.500	1.500	2.105	2.9846
		Je	4.070	1.500	1.400	0.400	0.770	1.5636
		Total	32.324					
0.6	2.9279	Nt	3.980	1.400	1.300	0.300	0.980	1.6528
		Xlz	7.260	2.100	2.100	1.200	1.860	2.6426
		Bd	7.903	2.500	2.400	1.100	1.903	3.2087
		Dt	8.860	2.600	2.500	1.500	2.260	3.0377
		Je	4.240	1.500	1.400	0.400	0.940	1.6288
		Total	32.243					

When γ changes from 1 to 0.9, we find that the Xlz and the Bd colliery are allocated a mining quota which allows them to conduct full load exploitation in each season. Meanwhile, the mining quotas for the Nt colliery and the Je colliery do not change. When γ is continuously relaxed to 0.8, the Dt colliery is the third to reach full load exploitation and, at the same time, the Je colliery also receives additional mining quota. When γ is continuously relaxed to 0.7, the Je colliery also reaches its full load exploitation and the Nt colliery receives a greater share of the quota compared with the situation when $\gamma = 0.8$. When the authority sets $\gamma = 0.6$, all the collieries are able to conduct full load exploitation. After their quota reaches the utmost mining capacity, the Dt colliery has the second priority and the Nt colliery will have the third priority, the Je colliery is the last one. The main reason for the above results derives from the proposed model. Note that, while this result is based on this regional case, in any other cases, this occurrence still takes place because of the physical logical and mathematical form of the model. When the authority sets a relatively relaxed environmental protection constraint, the amount of acceptable mining drainage increases at the same time. To maximize financial revenue, the total mining quota allocated to the collieries also increases and, the increased part

Table 4.9 Results of EP-MQC model under different satisfaction degree β, confidence levels $\gamma = 0.9$.

β	Total benefit: $F(X_{ij}, Y_i)$ (10^9 RMB)	Colliery	Total mining quota: Y_i (10^6 tonnes)	1st season: X_{i1} (10^6 tonnes)	2nd season: X_{i2} (10^6 tonnes)	3rd season: X_{i3} (10^6 tonnes)	4th season X_{i4} (10^6 tonnes)	Benefit: $G_i(X_{ij}, Y_i)$ (10^9 RMB)
0.5	3.1182	Nt	3.700	1.400	1.300	0.300	0.700	1.5365
		Xlz	8.100	2.100	2.100	1.900	2.000	2.9479
		Bd	9.300	2.500	2.400	2.100	2.300	3.7753
		Dt	9.350	2.600	2.500	1.650	2.600	3.2056
		Je	3.900	1.500	1.300	0.400	0.700	1.4983
		Total	34.35					
0.6	3.1210	Nt	3.980	1.400	1.300	0.300	0.980	1.6528
		Xlz	7.926	2.100	2.100	1.726	2.000	2.8847
		Bd	9.300	2.500	2.400	2.100	2.300	3.7753
		Dt	8.860	2.600	2.500	1.500	2.260	3.0377
		Je	4.240	1.500	1.400	0.400	0.940	1.6288
		Total	34.306					
0.7	3.1083	Nt	4.260	1.400	1.300	0.300	1.260	1.7691
		Xlz	7.470	2.100	2.100	1.270	2.000	2.7190
		Bd	8.704	2.500	2.400	1.504	2.300	3.5337
		Dt	9.170	2.600	2.500	1.500	2.570	3.1440
		Je	4.580	1.500	1.400	0.400	1.280	1.7594
		Total	34.184					

is first allocated to the collieries with smaller per unit mining drainage. This result also shows the guidance role of our model in reducing coal–water conflict to achieve sustainable development because no matter how the environmental protection confidence changes, and finally the collieries with the smaller unit mining drainage always have priority when competing for mining quotas.

Third, relaxed environmental protection constraints increase revenue, but cause more damage to the groundwater environment, which does more harm to the sustainable development. Table 4.7 shows that when the authority sets the environmental protection confidence at $\gamma = 0.9$ (take the situation of $\beta = 0.5$ as an example), the total financial revenue of the regional coal mining industry is 3.1182×10^9 RMB, which is a 5.7% growth compared to the situation when $\gamma = 1$. However, simultaneously, the total mining drainage in the coal field reaches 40.3135×10^6 m^3, an increase of about 6%. The increase in the financial revenue (5.7%) is lower than the total mining drainage (6%). When γ is changed to 0.8, the increasing ratio of the financial revenue and the total mining drainage is 8.4% and 9.1%, respectively, so the growth gap between them grows, from 0.3% to 0.7%. If the authority sets $\gamma = 0.7$,

Table 4.10 Results of EP-MQC model under different satisfaction degree β, confidence levels $\gamma = 0.8$.

β	Total benefit: $F(X_{ij}, Y_i)$ (10^9 RMB)	Colliery	Total mining quota: Y_i (10^6 tonnes)	1st season: X_{i1} (10^6 tonnes)	2nd season: X_{i2} (10^6 tonnes)	3rd season: X_{i3} (10^6 tonnes)	4th season X_{i4} (10^6 tonnes)	Benefit: $G_i(X_{ij}, Y_i)$ (10^9 RMB)
0.55	3.2005	Nt	3.840	1.400	1.300	0.300	0.840	1.5946
		Xlz	8.100	2.100	2.100	1.900	2.000	2.9479
		Bd	9.300	2.500	2.400	2.100	2.300	3.7753
		Dt	9.979	2.600	2.500	2.279	2.600	3.4210
		Je	4.070	1.500	1.400	0.400	0.770	1.5636
		Total	35.289					
0.65	3.2049	Nt	4.120	1.400	1.300	0.300	1.120	1.7109
		Xlz	8.100	2.100	2.100	1.900	2.000	2.9479
		Bd	9.300	2.500	2.400	2.100	2.300	3.7753
		Dt	9.325	2.600	2.500	1.625	2.600	3.1970
		Je	4.410	1.500	1.400	0.400	1.110	1.6941
		Total	35.255					
0.75	3.2007	Nt	4.400	1.400	1.300	0.400	1.300	1.8271
		Xlz	7.575	2.100	2.100	1.400	1.975	2.7571
		Bd	9.121	2.500	2.400	1.504	2.300	3.5337
		Dt	9.325	2.600	2.500	1.625	2.600	3.1970
		Je	4.750	1.500	1.400	0.450	1.400	1.8247
		Total	35.171					

this occurs again, and the financial revenue increases to 13.5% and the total mining drainage increases to 14.5%, so the growth gap between them increases again. A similar result can also be found when $\gamma = 0.6$. What is shown in Figure 4.3 is the changing trend in the total mining drainage and the total benefit under different confidence levels γ. Therefore, it can be surmised that when the authority has a relatively relaxed environmental protection policy, both revenue and total mining drainage increase with the growth proportion of the former being lower than the latter. We analyze the reason is that when the authority establishes the mining quota allocation mechanism, the collieries engage in full load exploitation in the most environmental friendly seasons which had already been discussed above and if the environmental protection constraints are relaxed, the mining quotas at each colliery increase as shown in Eq. (4.9) and the situation that collieries will try their best to exploit in more environmental friendly seasons will not change. As a result, the increased part is allocated to those seasons with a higher drainage coefficient which lead to higher revenue; however, due to the larger drainage coefficient, the total mining drainage has a higher ratio increase (Figure 4.4).

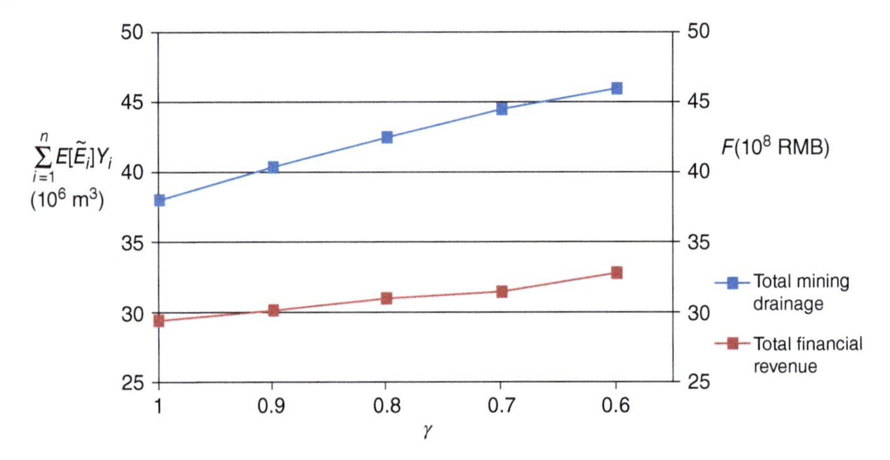

Figure 4.3 The changing trends for the total mining drainage and the total benefit under different confidence level γ, $\beta = 0.5$.

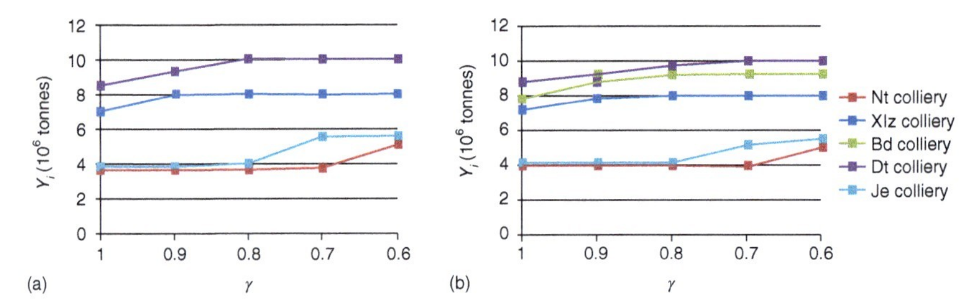

Figure 4.4 Mining quota allocation scheme under different confidence level γ, $\beta = 0.5$ (a) and $\beta = 0.6$ (b).

Forth, satisfaction degree does not significantly impact financial revenue. In this chapter, satisfaction degree is used to describe each colliery is allocated a relatively satisfactory quota, to be specific, parameter β in Eq. (4.12) is a subjective variable which represents the lowest satisfaction degree set by the regional authority. The results of sensitivity analysis are shown in Tables 4.8–4.10. Take the situation of $\gamma = 0.8$ as an example, as shown in Table 4.10, when the authority sets $\beta = 0.55$, the total financial revenue is 3.2005×10^9 RMB and as β changes to 0.65, the revenue is 3.2049×10^9 RMB, an increase of only 0.13%. When β is set at 0.75, the revenue is 3.2007×10^9 RMB, a slight decrease. When the authority has different confidence levels γ, as shown in Figure 4.5, similar conclusions can also be seen. These results show that under a certain confidence level γ, a change in the satisfaction degree β does not significantly impact financial revenue. We analyze that when the satisfaction degree changes, the allocated total mining quota to the collieries changes slightly.

4.4.2 Policy Recommendations

Based on the discussion above, in the following, some policy suggestions regarding coal field sustainability are given.

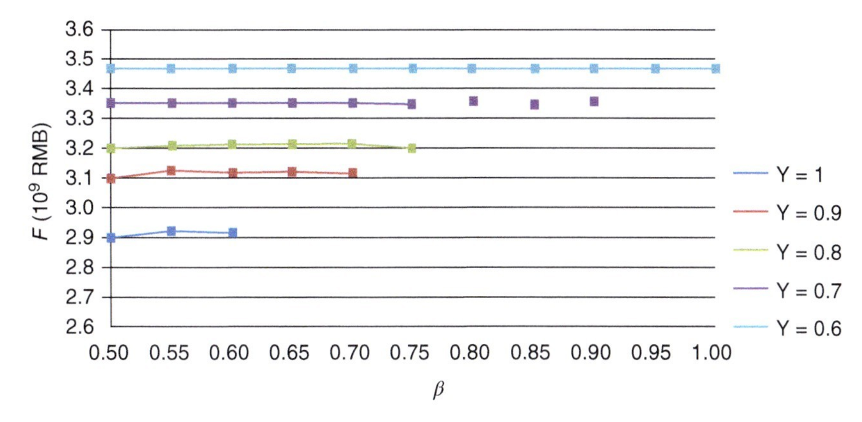

Figure 4.5 Changing trends for total financial revenue at different satisfaction degree β.

First of all, it is suggested to establish an environmental protection-based mining quota encouraged mechanism. If there was not a such mechanism, collieries, who are encouraged by the high potential economic benefits, would extensively exploit without any regard for the serious damage being caused to the environment. On the contrary, using the proposed method in this chapter, such a mining quota competition mechanism could be established and, as discussed before, this method could impel the collieries to conduct environmental friendly exploitation.

Second, it is recommended that when an authority establishes an environmental protection-based mining quota allocation scheme to achieve sustainable development, the satisfaction degree at each colliery should be emphasized. To comply with equity principle, one of the resource allocation principles proposed by the Chinese government, that is all the collieries' satisfaction degrees should be considered. In this chapter, Eqs. (4.11) and (4.12) were used to measure the satisfaction degree. As shown, a change in the required minimal satisfaction degree does not significantly impact financial revenue.

Third, there needs to be a dynamic environmental protection attitude. As China is still a developing country in which energy consumption is largely dependent on coal, when the authority selects its decision strategy, for example, by choosing the parameters such as γ and β in this chapter, a dynamic attitude should be taken. It is recommended that the authority fully consider the actual situation and set its own sustainable development targets as well as its own mining quota allocation schemes (i.e. choosing a specific γ and β). Further, to be dynamic, at the commencement of a mining quota competition mechanism that uses the proposed method, the authority can set a relatively loose environmental protection attitude (i.e. $\gamma = 0.7$) in consideration of economic development. However, after a period of time, the authority should upgrade this attitude to a higher level (i.e. $\gamma = 0.8$) and in the following period this degree should be upgraded again and such a process should continue over time until the strictest environmental protection requirements (i.e. $\gamma = 1$) are attained.

References

Angulo, E., Castillo, E., García-Ródenas, R., and Sánchez-Vizcaíno, J. (2014). A continuous bi-level model for the expansion of highway networks. *Computers & Operations Research* 41: 262–276.

Baker, K.A., Fennessy, M.S., and Mitsch, W.J. (1991). Designing wetlands for controlling coal mine drainage: an ecologic-economic modelling approach. *Ecological Economics* 3 (1): 1–24.

Ben-Ayed, O. and Blair, C.E. (1990). Computational difficulties of bilevel linear programming. *Operations Research* 38 (3): 556–560.

Bialas, W.F. and Karwan, M.H. (1984). Two-level linear programming. *Management Science* 30 (8): 1004–1020.

Bian, Z., Dong, J., Lei, S. et al. (2009). The impact of disposal and treatment of coal mining wastes on environment and farmland. *Environmental Geology* 58 (3): 625–634.

Boyd, S. and Vandenberghe, L. (2004). *Convex Optimization*. Cambridge University Press.

Chen, W. and Xu, R. (2010). Clean coal technology development in China. *Energy Policy* 38 (5): 2123–2130.

Cho, J.H. and Lee, J.H. (2014). Multi-objective waste load allocation model for optimizing waste load abatement and inequality among waste dischargers. *Water, Air, & Soil Pollution* 225 (3): 1892.

Colson, B., Marcotte, P., and Savard, G. (2005). Bilevel programming: a survey. *4OR* 3 (2): 87–107.

Dantzig, G. (2016). *Linear Programming and Extensions*. Princeton University Press.

Feng, T., Sun, L., and Zhang, Y. (2009). The relationship between energy consumption structure, economic structure and energy intensity in China. *Energy Policy* 37 (12): 5475–5483.

Ganesan, K. and Veeramani, P. (2006). Fuzzy linear programs with trapezoidal fuzzy numbers. *Annals of Operations Research* 143 (1): 305–315.

Glauser, S., McAllister, M.L., and Milioli, G. (2005). The challenges of sustainability in mining regions: the coal mining region of Santa Catarina, Brazil. In: *Natural Resources Forum*, vol. 29, 1–11. Wiley Online Library.

Golany, B. and Tamir, E. (1995). Evaluating efficiency-effectiveness-equality trade-offs: a data envelopment analysis approach. *Management Science* 41 (7): 1172–1184.

Grzegorzewski, P. (2008). Trapezoidal approximations of fuzzy numbers preserving the expected interval-algorithms and properties. *Fuzzy Sets and Systems* 159 (11): 1354–1364.

He, L., Huang, G.H., and Lu, H.W. (2008). A simulation-based fuzzy chance-constrained programming model for optimal groundwater remediation under uncertainty. *Advances in Water Resources* 31 (12): 1622–1635.

Hopwood, B., Mellor, M., and O'Brien, G. (2005). Sustainable development: mapping different approaches. *Sustainable Development* 13 (1): 38–52.

Hu, Z. and Xiao, W. (2013). Optimization of concurrent mining and reclamation plans for single coal seam: a case study in northern Anhui, China. *Environmental Earth Sciences* 68 (5): 1247–1254.

Jarvis, A.P. and Younger, P.L. (2000). Broadening the scope of mine water environmental impact assessment: a UK perspective. *Environmental Impact Assessment Review* 20 (1): 85–96.

Kalashnikov, V.V., Pérez-Valdés, G.A., Tomasgard, A., and Kalashnykova, N.I. (2010). Natural gas cash-out problem: bilevel stochastic optimization approach. *European Journal of Operational Research* 206 (1): 18–33.

Labbé, M., Marcotte, P., and Savard, G. (1998). A bilevel model of taxation and its application to optimal highway pricing. *Management Science* 44 (12-part-1): 1608–1622.

Li, R. and Leung, G.C.K. (2012). Coal consumption and economic growth in China. *Energy Policy* 40: 438–443.

Liu, B. and Iwamura, K. (1998). Chance constrained programming with fuzzy parameters. *Fuzzy Sets and Systems* 94 (2): 227–237.

Liu, B. and Liu, Y.K. (2002). Expected value of fuzzy variable and fuzzy expected value models. *IEEE Transactions on Fuzzy Systems* 10 (4): 445–450.

Lu, J., Shi, C., and Zhang, G. (2006). On bilevel multi-follower decision making: general framework and solutions. *Information Sciences* 176 (11): 1607–1627.

Marsh, M.T. and Schilling, D.A. (1994). Equity measurement in facility location analysis: a review and framework. *European Journal of Operational Research* 74 (1): 1–17.

Mudd, G.M. (2008). Sustainability reporting and water resources: a preliminary assessment of embodied water and sustainable mining. *Mine Water and the Environment* 27 (3): 136.

Qiao, X., Li, G., Li, M. et al. (2011). Influence of coal mining on regional karst groundwater system: a case study in West Mountain area of Taiyuan City, northern China. *Environmental Earth Sciences* 64 (6): 1525–1535.

Rapantova, N., Grmela, A., Vojtek, D. et al. (2007). Ground water flow modelling applications in mining hydrogeology. *Mine Water and the Environment* 26 (4): 264–270.

Robert, K.W., Parris, T.M., and Leiserowitz, A.A. (2005). What is sustainable development: goals, indicators, values, and practice. *Environment: Science and Policy for Sustainable Development* 47 (3): 8–21.

Rong, A. and Lahdelma, R. (2008). Fuzzy chance constrained linear programming model for optimizing the scrap charge in steel production. *European Journal of Operational Research* 186 (3): 953–964.

Shen, L., Gao, T.-m., and Cheng, X. (2012). China's coal policy since 1979: a brief overview. *Energy Policy* 40: 274–281.

Shi, C., Lu, J., and Zhang, G. (2005). An extended Kuhn–Tucker approach for linear bilevel programming. *Applied Mathematics and Computation* 162 (1): 51–63.

Silva, L.F.O., Vallejuelo, S.F.-O., Martinez-Arkarazo, I. et al. (2013). Study of environmental pollution and mineralogical characterization of sediment rivers from Brazilian coal mining acid drainage. *Science of the Total Environment* 447: 169–178.

Sun, W., Wu, Q., Dong, D., and Jiao, J. (2012). Avoiding coal–water conflicts during the development of China's large coal-producing regions. *Mine Water and the Environment* 31 (1): 74–78.

Tiwary, R.K. (2001). Environmental impact of coal mining on water regime and its management. *Water, Air, & Soil Pollution* 132 (1–2): 185–199.

Tsoutsos, T., Drandaki, M., Frantzeskaki, N. et al. (2009). Sustainable energy planning by using multi-criteria analysis application in the island of Crete. *Energy Policy* 37 (5): 1587–1600.

Wolkersdorfer, C. and Bowell, R. (2005). Contemporary reviews of mine water studies in Europe, Part 2. *Mine Water and the Environment* 24 (1): 2–37.

Xu, J. and Zhou, X. (2011). *Fuzzy-Like Multiple Objective Decision Making*, vol. 263. Springer.

Xu, J., Lv, C., Zhang, M. et al. (2015a). Equilibrium strategy based optimization method for the coal–water conflict: a perspective from China. *Journal of Environmental Management* 160: 312–323.

Xu, J., Yang, X., and Tao, Z. (2015b). A tripartite equilibrium for carbon emission allowance allocation in the power-supply industry. *Energy Policy* 82: 62–80.

You, C.F. and Xu, X.C. (2010). Coal combustion and its pollution control in China. *Energy* 35 (11): 4467–4472.

Younger, P.L. (2001). Mine water pollution in Scotland: nature, extent and preventative strategies. *Science of the Total Environment* 265 (1–3): 309–326.

Younger, P.L. and Wolkersdorfer, C. (2004). Mining impacts on the fresh water environment: technical and managerial guidelines for catchment scale management. *Mine Water and the Environment* 23: s2–s80.

Yuan, J.-H., Kang, J.-G., Zhao, C.-H., and Hu, Z.-G. (2008). Energy consumption and economic growth: evidence from china at both aggregated and disaggregated levels. *Energy Economics* 30 (6): 3077–3094.

Yuan, J., Xu, Y., Zhang, X. et al. (2014). China's 2020 clean energy target: consistency, pathways and policy implications. *Energy Policy* 65: 692–700.

Zadeh, L.A. (1965). Fuzzy sets. *Information and Control* 8 (3): 338–353.

5

GIS-Oriented Equilibrium Strategy Toward Coal Gangue Contamination Mitigating

Coal is one of the main resources in the world, especially in developing countries [Choudhary and Shankar, 2012, Wu et al., 2014, Nicky M et al., 2016, Blumberg et al., 2015]. For example, China is one of the largest coal-producing countries, with coal accounting for 67.5% of the primary energy supply [You and Xu, 2010]. Coal gangue, a byproduct of coal mining and washing, is a major solid waste [Xiao et al., 2010, Zhou et al., 2014, Liu et al., 2012, Cardoso et al., 2016, Liu and Liu, 2015] that not only occupies plenty of land but also results in series environmental problems, such as the water, soil, and atmospheric pollution [Xiao et al., 2015, Wang et al., 2016b, Guo et al., 2015, Lü et al., 2014, Qian and Li, 2015]. Therefore, under pressure from increased coal consumption and the subsequent increased environmental pollution, there is an urgent need to control coal gangue contamination.

5.1 Review of Background

As the widest distribution and the largest reserves conventional energy in the world, coal is also an important strategic resource. It is widely used in iron and steel, electric power, chemical industry, and other industrial production and living areas. As one of main resources in the world, coal plays an important role in the energy supply of countries, especially in the developing countries [Choudhary and Shankar, 2012, Wu et al., 2014, Nicky M et al., 2016, Blumberg et al., 2015]. For example, coal accounts for 67.5% of the primary energy supply in China, which is one of the largest coal-producing countries. With the coal output increasing, a great deal of industrial residues is produced, especially the coal gangue, which is one of the main byproduct in the coal mining and washing process [Xiao et al., 2010, Zhou et al., 2014, Liu et al., 2012, Cardoso et al., 2016, Liu and Liu, 2015]. Because of the directly emitting without any treatment, huge coal gangue occupies plenty of land and causes series environmental pollution [Xiao et al., 2015, Wang et al., 2016b, Guo et al., 2015, Lü et al., 2014, Qian and Li, 2015]. Thus, along with the coal consumption and demand of environmental promotion increasing, it is extremely necessary to reduce the coal gangue pollution.

Innovative Approaches towards Ecological Coal Mining and Utilization, First Edition.
Jiuping Xu, Heping Xie, and Chengwei Lv.

To reduce the series environmental influence caused by the coal gangue, scholars from all over the word have carried out many focusing studies. For example, Zhang et al. Zhang et al. [2015a] proposed to backfill mining using coal gangue, which has been proved to be an effective method. In the glass ceramics production process, coal gangue accounts for 70% of the raw materials [Yang et al., 2012]. Wang et al. Wang et al. [2016a] developed to use coal gangue to manufacture a new type of auto-claved aerated concrete. Meanwhile, as an alternative raw material, coal gangue can be used for clay in the cement industry [Wu et al., 2015a]. To mitigate the series impacts of coal gangue, governments and institutions have made a variety of policies and laws, such as the Pollution Prevention Law of the United States, the German's Dual Recovery System (DSD) Mode, and the Chinese Management Measures for the Comprehensive Utilization of Coal Gangue. Obviously, the government policies and recycling technologies have reduced the coal gangue contamination to some extent. However, the utilization rate of coal gangue in China is lower than 15% and the coal gangue-related environmental pollution is still serious [Gao et al., 2015]. Therefore, there is an urgent need to find new method to promote the coal gangue utilization [Zhang et al., 2015b, Wu et al., 2015b, Li et al., 2010]. The coal gangue foundation (CGF), which is constructed based on the gangue-by-gangue characteristics, has been advised to increase the coal gangue utilization rate and realize coal gangue stacks reduction [Li et al., 2015, Chugh and Patwardhan, 2004, Zhou et al., 2012]. However, few studies considered the management optimization problem in the recycling process. In fact, the CGF location has a great deal of influence on many aspects, such as coal gangue stack amounts, coal production quantity, and the coal gangue transportation quantity, all of which further impacts the coal gangue utilization rate. Therefore, identifying and selecting suitable sites is the basis of CGF construction.

The CGF site selection should consider various aspects, such as costs of constructing CGF, local development planning, local geographical features, traffic conditions of sites, and the environmental pollution, which is a complicated decision process. Many studies focusing on site selection have been proposed. Under different sales tax structures, a nonlinear mixed integer programming model was established by Avittathur et al. Avittathur et al. [2005] to select the optimal location of a distribution center. He et al. He et al. [2016] developed a mean-shift algorithm to solve large-scale planar maximal covering location problems. Gołębiewski et al. Gołębiewski et al. [2013] defined the most suitable site for dismantling facilities and developed a nonlinear model for the site selection of vehicle recycling facilities. A stochastic model was formulated to decide the location and capacities of distribution centers by Paul and MacDonald for emergency stockpiles [Paul and MacDonald, 2016]. Although these studies have a great impact on determining the location of the predetermined construction project, most of them are not suitable for CGF location problems. First, the influence of geospatial data on candidate site identification is ignored by many studies; second, the colliery behavior, such as coal output and coal gangue transportation plan, also affects the final CGF site selection. However, the previous studies only consider problems from the perspective of the local authority and ignore the interaction relationships among multiple stakeholders. Further, there are many uncertainties in the CGF site selection, which also should be considered. According to the above analysis, the existing studies cannot directly apply to the CGF site selection problem. In fact, some scholars proposed geographic information system (GIS) technique to deal with the geospatial data in the candidate identification process [Yang et al., 2007, Awasthi et al., 2011]. To solve the hybrid wind solar-photovoltaic (PV) renewable energy systems location problem, a decision tool integrating GIS was developed by Aydin et al. Aydin et al. [2013]. To select waste disposal site, Eskandari et al. Eskandari et al. [2016] used GIS to mask unsuitable areas by considering landslide exposure. A methodology which integrated GIS and the spatial multi-criteria decision analysis was developed by Latinopoulos and Kechagia Latinopoulos and Kechagia [2015] for wind farm planning. These studies show that GIS technique was efficient for ensuring candidate sites, which can be used to solve the CGF site identification. In addition, considering the complex relationship among the multiple stakeholders in site selection process, some scholars have applied the bilevel programming to deal with such problems. To select the appropriate site for logistics distribution centers, Sun et al. Sun et al. [2008] established a bilevel programming model by considering logistics planning departments and benefits of customers. A bilevel programming model described the hierarchical structure among multiple decision-makers was formulated by Gang et al. to find a suitable stone industrial park site [Gang et al., 2015]. It is easy to conclude that the bilevel programming can efficiently deal with the complex interaction among the multiple stakeholders. In a word, an integrated method by combing a GIS technique and a bilevel programming model was proposed in this chapter to identify candidate CGF sites and to select the most suitable one. Obviously, the new methodology is more suitable for the practical CGF site selection problem, which consider the geospatial data and inherent relationship among multi-decision-makers at the same time.

Based on the CGF site selection process, there are two parts for a CGF location problem: identifying alternative sites and selecting the feasible site. First, to integrate geospatial data, GIS technology is used to screen and identify alternative CGF sites, which has been successfully applied to kinds of sites identification problems [Joachim and Achim, 2014]. Then, the conflicts between the government and each colliery are dealt by a bilevel model. To ensure the optimal feasible site, this chapter proposed a bilevel multiobjective programming (BLMOP) model with fuzzy variables. In the decision-making process, to achieve their own goals, the government and each colliery make their own decisions based on the practical constraints, the interaction between themselves and the other decision-makers. Further, the equilibrium between environmental protection and economic development is ensured by a multiobjective technique, and uncertain decision information is dealt by the fuzzy theory in the CGF site selection problem [Zadeh, 1965].

5.2 Key Problem Statement

CGF is crucial to effectively utilize coal gangue, improve the coal gangue recycling efficiency, and control the environmental pollution. The construction of CGF greatly affects both the coal gangue stack quantity and finance revenues of the government. An inappropriate CGF site may fail to reduce the coal gangue emission. The failure of the coal gangue reduction targets will lead to a coal supply shortage in the local area, thereby causing serious environmental pollution and impeding the development of local economy. Therefore, from the perspective of development, it is necessary to effectively identify the alternative sites and develop a rational site selection strategy for the optimal CGF site.

The CGF site identification system requires important geological data, such as geological fault data, climatic conditions, and soil data, etc. The GIS technology is an effective tool to access these data. Without the GIS technology, it is difficult to quantify the spatial relationship of geographical data in practical situations. Therefore, before determining the optimal CGF site, the GIS technology should be used to identify appropriate candidate sites. To our knowledge, Joachim and Achim Joachim and Achim [2014] and Qaddah and Abdelwahed Qaddah and Abdelwahed [2015] have introduced spatial characteristics required for CGF cite identification and it has been proved that these features are feasible in many large construction projects. In general, different sources of geospatial data have different scales. These data with different sales will increase the complexity of the site identification process. Therefore, a tool to model spatial data and a spatial analysis method should be included in the GIS technology.

There are two kinds of decision-makers in the CGF site selection system, one is the local government and the other is the owner of each coal mine. Their decisions affect the amount of piled coal gangue. The government, on behalf of the public, is responsible for environmental protection and economic development. The purpose of the government to decide the site selection is to reduce the total amount of coal gangue and to increase the total social benefits while ensuring the minimum satisfaction of the colliery. The government has a priority to determine the best CGF location to achieve the balance between the environment and the economy. Meanwhile, each colliery makes its own coal output and coal gangue transportation plan according to the government's site selection strategy. In fact, collieries are independent and each one tries to maximize profits. The conflict of CGF is obvious in the siting system, where the government tries to reduce the level of coal gangue heap and increase fiscal revenue by constructing a coal gangue facility, while each coal mine tries to maximize its profits by increasing the coal output and reducing the quantity of coal gangue transportation. In the CGF site selection system, the conflicts are obvious: the government tends to reduce the level of coal gangue stack and increase financial revenues through constructing a coal gangue facility; however, each coal colliery tends to maximize its own profits by increasing the coal output and reducing the quantity of coal gangue transportation.

In this case, the government is regarded as the leader who first determines the CGF site from the historical information of the coal field. Each colliery then makes plans

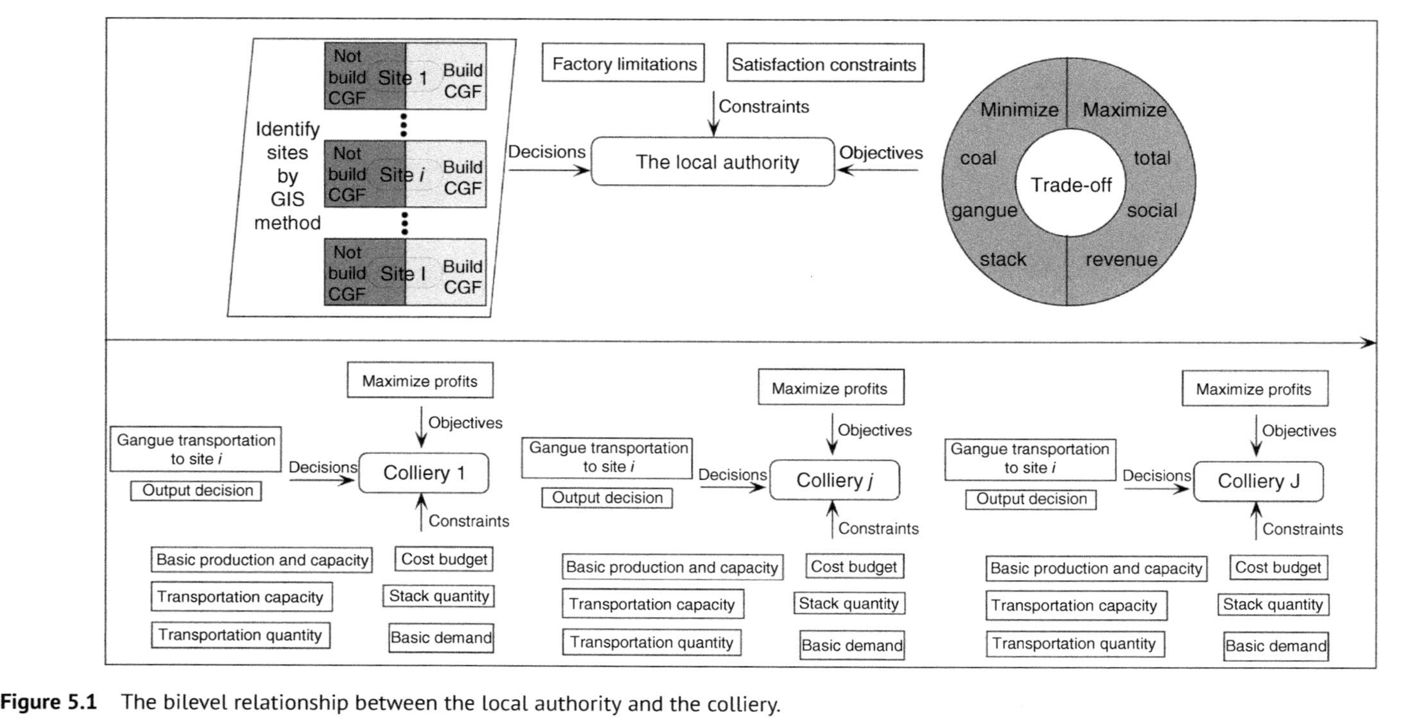

Figure 5.1 The bilevel relationship between the local authority and the colliery.

for the coal output and coal gangue transportation according to the decisions of the local authority. The authority and the colliery are both independent but there is also a mutual influence between them. From the perspective of competition, this influence plays a crucial role in the CGF site selection problem. Thus, a decision-making process with a leader–follower relationship can be considered a Stackelberg game. In order to solve conflicts between the government and the collieries, a bilevel model is introduced to present the Stackelberg game in the CGF site selection process, in which the local authority is the leader level and the collieries are the follower level. The structure of the bilevel relationship between the local authority and the colliery is illustrated in Figure 5.1.

5.3 Coal Gangue Facility Siting Method

The selection of the coal gangue recycling facility includes two parts: identifying candidate sites and selecting the optimal site. Specifically, a GIS technique is used to obtain the geospatial data and then a bilevel programming method is developed to depict the complex relationship between the local authority and the colliery owner and further select the most appropriate CGF site. To solve the CGF siting problem, we propose a GIS-based bilevel programming model and employ a Karush–Kuhn–Tucker (KKT) condition to degrade the bilevel model to a single-level model, which is discussed in detail in the following.

5.3.1 Identifying Candidate Sites Using GIS Technique

The GIS technology is an effective tool to quantify the spatial relationship of geographical data and it provides a framework for collecting, storing, analyzing, transforming, and displaying the spatial and nonspatial data. In practice, this technique has been widely used to determine candidate sites for large construction projects. In particular, it has been successfully applied to filter the appropriate site for seismic stations [Joachim and Achim, 2014] and identify potential sites for soil and water conservation technologies [Qaddah and Abdelwahed, 2015]. Based on these achievements, we attempt to use spatial data modeling tools and spatial analysis methods to screen for viable CGF sites based on GIS technology of geospatial data. For screening for viable CGF sites, we take ArcGIS software and MapInfo as the core spatial data analysis GIS platform to select feasible CGF sites. Joachim and Achim Joachim and Achim [2014] developed a GIS framework for seismic station identification, which serves as the basis for the identification of potential CGF sites by the two-layer modeling method based on GIS proposed. In this chapter, the GIS framework for seismic stations identification that is introduced by Joachim and Achim Joachim and Achim [2014] is used as the basis of the proposed GIS-based bilevel modeling method, which can identify the potential CGF sites. The flowchart of the GIS-based bilevel modeling method is illustrated in Figure 5.2.

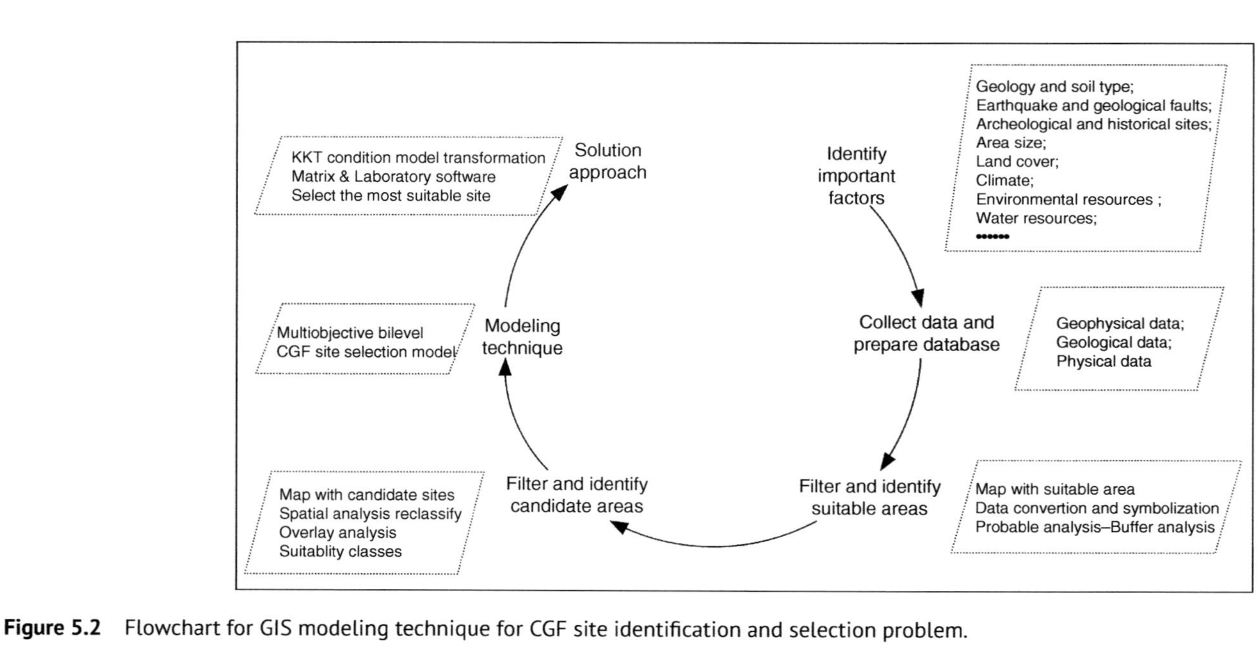

Figure 5.2 Flowchart for GIS modeling technique for CGF site identification and selection problem.

5.3.2 Selecting the Optimal Site Using the Modeling Technique

This section first give a description of the relevant assumptions and then constructs a bilevel multiobjective model under the uncertain environment for selecting the optimal CGF site based on the identification results. The mathematical description of the problem is presented as follows.

5.3.2.1 Assumptions

To construct a model for the CGF site selection problem, the following assumptions were adopted:

1. The CGF's construction costs are undertaken by the local authority.
2. The tax rate is the same for each colliery.
3. The CGF capacity is the same at each alternative site.

5.3.2.2 Notations

The following mathematical notations are used to describe the CGF site selection problem.

Indices

i = potential site index, where $i = 1, 2, \dots, I$;
j = colliery index, where $j = 1, 2, \dots, J$;

Certain parameters

I = number of potential CGF sites;
J = umber of collieries;
C_j^{basic} = basic production of the colliery j;
$C_j^{capacity}$ = maximum production capacity of the colliery j;
T_{ij}^{cg} = volume of transportation from the colliery j to CGF i;
ϑ = economic revenue coefficient of unit tonne coal gangue in the CGF;
P_j^c = sale price of unit tonne coal of the colliery j;
D^{basic} = basic amount of coal the coal field should supply;
β = satisfaction level of the colliery;
α = the excess stack level permitted;
H_j = historical coal gangue stack data at colliery j;
d_{ij} = distance between the CGF i and colliery j;
C_j^m = budget for each colliery.

Uncertain parameters

$\widetilde{C_j^o}$ = unit operating and maintenance costs for colliery j;
$\widetilde{C_j^t}$ = unit coal gangue transportation costs for the colliery j;
$\widetilde{E_j^g}$ = coal gangue emission coefficient for a unit tonne of coal at colliery j;

Decision variables
X_i = binary variable, if site i selected to build CGF, $X_i = 1$, otherwise $X_i = 0$;
Y_j = output decision of colliery j;
R_{ij} = coal gangue transportation quantity from colliery j to CGF i;

5.3.2.3 Model Formulation

To protect the environment and make full use of the coal gangue, the local authority attempts to build a coal gangue facility based on the 3R principle. However, as this is a large project, it is very important to control construction costs and reduce the total coal gangue at the colliery. For the colliery, an optimal CGF site means greater profits. Therefore, the CGF site has a significant effect on the colliery's behavior. To solve the complex coal gangue facility site selection problem, a bilevel multiobjective model is constructed from an environmental protection and economic development view point. The local authority on the upper level selects the CGF site and the colliery's owner on the lower level decides on the coal output and coal gangue transportation plan.

5.3.2.3.1 Coal Gangue Facility Site Selection

In the CGF-SS (site selection) system, there are various factors (i.e. environmental protection and social revenues) that affect the local authority's behavior, so the authority seeks to try to achieve multiple objectives. In this chapter, two objectives minimize the total coal gangue stack level and maximize the total social revenues.

Coal Gangue Stack Minimization Because the amount of coal gangue at the colliery can have the adverse bad effects on the environment and human health, it is necessary to control the coal output and reduce the coal gangue stack quantity. In the real world, reducing or mitigating the pollution effect on environmental and human health is one of the main objectives of the local authority; therefore, minimizing the total coal gangue stack levels at the colliery is the first objective.

Significant research has been conducted on the coal gangue pollution problem, from which the establishment of a coal gangue facility based on the 3R principle has been found to be the best scientific reference for policymakers. According to the above assumptions, when a CGF site is identified, collieries j have a fixed transportation capacity, so each colliery decides on their respective transportation coal gangue quantity, i.e. $\sum_{i=1}^{I} X_i R(ij)$. However, the coal quality is different in each colliery, the coal gangue production coefficient for a unit tonne of coal is also different, meaning that the production coal gangue quantity from a colliery is $\widetilde{E_j^g} Y_j$. Therefore, the coal gangue stack at the colliery j is $\left(\widetilde{E_j^g} Y_j - \sum_{i=1}^{I} X_i R(ij)\right)$.

There are in fact many factors such as mining methods, geographical structures, and processing methods that can have a significant effect on the coal gangue emissions coefficient for a unit tonne of coal at colliery j (i.e. $\widetilde{E_j^g}$); therefore, it is difficult to estimate $\widetilde{E_j^g}$ as a certain parameter. Consistent with reality, experienced experts and engineers are asked to estimate the most likely coal gangue emission coefficient, from which a value range is determined. The minimum value is regarded as the left border (i.e. a_{1j}), the maximum value is regarded as the right border (i.e. a_{3j}), and the most possible value is a_{2j}, $(a_{1j} \leq a_{2j} \leq a_{3j})$. It is reasonable to deal this uncertainty using fuzzy theory [27], which is a triangular fuzzy number $\widetilde{E_j^g} = (a_{1j}, a_{2j}, a_{3j})$, where $a[.]$ are the parameters in the membership function for the triangular fuzzy number $\widetilde{E_j^g}$.

Because the parameter $\widetilde{E_j^g}$ is fuzzy, the objective function Z_1 is also a fuzzy goal which can be described precisely as fuzzy sets in the spaces of the alternatives. An optimistic-pessimistic parameter is imported using the fuzzy measure Me to obtain the decision-maker attitudes [Xu and Zhou, 2011] (see Figure 5.3). From Xu's methods, the fuzzy variable $\widetilde{E_j^g}$ can be converted into a crisp form as follows:

$$\widetilde{E_j^g} \to E\left[\widetilde{E_j^g}\right] = \frac{1-\lambda}{2}(a_{1j} + a_{2j}) + \frac{\lambda}{2}(a_{2j} + a_{3j}) = \frac{1}{2}\left[\left(a_{2j} + a_{1j}\right) + \lambda\left(a_{3j} - a_{1j}\right)\right] \tag{5.1}$$

From a system's point of view, the optimal CGF site minimizes the total piled coal gangue quantity and the total coal gangue stacks, which can be expressed as follows:

$$\min_{X_i} Z_1 = \sum_{j=1}^{J}\left(E\left[\widetilde{E_j^g}\right] Y_j - \sum_{i=1}^{I} X_i R(ij)\right) \tag{5.2}$$

Total Social Revenue Maximization At the same time as seeking to reduce the adverse effects from the coal gangue piled at different collieries, the local authority is also seeking to maximize total revenues, which includes taxes revenues and the CGF revenues. Further, the taxes noticeably influence the collieries' output decisions. Let ψ be the tax rate, P_j^c be the sale price of the colliery j, and Y_j be the coal output of colliery decided by the collieries' owner, thus the taxes revenues is $\sum_{j=1}^{J} \psi P_j^c Y_j$. As a possible renewable resource, the coal gangue processed at the CGF is also revenue producing, which is an important factor for the local authority. The economic revenue coefficient for a unit of coal gangue at the CGF is ϑ, so total revenues can be expressed as: $\vartheta \sum_{j=1}^{J} \sum_{i=1}^{I} X_i R_{ij}$. As the aim of the local authority is to maximize total revenues, the objective function is described as:

$$\max_{X_i} Z_2 = \sum_{j=1}^{J} \psi P_j^c Y_j + \vartheta \sum_{j=1}^{J} \sum_{i=1}^{I} X_i R_{ij} \tag{5.3}$$

Satisfaction Constraints Apart from protecting the environment and increasing total social revenues, the colliery production plan can have a significant impact on the local authority's behavior. Therefore, the site selection decisions need to ensure a certain colliery satisfaction degree. Depending on the coal gangue stack allowance, the colliery requires a greater transportation capacity, but the local authority's CGF site selection means that the transportation capacity is fixed. However, as the transportation capacity is closely related to the colliery's satisfaction, the local authority defines

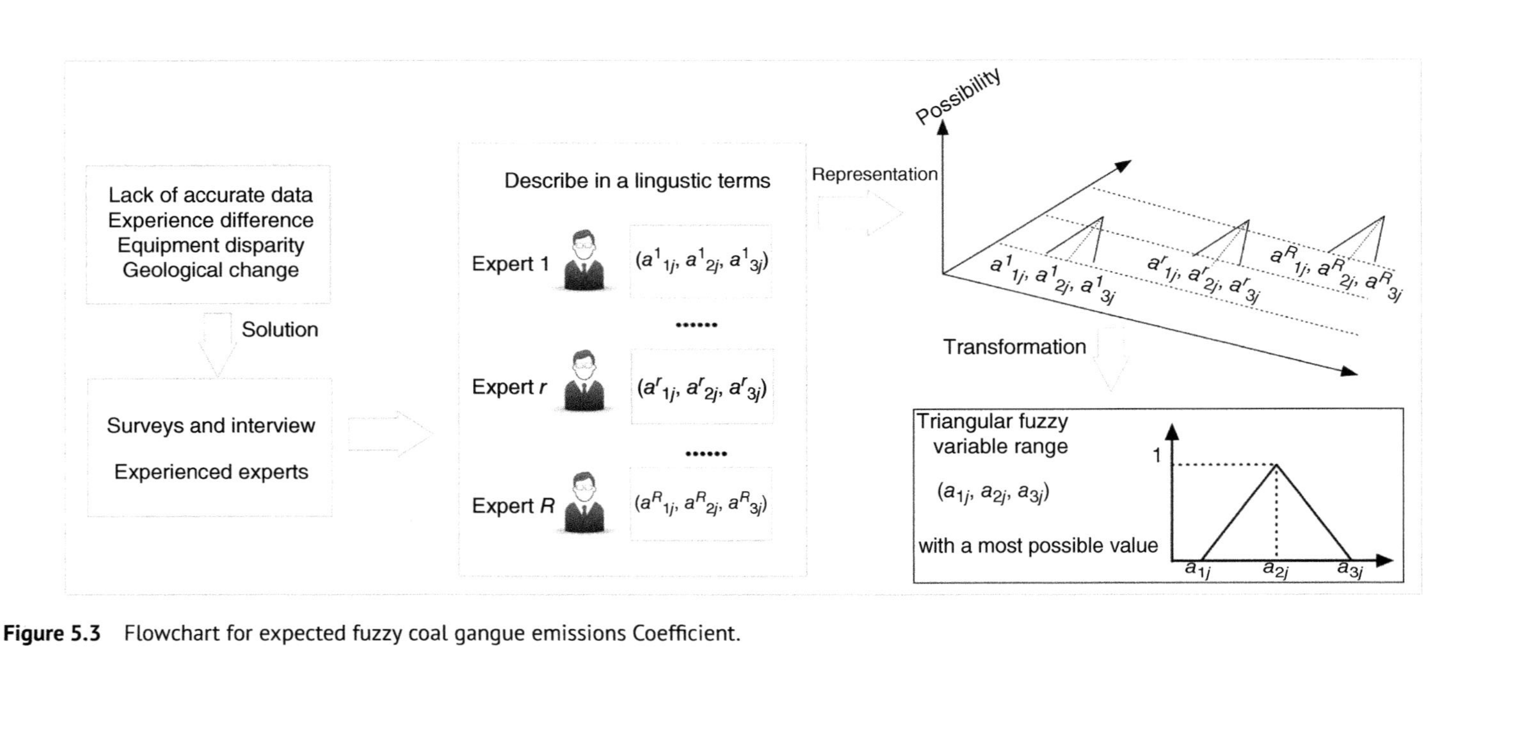

Figure 5.3 Flowchart for expected fuzzy coal gangue emissions Coefficient.

the location satisfaction degree for each colliery using the following [Xu, 2014]:

$$g_j(X_i) = \begin{cases} 0, & \sum_{i=1}^{I} X_i T_{ij}^{cg} < E\left[\widetilde{E_j^g}\right] Y_j^{basic} \\ \frac{\sum_{i=1}^{I} X_i T_{ij}^{cg} - E\left[\widetilde{E_j^g}\right] Y_j^{basic}}{E\left[\widetilde{E_j^g}\right]\left(Y_j^{capacity} - Y_j^{basic}\right)}, & E\left[\widetilde{E_j^g}\right] Y_j^{basic} < \sum_{i=1}^{I} X_i T_{ij}^{cg} < E\left[\widetilde{E_j^g}\right] Y_j^{capacity} \\ 1, & \sum_{i=1}^{I} X_i T_{ij}^{cg} > E\left[\widetilde{E_j^g}\right] Y_j^{capacity} \end{cases} \tag{5.4}$$

This function means that there is a basic output ($E\left[\widetilde{E_j^g}\right] Y_j^{basic}$) and a capacity output ($E\left[\widetilde{E_j^g}\right] Y_j^{capacity}$) for the coal gangue at each colliery. If $\sum_{i=1}^{I} X_i T_{ij}^{cg}$ is less than $E\left[\widetilde{E_j^g}\right] Y_j^{basic}$, the satisfaction degree is zero; if $\sum_{i=1}^{I} X_i T_{ij}^{cg}$ is more than $E\left[\widetilde{E_j^g}\right] Y_j^{capacity}$, the satisfaction degree is 1. To ensure the policy successful, the local authority needs to ensure that the satisfaction degree at each colliery is more than β, which can be expressed as follows:

$$\min\left(g_j(X_i)\right) \geq \beta \tag{5.5}$$

Factory Limitations Considering the construction costs and efficiency, the local authority can select only one site from the total alternative sites to built the CGF.

$$\sum_{i=1}^{I} x_i = 1 \tag{5.6}$$

5.3.2.3.2 Colliery Coal Production and Gangue Transportation Plan

As the main coal gangue producer, the coal mine company owner seeks to maximize profits consideration of the taxes. Therefore, an efficient production plan is essential.

Profits Maximization The local authority site selection strategy restrains colliery behavior; however, colliery plans impacts the local authority's achievement of their own objectives to minimize the total coal gangue stacks and maximize taxes revenues. Each colliery, therefore, requires an optimal production and transportation plan to maximize the profits.

As a scarce resource, coal supply is generally less than demand. Colliery revenues, therefore, depends mainly on the coal output, which is expressed as $Y_j P_j^c$. Total costs are made up of operating costs, taxes, and coal gangue-related costs. Operating costs are expressed as $Y_j \widetilde{C_j^o}$ and taxes are expressed as $\psi P_j^c Y_j$. The coal gangue-related costs are transportation costs and taxes. Each colliery's coal gangue transportation quantity is decided by the colliery i.e. R_{ij}. With the transportation costs at each colliery expressed as $\left(\sum_{i=1}^{I} X_i R_{ij}\right) \widetilde{C_j^t}$. So, the total costs are $Y_j \widetilde{C_j^o} + \psi P_j^c Y_j + \left(\sum_{i=1}^{I} X_i R_{ij}\right) \widetilde{C_j^t}$. Similar to parameter $\widetilde{E_j^g}$, the variable costs (i.e. $\widetilde{C_j^o}, \widetilde{C_j^t}$) of the CGF-SS system are also triangular fuzzy numbers. That is, $\widetilde{C_j^o} = (b_{1j}, b_{2j}, b_{3j})$, $\widetilde{C_j^t} = (c_{1j}, c_{2j}, c_{3j})$, and $b[.], c[.]$, are the respective membership function parameters for the above triangular fuzzy numbers. Using the fuzzy measure Me, the fuzzy variables $\widetilde{C_j^o}$ and $\widetilde{C_j^t}$ can be transformed into crisp forms $E\left[\widetilde{C_j^o}\right], E\left[\widetilde{C_j^t}\right]$, so the total profits at colliery j are established as:

$$\max_{Y_j} F_j = Y_j P_j^c - E\left[\widetilde{C_j^o}\right] Y_j - \psi P_j^c Y_j - E\left[\widetilde{C_j^t}\right]\left(\sum_{i=1}^{I} X_i d_{ij} R_{ij}\right) \tag{5.7}$$

Basic Production and Capacity Limitations As the CGF project investment cycle is long, there are many fixed costs for each colliery in the construction process, even when the output is 0, so the collieries need to maintain a certain basic output. In other words, the coal output cannot less than a basic line, C_j^{basic}. Further, based on the economic development plan and the productivity restrictions, coal output cannot be greater than the capacity, $C_j^{capacity}$.

$$C_j^{basic} < Y_j \leq C_j^{capacity} \tag{5.8}$$

Cost Budget Constraints At the CGF site selection problem, total costs are strictly controlled by the colliery decision-makers and cannot be more than the allocated budget C_m.

$$E\left[\widetilde{C_j^o}\right] Y_j + E\left[\widetilde{C_j^t}\right]\left(\sum_{i=1}^{I} X_i d_{ij} R_{ij}\right) \leq C_j^m \tag{5.9}$$

Stack Quantity Constraints Even if the local authority allows coal gangue piles at the colliery, but there are stack level limitations α which are based on historical data H_j, set by the local authority to avoid unrecoverable effects on the environment, so under these restrictions, the collieries cannot generate additional coal gangue:

$$E\left[\widetilde{E_j^g}\right] Y_j - \sum_{i=1}^{I} X_i R_{ij} \leq \alpha H_j \tag{5.10}$$

Basic Demand Constraint One of the main targets for the local authority is to maintain a stable and sustainable development of the coal fields, so all the collieries' coal outputs should be more than the basic market demand D^{basic}.

$$\sum_{j=1}^{J} Y_i \geq D^{basic} \tag{5.11}$$

Transportation Quantity Constraint Under the local authority, each colliery attempts to transport as great a coal gangue quantity as possible to the CGF. However, the transportation quantity is limited by the coal production plan. Namely, the total coal gangue transportation quantity at each colliery cannot be more than the production quantity $E\left[\widetilde{E_j^g}\right] Y_j$.

$$\sum_{i=1}^{I} X_i R_{ij} \leq E\left[\widetilde{E_j^g}\right] Y_j \tag{5.12}$$

Transportation Capacity Constraint At the same time, the transportation quantity is limited by transportation capacity. In other words, the transportation quantity decision R_{ij} cannot be more than the transportation capacity T_{ij}^{cg}.

$$R_{ij} \leq T_{ij}^{cg} \tag{5.13}$$

5.3.2.3.3 Global Model for CGF Site Selection

To reduce coal gangue stack, the local authority establishes a coal gangue facility based on the 3R principle policy, for which CGF site selection is particularly important. The government is responsible for the selection of the CGF site, but is constrained by the site decision, local policies, and the coal output and gangue transportation plan from each colliery. When selecting a CGF site, the local authority first ensures the collieries' satisfaction degree and then seeks to minimize the total coal gangue stacks, as expressed in Eq. (5.2). The site selection decision should also maximize total social revenues, as expressed in Eq. (5.3), which is dependent on the collieries' decisions. To achieve profit maximization, as expressed in Eq. (5.7), colliery owners decide on the coal output and gangue transportation plan based on the selected CGF site. There are, therefore, unavoidable conflicts within the CGF-SS system, so it can be described as a typical Stackelberg game between the local authority and collieries. In this situation, it is too difficult to describe the complex inherent relationship using a simple site selection model. Therefore, based on Eqs. (5.1)–(5.13), a bilevel model is formulated to deal with these conflicts as follows:

In consideration of the conflict between the local authority and the colliery objectives, the model in Eq. (5.14) is proposed optimal CGF site selection, within which the local authority considers each colliery's owner satisfaction degree. The model can also be applied to an analogical situation when adding, reducing, or altering the objective functions or constraints. For example, if the government prioritizes gangue stack reduction without considering social revenues, then the objective function in Eq. (5.7) focus only on total coal gangue stack minimization; if the colliery wished to prioritize the trade-off between increased profits and total costs, the cost budget constraint should be removed. In this model, parameter β expresses the colliery cooperation degree to the CGF-SS policy and parameter α indicates the local authority's lowest limits for the gangue stacks. Intuitively, the Stackelberg equilibrium model established in Eq. (5.7) is effectively describing the CGF-SS problem based on the 3R principle.

$$\begin{cases} \min\limits_{X_i} Z_1 = \sum\limits_{j=1}^{J} \left(E\left[\widetilde{E_j^g}\right] Y_j - \sum\limits_{i=1}^{I} X_i R(ij) \right) \\ \max\limits_{X_i} Z_2 = \sum\limits_{j=1}^{J} \psi P_j^c Y_j + \vartheta \sum\limits_{j=1}^{J} \sum\limits_{i=1}^{I} X_i R_{ij} \\ \text{s.t.} \begin{cases} \min\ (g_j(X_i)) \geq \beta \\ \sum\limits_{i=1}^{I} X_i = 1 \\ X_i = 0 \text{ or } 1 \\ \forall i = 1, 2, \ldots, I \\ \forall j = 1, 2, \ldots, J \\ \max\limits_{Y_j, R_{ij}} F_j = Y_j P_j^c - E\left[\widetilde{C_j^o}\right] Y_j - \psi P_j^c Y_j - E\left[\widetilde{C_j^t}\right] \left(\sum\limits_{i=1}^{I} X_i d_{ij} R_{ij} \right) \\ \text{s.t.} \begin{cases} C_j^{basic} < Y_j \leq C_j^{capacity} \\ E\left[\widetilde{C_j^o}\right] Y_j + E\left[\widetilde{C_j^t}\right] \left(\sum\limits_{i=1}^{I} X_i d_{ij} R_{ij} \right) \leq C_j^m \\ E\left[\widetilde{E_j^g}\right] Y_j - \sum\limits_{i=1}^{I} X_i R_{ij} \leq \alpha H_j \\ \sum\limits_{j=1}^{J} Y_i \geq D^{basic} \\ \sum\limits_{i=1}^{I} X_i R_{ij} \leq E\left[\widetilde{E_j^g}\right] Y_j \\ R_{ij} \leq T_{ij}^{cg} \\ \forall i = 1, 2, \ldots, I \\ \forall j = 1, 2, \ldots, J \end{cases} \end{cases} \end{cases} \tag{5.14}$$

5.3.3 Model Transformation

The proposed model in Eq. (5.7) is a bilevel multiobjective decision-making problem, which represents the government's targets and reflects the complex relationship between the local authority and the colliery. However, there are many difficulties in multiple objective problems because the dimensionality of each objective function is generally different. Therefore, it is not suitable to simply reduce the multiple objectives to a single objective using weight coefficients. Assuming that the local authority has an acceptable stack level limit (i.e. γ), the objective function for the total coal gangue stack $\min\limits_{X_i} Z_1$ can be transformed into an inequality as follows:

$$\sum_{j=1}^{J} \left(E\left[\widetilde{E_j^g}\right] Y_j - \sum_{i=1}^{I} X_i R_{ij} \right) \leq \gamma \sum_{j=1}^{J} H_j \tag{5.15}$$

In addition, it is also particularly difficult to handle the bilevel programming model because it is a NP-hard problem [Ben-Ayed and Blair, 1990]. Research on bilevel programming problems has found that the KKT approach is an efficient method for converting the bilevel problem into a single-level problem [Hanson, 1981]. The core theory for the KKT approach is that the game between the upper level and the lower level is reduced to a single decision-making problem by constraining the upper level using the KKT conditions [Liu et al., 2007]. Therefore, based on previous work by Shi et al. Shi et al. [2005], the KKT optimality conditions are formulated as follows:

$$
\begin{cases}
C_j^{basic} < Y_j \leq C_j^{capacity} \\
E\left[\widetilde{C_j^o}\right] Y_j + E\left[\widetilde{C_j^t}\right] \left(\sum_{i=1}^{I} X_i R_{ij}\right) \leq C_j^m \\
E\left[\widetilde{E_j^g}\right] Y_j - \sum_{i=1}^{I} X_i R_{ij} \leq \alpha H_j \\
\sum_{j=1}^{J} Y_i \geq D^{basic} \\
\sum_{i=1}^{I} X_i R_{ij} \leq E\left[\widetilde{E_j^g}\right] Y_j \\
R_{ij} \leq T_{ij}^{cg} \\
(1 - \upsilon_3) E\left[\widetilde{C_j^t}\right] \sum_{i=1}^{I} X_i d_{ij} + \upsilon_4 - \upsilon_6 - \upsilon_7 = 0 \\
(1 - \psi) P_j^c - \upsilon_1 + \upsilon_2 - (1 + \upsilon_3) E\left[\widetilde{C_j^o}\right] + \upsilon_5 + (\upsilon_6 - \upsilon_4) E\left[\widetilde{E_j^g}\right] = 0 \\
\upsilon_1 f_1(X_i, Y_j, R_{ij}) + \upsilon_2 f_2(X_i, Y_j, R_{ij}) + \upsilon_3 f_3(X_i, Y_j, R_{ij}) + \upsilon_4 f_4(X_i, Y_j, R_{ij}) \\
\quad + \upsilon_5 f_5(X_i, Y_j, R_{ij}) + \upsilon_6 f_6(X_i, Y_j, R_{ij}) + \upsilon_7 f_7(X_i, Y_j, R_{ij}) = 0 \\
f_1(X_i, Y_j, R_{ij}) = C_j^{capacity} - Y_j \geq 0 \\
f_2(X_i, Y_j, R_{ij}) = Y_j - C_j^{basic} \geq 0 \\
f_3(X_i, Y_j, R_{ij}) = C_j^m - E\left[\widetilde{C_j^o}\right] Y_j - E\left[\widetilde{C_j^t}\right] \left(\sum_{i=1}^{I} X_i d_{ij} R_{ij}\right) \geq 0 \\
f_4(X_i, Y_j, R_{ij}) = \alpha H_j - E\left[\widetilde{E_j^g}\right] Y_j + \sum_{i=1}^{I} X_i R_{ij} \geq 0 \\
f_5(X_i, Y_j, R_{ij}) = \sum_{j=1}^{J} Y_i - D^{basic} \geq 0 \\
f_6(X_i, Y_j, R_{ij}) = E\left[\widetilde{E_j^g}\right] Y_j - \sum_{i=1}^{I} X_i R_{ij} \geq 0 \\
f_7(X_i, Y_j, R_{ij}) = T_{ij}^{cg} - R_{ij} \geq 0 \\
\forall i = 1, 2, \dots, I \\
\forall j = 1, 2, \dots, J
\end{cases}
\tag{5.16}
$$

From the above introduction, the bilevel multiobjective model is transformed into an equivalent crisp model as shown in Eq. (5.17):

In the objective conversion process, the stack level limit allows the government to achieve social revenues maximization and ensures the sustainable development of the local environment. In the KKT transformation, each local authority decision corresponds to an optimal colliery behavior. Therefore, the complex inherent conflicts between the government and colliery owner are integrated in the KKT conditions, which changes the CGF site selection problem from a game between the local authority and colliery to a single decision-making problem by the local authority. Though a single-level programming with only one objective is also complicated, it can be solved using the existing algorithms [Shi et al., 2005].

5.4 Case Study

In this section, a case study for a practical CGF site identification and selection problem in the Yanzhou coal field in China is introduced to demonstrate the effectiveness of the proposed optimal methodology.

5.4.1 Case Region Presentation

With the growth of the populations all over the world, the demand for coal is increasing in a rapid speed. According to the US International Energy Statistics (2014), coal consumption accounted for approximately 69% of total energy consumption in China, leading to a large amount of coal gangue plied in the coal field causing negative influences to the environment, and put pressure on the Chinese government to reduce the accumulation of coal gangue. Therefore, the optimal site of CGF has a significant impact on the coal gangue reduction target, and the local authority needs to consider geospatial data and the colliery when they identify and select the site. In the following, taking the Yanzhou coal field as an application case, the applicability of the proposed method is discussed.

Yanzhou coal field is located in the southwest of Shandong province, covering an area of 3400 km^2. It consists of two parts, Yanzhou and Zaozhuang, and spans 12 counties and cities, including Jining, Zou county, Yanzhou, Qufu, Juye, and Zaozhuang. The total coal reserves are estimated at about 9.1 billion tonnes. There are six major collieries, the South Tuen colliery, the Xinglongzhuang colliery, the Baodian colliery, the Dongtan colliery, the Jining no.2 colliery, the Jining no.3 colliery (i.e. ST-South Tuen, XLZ-Xing longzhuang, BD-Baodian, DT-Dongtan, JN2-Jining no.2, JN3-Jining no.3). Yanzhou coal field produces and the carbon in gangue is greater than 20%, which means that it can be used for the electricity generation [Li et al., 2007]. Thus, the local authority wants to build a coal gangue power facility (CGPF) to accomplish the goal of reducing gangue. Furthermore, taking into account the cost constraints and economic development, the coal gangue stack reduction target is set at 10% (i.e. $\gamma = 0.9$), which shows the government's attitude toward environmental protection.

$$
\begin{cases}
\max\limits_{X_i} Z_2 = \sum\limits_{j=1}^{J} \psi P_j^c Y_j + \vartheta \sum\limits_{j=1}^{J} \sum\limits_{i=1}^{I} X_i R_{ij} \\
\text{s.t.} \begin{cases}
\sum\limits_{j=1}^{J} \left(E\left[\widetilde{E_j^g}\right] Y_j - \sum\limits_{i=1}^{I} X_i R_{ij} \right) \leq \gamma \sum\limits_{j=1}^{J} H_j \\
\min\left(g_j(X_i)\right) \geq \beta \\
\sum\limits_{i=1}^{I} X_i = 1 \\
X_i = 0 \text{ or } 1 \\
C_j^{basic} < Y_j \leq C_j^{capacity} \\
E\left[\widetilde{C_j^o}\right] Y_j + E\left[\widetilde{C_j^t}\right] \left(\sum\limits_{i=1}^{I} X_i R_{ij} \right) \leq C_j^m \\
E\left[\widetilde{E_j^g}\right] Y_j - \sum\limits_{i=1}^{I} X_i R_{ij} \leq \alpha H_j \\
\sum\limits_{j=1}^{J} Y_i \geq D^{basic} \\
\sum\limits_{i=1}^{I} X_i R_{ij} \leq E\left[\widetilde{E_j^g}\right] Y_j \\
R_{ij} \leq T_{ij}^{cg} \\
(1 - \upsilon_3) E\left[\widetilde{C_j^t}\right] \sum\limits_{i=1}^{I} X_i d_{ij} + \upsilon_4 - \upsilon_6 - \upsilon_7 = 0 \\
(1 - \psi) P_j^c - \upsilon_1 + \upsilon_2 - (1 + \upsilon_3) E\left[\widetilde{C_j^o}\right] + \upsilon_5 + (\upsilon_6 - \upsilon_4) E\left[\widetilde{E_j^g}\right] = 0 \\
\upsilon_1 f_1(X_i, Y_j, R_{ij}) + \upsilon_2 f_2(X_i, Y_j, R_{ij}) + \upsilon_3 f_3(X_i, Y_j, R_{ij}) + \upsilon_4 f_4(X_i, Y_j, R_{ij}) \\
\quad + \upsilon_5 f_5(X_i, Y_j, R_{ij}) + \upsilon_6 f_6(X_i, Y_j, R_{ij}) + \upsilon_7 f_7(X_i, Y_j, R_{ij}) = 0 \\
f_1(X_i, Y_j, R_{ij}) = C_j^{capacity} - Y_j \geq 0 \\
f_2(X_i, Y_j, R_{ij}) = Y_j - C_j^{basic} \geq 0 \\
f_3(X_i, Y_j, R_{ij}) = C_j^m - E\left[\widetilde{C_j^o}\right] Y_j - E\left[\widetilde{C_j^t}\right] \left(\sum\limits_{i=1}^{I} X_i d_{ij} R_{ij} \right) \geq 0 \\
f_4(X_i, Y_j, R_{ij}) = \alpha H_j - E\left[\widetilde{E_j^g}\right] Y_j + \sum\limits_{i=1}^{I} X_i R_{ij} \geq 0 \\
f_5(X_i, Y_j, R_{ij}) = \sum\limits_{j=1}^{J} Y_i - D^{basic} \geq 0 \\
f_6(X_i, Y_j, R_{ij}) = E\left[\widetilde{E_j^g}\right] Y_j - \sum\limits_{i=1}^{I} X_i R_{ij} \geq 0 \\
f_7(X_i, Y_j, R_{ij}) = T_{ij}^{cg} - R_{ij} \geq 0 \\
\forall i = 1, 2, \ldots, I \\
\forall j = 1, 2, \ldots, J.
\end{cases}
\end{cases}
\tag{5.17}
$$

5.4.2 GIS Technique

Here in this section, we applied the ArcGIS software to produce a CGPF map for the Yanzhou coal field. Based on the buffer zone dimensions proposed by

Argyriou et al. [2016], the buffer zone characteristics in this chapter are as follows; namely, 0.25 km from electricity networks and vehicular roads, 1 km from farm pumps, water wells, residential areas, railways, and paved roads, and 5 km from pipelines and airports. Then, we divide the suitable area into 5 km × 7 km units using ArcGIS in Environmental Systems Research Institute (ESRI). Thirdly, we use the Global Digital Elevation Model to analyze the topographic characteristics [Rahmati et al., 2016], the GRID module in ARCInfo to determine slope information [Srivastavaa and Singhb, 2016], and the overlay analysis to ensure the location of the alternative sites [Kerry et al., 2016]. Finally, 10 candidate sites are finally identified.

5.4.3 Modeling Technique

Based on the GIS technique results, the modeling was used to select the most appropriate site for the CGPF construction project.

5.4.4 Data Collection

In this case, six major collieries were considered. Certain parameter shown in Table 5.1 were obtained from the Statistical Yearbook of Chinese Coal Industry 53N [2013] and website information published by these companies. Distance and transportation capacity data between each colliery and the CGPF candidate site shown in Tables 5.2 and 5.3 were collected from field research. Uncertain parameters in fuzzy form and shown in Table 5.4 were estimated from expert's experience. Coal field parameters shown in Table 5.5 were presented to ensure the project effectiveness. Reduction target was collected from the Chinese Waste Recycling Technology Engineering Twelfth Five-Year Special Plan 54N [2012], and the basic coal demand were collected from the Chinese Coal Industry Development Plan for the Twelfth Five-Year Plan 55N [2010]. The coefficient of economic revenue for a unit of gangue was obtained from an existing CGPF, the associated satisfaction degree was set at 0.8.

Table 5.1 Basic information for each colliery.

	Basic output C_j^{basic} (10^6 tonnes)	Production capacity $C_j^{capacity}$ (10^6 tonnes)	Sale price P_j^c (yuan/tonne)	History output H_j (10^6 tonne)	Cost budget C_j^m (10^8 yuan)
ST	2.4	4	808	3.9	20
XLZ	3	7.5	776	6.9	46
BD	3	6.5	770	6.2	30
DT	4	9	798	8.1	53
JN2	4	6	810	4.4	29
JN3	5	7	773	6.5	32

Table 5.2 Distance between the colliery and the CGPF candidate site (km).

	Site 1	Site 2	Site 3	Site 4	Site 5	Site 6	Site 7	Site 8	Site 9	Site 10
ST	46	52	101	52	68	147	89	85	61	76
XLZ	72	55	85	97	98	62	111	95	94	148
BD	55	75	98	49	54	78	94	94	64	74
DT	98	69	64	152	49	94	67	76	130	83
JN2	62	42	88	71	152	126	82	82	109	119
JN3	53	90	76	68	132	66	90	67	80	78

Table 5.3 Transportation capacity between the colliery and the CGPF candidate site (10^6 tonnes).

	Site 1	Site 2	Site 3	Site 4	Site 5	Site 6	Site 7	Site 8	Site 9	Site 10
ST	0.3691	0.3356	0.1025	0.3356	0.2034	0.0638	0.1271	0.1402	0.2445	0.1534
XLZ	0.3168	0.5658	0.2809	0.2625	0.2621	0.4873	0.2328	0.2642	0.2651	0.1973
BD	0.5007	0.2478	0.1898	0.5219	0.5114	0.2415	0.2128	0.2128	0.3876	0.2513
DT	0.3268	0.3632	0.3679	0.1872	0.5743	0.3306	0.3653	0.3589	0.2105	0.3412
JN2	0.4375	0.6261	0.3629	0.4198	0.1918	0.2371	0.3811	0.3811	0.3363	0.2979
JN3	0.5837	0.2051	0.2753	0.3905	0.1873	0.4015	0.2383	0.3984	0.2607	0.2692

Table 5.4 Uncertainty parameter for each colliery.

	Gangue emission coefficient $\widetilde{E}_j^g$	Unit operating cost $\widetilde{C}_j^o$ (yuan/tonne)	Unit transportation cost $\widetilde{C}_j^t$ (yuan/tonne)
ST	(18.21%,19.86%,20.13%)	(221,223,225)	(56,58,60)
XLZ	(13.71%,16.35%,18.29%)	(199,207,214)	(49,52,55)
BD	(15.36%,18.07%,19.74%)	(211,219,232)	(53,55,58)
DT	(12.92%,15.03%,16.08%)	(208,220,224)	(52,56,57)
JN2	(15.74%,18.62%,19.91%)	(207,218,227)	(51,54,57)
JN3	(16.12%,17.49%,19.25%)	(198,213,226)	(50,53,56)

Table 5.5 Input parameters in the CGPF site selection system.

γ (%)	ϑ (yuan/tonne)	D^{basic} (10^6 tonnes)	Satisfaction degree	Tax rate (%)
90	280	30	0.8	20

5.4.5 Computational Results and Analysis

Coal gangue byproduct from the coal mining and washing has resulted in significant environmental pollution. Based on the 3R principle, the government seeks to build a coal gangue facility and limits the coal gangue stack levels to reduce the adverse effects and increase social revenues. After transformation using the reality-oriented meaning and KKT conditions, the single-level programming with only one objective was run on the Matlab 2007, and the results were calculated based on $\theta_1 = \theta_2 = \theta_3 = \theta_4 = \theta_5 = \theta_6 = 0.1$, and $\theta_7 = 0.4$, where $\theta_1 + \theta_2 + \theta_3 + \theta_4 + \theta_5 + \theta_6 + \theta_7 = 1$. The stack level significantly influences colliery owner behavior and the government social revenues. Consequently, to ensure the efficiency of the proposed model, the coal gangue stack limit α was considered the following five scenarios.

5.4.5.1 Scenario 1: $\alpha = 1.0$

When the stack limit level is 1.0, this indicates that the coal gangue stack quantity at each colliery cannot be more than in the previous year. In this situation, the colliery needs to transport the extra coal gangue to the coal gangue facility, if it wishes to expand scale. The objective values for both the local authority and the colliery owner and the integrated value for each candidate site were calculated, as shown in Table 5.6. From Table 5.6, Site 2 was seen to be the most suitable site for the CGPF as it ranked first for integrated value. Site 2 was satisfactory to the local authority, the colliery owners, and the local consumers. The objective values for the local authority were $Z_2 = 71.8963 \times 10^8$ yuan, indicating that the social revenues was 71.8963×10^8 yuan from the colliery and the coal gangue recycling facility per year. For the colliery owner, the optimal profits were determined as $F_1 = 11.6640 \times 10^8$ yuan, $F_2 = 21.7772 \times 10^8$ yuan, $F_3 = 14.9135 \times 10^8$ yuan, $F_4 = 23.2036 \times 10^8$ yuan, $F_5 = 15.3264 \times 10^8$ yuan, and $F_6 = 13.2254 \times 10^8$ yuan, respectively. The coal output and coal gangue transportation plan at each colliery were then computed, as shown in Table 5.7. When the stack level was $\alpha = 1.0$, the coal gangue stack was not reduced. However, the strategy based on the 3R principle to build gangue recycling facility increased government social revenue and it was obvious that the appropriate site played an important role in the maximization of revenue target.

5.4.5.2 Scenario 2: $\alpha = 0.9$

In this scenario, the coal gangue stack quantity cannot exceed the limit. Each colliery develops their respective plan to maximize production and minimize coal gangue transportation costs, the optimal production plan, and transportation strategy for each colliery is shown in Figure 5.4. Table 5.8 shows that the local authority's optimal decision is Site 2 when the allowed stack level α is 0.9. The total coal gangue stack quantity is 4.5474×10^6 tonnes and the total social revenues is 69.8002×10^8 yuan. Therefore, it can be seen that the government revenues increased by 4.23% and the total coal gangue stack quantity declines 10% compared with when there is no stack limit.

Table 5.6 Results for all CGPF candidate sites at stack limit is $\alpha = 1.0$ (10^8 RMB).

	Site 1	Site 2	Site 3	Site 4	Site 5	Site 6	Site 7	Site 8	Site 9	Site 10
Z2	67.4481	66.9668	63.4400	63.5672	61.6511	62.4032	63.0563	64.4938	61.8579	62.2816
ST	10.7432	10.0063	10.6526	10.0063	10.1504	10.9947	10.4351	11.4563	10.1257	10.2026
XLZ	14.8367	18.8290	19.0620	19.1226	19.1239	18.8866	19.2203	18.6394	19.1140	19.3372
BD	15.7514	12.4522	13.5704	11.8083	11.9259	12.5737	13.1270	14.0757	12.2073	12.3847
DT	12.8882	19.7998	19.7491	21.7010	19.5816	20.1520	19.7771	19.4985	21.4493	20.0375
JN2	7.2736	13.4425	13.3940	13.4175	13.3233	13.3420	13.4015	13.8143	13.3830	13.3671
JN3	20.2477	10.5901	7.9014	7.3192	11.2719	6.8979	9.3185	8.4368	8.4606	8.1350
Integrated value	34.3896	**35.2987**	33.8090	33.2460	33.2171	33.7644	33.7505	35.1533	33.1981	33.2590
Rankings	Site 2 > Site 8 > Site 1 > Site 3 > Site 6 > Site 7 > Site 10 > Site 4 > Site 5 > Site 9									

Underlined bold indicates calculation results of site 2.

Table 5.7 Coal output and coal gangue transportation plan for each colliery when the stack limit is $\alpha = 1.0$ (10^6 tonnes).

	Site 1 Output	Quantity	Site 2 Output	Quantity	Site 3 Output	Quantity	Site 4 Output	Quantity	Site 5 Output	Quantity
ST	4.0000	0.1747	**<u>4.0000</u>**	**<u>0.1747</u>**	3.6361	0.1025	4.0000	0.1747	3.9221	0.1593
XLZ	7.4576	0.3168	**<u>7.5000</u>**	**<u>0.3237</u>**	7.238	0.2809	7.1255	0.2625	7.1231	0.2621
BD	6.4758	0.2739	**<u>6.3313</u>**	**<u>0.2478</u>**	6.0104	0.1898	6.5	0.2783	6.4824	0.2751
DT	8.6543	0.3268	**<u>8.8965</u>**	**<u>0.3632</u>**	8.9278	0.3679	7.7255	0.1872	9.0000	0.3788
JN2	5.8696	0.4375	**<u>6.0000</u>**	**<u>0.4617</u>**	5.4689	0.3629	5.7745	0.4198	4.5500	0.1918
JN3	7.0000	0.3148	**<u>6.3726</u>**	**<u>0.2051</u>**	6.7740	0.2753	6.8909	0.2905	6.2708	0.1873

	Site 6 Output	Quantity	Site 7 Output	Quantity	Site 8 Output	Quantity	Site 9 Output	Quantity	Site 10 Output	Quantity
ST	3.4412	0.0638	3.7599	0.1271	3.8259	0.1402	3.9362	0.1621	3.8924	0.1534
XLZ	7.5000	0.3237	6.9439	0.2328	7.1359	0.2642	7.1414	0.2651	6.7267	0.1973
BD	6.2965	0.2415	6.1376	0.2128	6.1376	0.2128	6.4016	0.2605	6.3507	0.2513
DT	8.6796	0.3306	8.9105	0.3589	8.8679	0.3589	7.8805	0.2105	8.7501	0.3412
JN2	4.7933	0.2371	5.5667	0.3811	5.5667	0.3811	5.3261	0.3363	5.1198	0.2979
JN3	6.9238	0.3015	6.5624	0.2984	6.9061	0.2948	6.6905	0.2607	6.7391	0.2692

Underlined bold indicates calculation results of site 2.

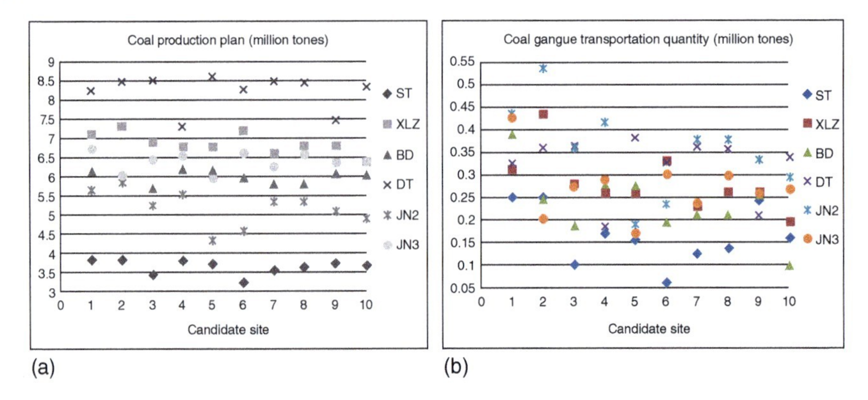

Figure 5.4 Reaction of each colliery when the reduction target is 0.9.

Table 5.8 Coal output and gangue transportation quantity from each colliery when the CGPF is located at Site 2 (10^6 tonnes).

α	Description	ST	XLZ	BD	DT	JN2	JN3
0.9	Coal output	3.8092	7.3261	6.0213	8.4915	5.8784	6.0477
	Gangue transportation quantity	0.1756	0.3517	0.2478	0.3632	0.4801	0.2051
0.8	Coal output	3.6142	6.9811	5.7113	8.0865	5.6584	5.7227
	Gangue transportation quantity	0.1756	0.3517	0.2478	0.3632	0.4801	0.2051
0.7	Coal output	3.4192	6.6361	5.4013	7.6815	5.4384	5.3977
	Gangue transportation quantity	0.1756	0.3517	0.2478	0.3632	0.4801	0.2051
0.6	Coal output	2.4242	6.2911	5.0913	7.2765	5.2184	5.0727
	Gangue transportation quantity	0.1756	0.3517	0.2478	0.3632	0.4801	0.2051

5.4.5.3 Scenario 3: $\alpha = 0.8$

Figure 5.5 shows each colliery's optimal production plan and transportation strategy when a 0.8 stack level is allowed. Table 5.8 shows that the local authority optimal decision is also Site 2 under this condition. Limited by the allowed stack level, ST, XLZ, and JN2 collieries can also maximize output with transportation capacity. In turn, the other collieries have to decrease output to avoid excessive transportation costs. The total coal output is 39.1004×10^6 tonnes, the total coal gangue transportation quantity is 1.7762×10^6 tonnes, which improved the government social revenues by about 7.36%.

5.4.5.4 Scenario 4: $\alpha = 0.7$

Each colliery's optimal reactions to the changing stack limitation level when 0.7 is the maximum level are displayed in Figure 5.6. The local authority optimal decision is also Site 2 as shown in Table 5.8. However, each colliery is unable to realize maximum output because the allowed stack level is too low. Therefore, each colliery attempts to make full use of the stack level and transportation capability. The total

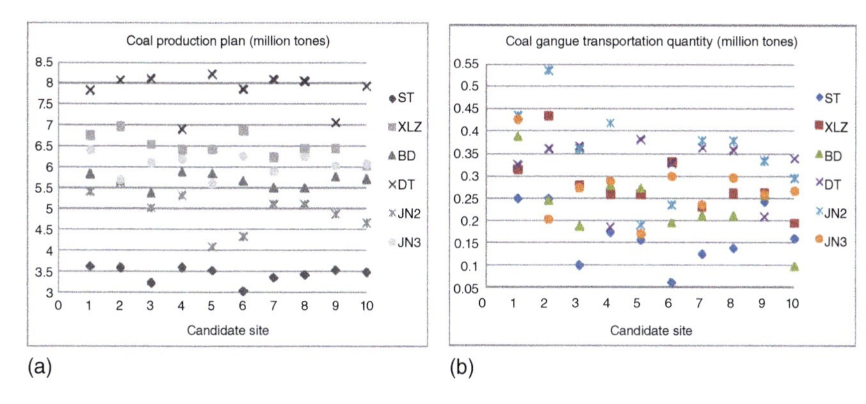

Figure 5.5 Reaction of each colliery when the reduction target is 0.8.

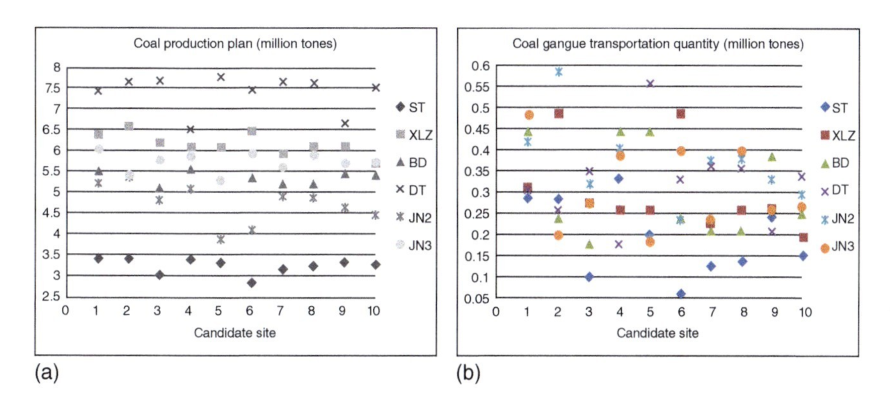

Figure 5.6 Reaction of each colliery when the reduction target is 0.7.

coal gangue stack quantity has significantly declined and the local authority's revenues also decreased by 4.08%.

5.4.5.5 Scenario 5: $\alpha = 0.6$

Figure 5.7 shows each colliery's optimal production plan, transportation strategy, and total profits at each candidate site when a 0.60 stack level is allowed. Under these conditions, Site 2 is the optimal decision as shown in Table 5.8. The suitable strategy for ST and JN3 collieries was to meet basic demand by making full use of the allowed stack level. Even if a lower reduction target were also suitable for the other collieries, the limitation stack level need to be more than 0.6 or equal to 0.6 to ensure each colliery's normal operations. The total coal gangue transportation quantity was 1.8235×10^6 and the social revenues decreased 8.46%

From the above analysis and results showed in the tables, the most appropriate selection for the local authority was Site 2 with a 0.8 limit, which resulted in maximum social revenue for the government and also maximized each colliery's profits. Nonetheless, if the main target is to reduce the coal gangue stack quantities to alleviate the environmental effect, a limit of 0.6 is the best strategy.

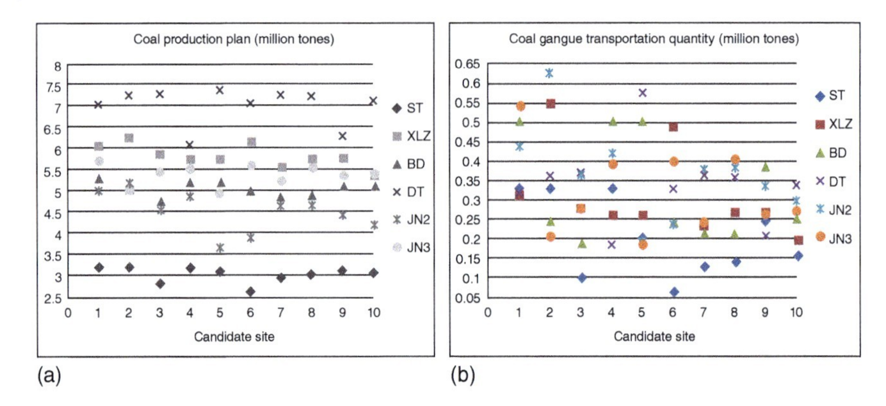

Figure 5.7 Reaction of each colliery when the reduction target is 0.6.

5.5 Discussion

The above results indicate that a coal gangue facility can play an important role in reducing gangue contamination and that efficient CGF site identification and selection is vital. Based on the results under different situations, we further do a deep discussion to verify the efficient of the proposed methodology.

5.5.1 Propositions

First, environmental protection could be enhanced by the facility recycling mechanism from the coal gangue contamination reduction perspective. Based on the formulated BLMOP model, a coal gangue facility recycling mechanism could be established. The test results which were shown from the Figures 5.4–5.7 have demonstrated that each colliery attempts to optimize their decisions to achieve the trade-off between transportation costs and stack cost. To negotiate for a larger transportation capacity, although each colliery gets a smaller unit revenue, it also transports more coal gangue to the CGF under the limited stack level. Meanwhile, to avoid paying for huge stack costs, the colliery owner still has to control coal, even if the revenue of unit coal production is higher output. In this situation, the facility recycling mechanism is conducive to achieving the coal gangue stack reduction target, thereby mitigating coal gangue contamination to promote environmental improvement. By applying this approach with changing some parameters, similar results also can be drawn in other coal fields. Under the facility recycling mechanism, each colliery will transport as much coal gangue as possible to the CGF and control coal output.

Then, the total coal gangue recycling quantity was significantly influenced by the allowed excess stack levels. The total coal gangue transportation is 0.3440 million tonnes and 1.1674 million tonnes when the allowed excess stack level changes from 40% from 0 with the basic degree is 0.7, which increases about 3.39 times. The total coal gangue transportation quantity gap between 40% and 0 has increased about 2.87

times and 4.86 times when $\beta = 0.8$ and $\beta = 0.9$. Although the total coal gangue also fluctuates with the changing β, the actual change margin in the allowed excess stack level is hugely higher than in the satisfaction degree. Therefore, it can be concluded that the allowed excess stack levels has huge impacts on the total coal gangue recycling quantity.

Further, the total coal gangue stack in the coal fields almost was not influenced by the basic satisfaction degree. A subjective parameter β is developed by the authority to ensure site selection equity in the formulated facility recycling mechanism. Taking $\gamma = 40\%$ as an example, the coal stack quantity is 6.1966×10^6 tonnes, 6.1895×10^6 tonnes, and 6.1905×10^6 tonnes when parameter β changes from 0.7 to 0.9. The result of total coal stack quantity when $\gamma = 40\%$, $\gamma = 30\%$, $\gamma = 20\%$, $\gamma = 10\%$, and $\gamma = 0$ draw the same conclusion can be obtained. The above results demonstrate that the basic satisfaction degree β almost does little impacts on the total coal gangue stack quantity when the allowed excess stack level γ is fixed. In fact, it is more efficient for the colliery to obtain more transportation capacity rather than undertaking huge stack costs. Thus, the allowed excess stack level is the main factor influencing the total coal gangue stack quantity (Figure 5.8).

Finally, collieries with smaller unit mining gangue coefficients have priority in the gangue facility recycling mechanism. From Figures 5.4–5.7 have shown that the reaction of each colliery is very different with the changing environmental protection target. Taking the situation $\beta = 0.7$ as an example, when the allowed excess stack level changes from 40% to 10%, ST and JN2 collieries have to decrease more coal output and can just meet the basic demand to realize the 30% stack reduction target. However, the XLZ and DT collieries can guarantee the coal output by adjusting the coal gangue transportation plan. Compared with ST and JN2 collieries, the change

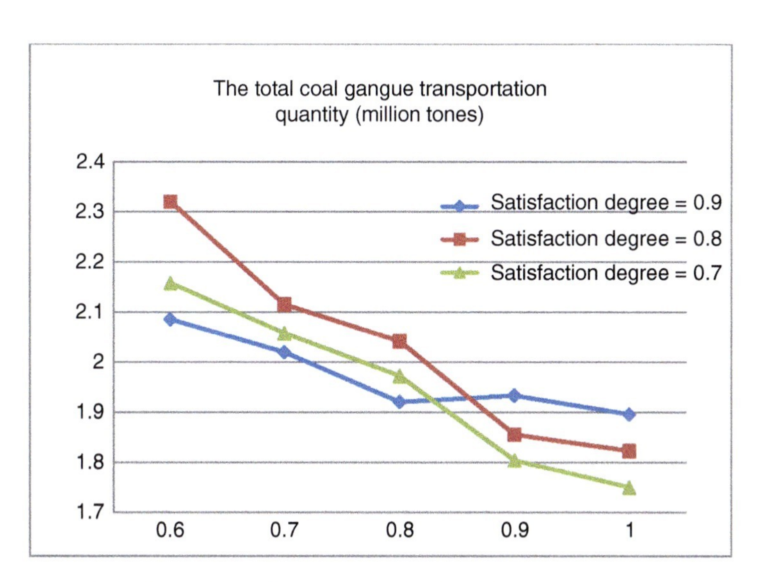

Figure 5.8 The total coal gangue transportation quantity with the limitation level and satisfaction degree changes.

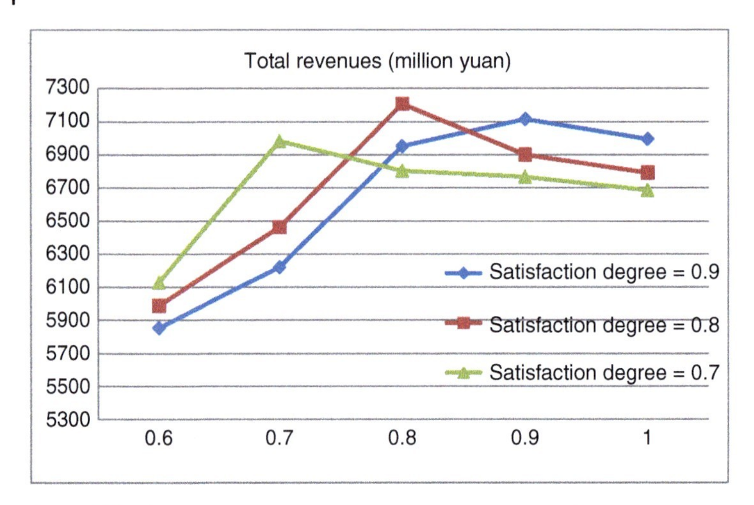

Figure 5.9 The government's total social revenues with the limitation level and satisfaction degree changes.

margin in coal output and coal gangue transportation quantity of XLZ and DT collieries is relatively small under a lower allowed excess stack level. That is to say, in the formulated coal gangue facility recycling mechanism, the collieries with smaller mining gangue coefficients which have priority to get larger transportation capacity and produce more coal to maximize profits (Figure 5.9).

5.5.2 Management Recommendations

Based on the above application results and discussions, some management recommendations are proposed. The appropriate CGF site selection not only mitigates the adverse effects caused by the coal gangue stack but also improves the local authority financial revenue. Furthermore, the GIS technique and the bilevel model used to screen for the optimal CGF site are proven to be an efficient method for the achievement of environmental economic development goals.

In addition, using the GIS technique-based bilevel model, the authority can combine the actual situation to set up a suitable allowed excess stack level γ and satisfaction degree β. When the allowed excess stack level γ is strict, set a lower value for β to ensure the colliery transportation capacity, which has a big gangue stack quantity in the previous year. In turn, when the allowed excess stack level γ is relaxed, set a higher value for β to ensure the colliery transportation capacity, which has a smaller mining gangue coefficient. In this situation, when the allowed stack level is very high, the colliery with smaller mining gangue coefficients can increase the coal output, which increases the government's tax revenue. Meanwhile, even though the allowed stack level is very low, the colliery with big historical stack quantities will make full use of the transportation capacity, and the government's financial revenue will not significantly decline. The total coal gangue stack quantity reduction target and the financial revenue target for the local authority also can be effectively achieved.

References

Argyriou, A.V., Teeuwa, R.M., Rust, D., and Sarris, A. (2016). GIS multi-criteria decision analysis for assessment and mapping of neotectonic landscape deformation: a case study from Crete. *Geomorphology* 253: 262–274.

Avittathur, B., Shah, J., and Gupta, O.K. (2005). Distribution centre location modelling for differential sales tax structure. *European Journal of Operational Research* 162: 191–205.

Awasthi, A., Chauhan, S.S., and Goyal, S.K. (2011). A multi-criteria decision making approach for location planning for urban distribution centers under uncertainty. *Mathematical and Computer Modelling* 53: 98–109.

Aydin, N.Y., Kentel, E., and Duzgun, H.S. (2013). GIS-based site selection methodology for hybrid renewable energy systems: a case study from western Turkey. *Energy Conversion and Management* 70: 90–106.

Ben-Ayed, O. and Blair, C.E. (1990). Computational difficulties of bilevel linear programming. *Operations Research* 38: 556–560.

Blumberg, T., Sorgenfrei, M., and Tsatsaronis, G. (2015). Design and assessment of an IGCC concept with CO_2 capture for the co-generation of electricity and substitute natural gas. *Sustainability* 7 (12): 16213–16225.

Cardoso, R., Silva, R.V., Brito, G.D., and Dhir, R. (2016). Use of recycled aggregates from construction and demolition waste in geotechnical applications: a literature review. *Waste Management* 46: 131–145.

Choudhary, D. and Shankar, R. (2012). An steep-fuzzy AHP-topsis framework for evaluation and selection of thermal power plant location: a case study from India. *Energy* 42: 510–521.

Chugh, Y.P. and Patwardhan, A. (2004). Mine-mouth power and process steam generation using fine coal waste fuel. *Resources, Conservation and Recycling* 40: 225–243.

Eskandari, M., Homaee, M., and Falamaki, A. (2016). Landfill site selection for municipal solid wastes in mountainous areas with landslide susceptibility. *Environmental Science and Pollution Research* 23: 1–12.

Gang, J., Tu, Y., Lev, B. et al. (2015). A multi-objective bi-level location planning problem for stone industrial parks. *Computers & Operations Research* 56: 8–21.

Gao, Y., Huang, H., Tang, W. et al. (2015). Preparation and characterization of a novel porous silicate material from coal gangue. *Microporous and Mesoporous Materials* 217: 210–218.

Gołębiewski, G., Trajer, J., Jaros, M., and Winiczenko, R. (2013). Modelling of the location of vehicle recycling facilities: a casestudy in Poland. *Resources, Conservation and Recycling* 80: 10–20.

Guo, Y.X., Lv, H.B., Yang, X., and Cheng, F.Q. (2015). $AlCl_3 \cdot 6H_2O$ recovery from the acid leaching liquor of coal gangue by using concentrated hydrochloric inpouring. *Separation and Purification Technology* 151: 177–183.

Hanson, M.A. (1981). On sufficiency of the Kuhn–Tucker conditions. *Journal of Mathematical Analysis and Applications* 80: 545–550.

He, Z., Fan, B., Cheng, T.C.E. et al. (2016). A mean-shift algorithm for large-scale planar maximal covering location problems. *European Journal of Operational Research* 250: 65–76.

Joachim, K. and Achim, S. (2014). GIS-based multi-criteria evaluation to identify potential sites for soil and water conservation techniques in the Ronquillo watershed, northern Peru. *Applied Geography* 51: 131–142.

Kerry, S., Olivier, B., Geoffrey, C., and Ulrich, L. (2016). GIS-based modelling of shallow geothermal energy potential for CO_2 emission mitigation in urban areas. *Renewable Energy* 86: 1023–1036.

Latinopoulos, D. and Kechagia, K. (2015). A GIS-based multi-criteria evaluation for wind farm site selection. A regional scale application in Greece. *Renew Energy* 78: 550–560.

Li, Q., Sun, G.N., Han, Y.F., and Chen, S.J. (2007). Analysis of the regeneration utilization ways of coal gangue resource in China. *Coal Conversion* 1: 018.

Li, C., Wan, J., Sun, H., and Li, L. (2010). Investigation on the activation of coal gangue by a new compound method. *Journal of Hazardous Materials* 179: 515–520.

Li, G.F., Yue, C.S., Zhang, M. et al. (2015). Facile and economical preparation of SiAlON-based composites using coal gangue: from fundamental to industrial application. *Energies* 8: 7428–7440.

Liu, H.B. and Liu, Z.L. (2015). Recycling utilization patterns of coal mining waste in China. *Resources, Conservation and Recycling* 54: 1331–1340.

Liu, J., Shi, C., Zhang, G., and Dillon, T. (2007). Model and extended Kuhn–Tucker approach for bilevel multi-follower decision making in a referential-uncooperative situation. *Journal of Global Optimization* 38: 597–600.

Liu, Y.G., Yu, L.N., and Wang, H.C. (2012). Study on countermeasures of coal gangue pollution prevention and regional sustainable development in China. *Applied Mechanics and Materials* 307: 510–513.

Lü, Q.K., Dong, X.F., Zhua, Z.W., and Dong, Y.C. (2014). Environment-oriented low-cost porous mullite ceramic membrane supports fabricated from coal gangue and bauxite. *Journal of Hazardous Materials* 273: 136–145.

National Development and Reform Commission (2010). *Chinese Coal Industry Development Plan in the Twelfth Five-year Plan*. National Energy Administration.

National Development and Reform Commission (2012). *Chinese Waste Recycling Technology Engineering Twelfth Five-Year Special Plan*. National Energy Administration.

National Bureau of Statistics (2013). *The Statistical Yearbook of Chinese Coal Industry*. National Bureau of Statistics of the Peoples Republic of China.

Mushia, N.M., Ramoelo, A., and Ayisi, K.K. (2016). The impact of the quality of coal mine stockpile soils on sustainable vegetation growth and productivity. *Sustainability* 8 (6): 546.

Paul, J.A. and MacDonald, L. (2016). Location and capacity allocations decisions to mitigate the impacts of unexpected disasters. *European Journal of Operational Research* 251: 252–263.

Qaddah, A.A. and Abdelwahed, M.F. (2015). GIS-based site-suitability modeling for seismic stations: case study of the northern Rahat volcanic field, Saudi Arabia. *Computers & Geosciences-UK* 83: 193–208.

Qian, T.T. and Li, J.H. (2015). Synthesis of Na-A zeolite from coal gangue with the in-situ crystallization technique. *Advanced Powder Technology* 26: 98–104.

Rahmati, O., Pourghasemi, H.R., and Melesse, A.M. (2016). Application of GIS-based data driven random forest and maximum entropy models for groundwater potential mapping: a case study at Mehran Region, Iran. *Catena* 137: 360–373.

Shi, C., Lu, J., and Zhang, G. (2005). An extended Kuhn–Tucker approach for linear bilevel programming. *Applied Mathematics and Computation* 162: 51–63.

Srivastavaa, P.K. and Singhb, R.M. (2016). GIS based integrated modelling framework for agricultural canal system simulation and management in Indo-Gangetic plains of India. *Agricultural Water Management* 163: 37–47.

Sun, H., Gao, Z., and Wu, J. (2008). A bi-level programming model and solution algorithm for the location of logistics distribution centers. *Applied Mathematical Modelling* 32 (4): 610–616.

Wang, C.L., Ni, W., Zhang, S.Q. et al. (2016a). Preparation and properties of autoclaved aerated concrete using coal gangue and iron ore tailings. *Construction and Building Materials* 104: 109–115.

Wang, S.B., Luo, K.L., Wang, X., and Sun, Y.Z. (2016b). Estimate of sulfur, arsenic, mercury, fluorine emissions due to spontaneous combustion of coal gangue: an important part of Chinese emission inventories. *Environmental Pollution* 209: 107–113.

Wu, D., Yang, B.J., and Liu, Y.C. (2015a). Transportability and pressure drop of fresh cemented coal gangue-fly ash backfill slurry in pipe loop. *Powder Technology* 284: 218–224.

Wu, D., Yang, B.J., and Liu, y.C. (2015b). Pressure drop in loop pipe flow of fresh cemented coal gangue-fly ash slurry: experiment and simulation. *Advanced Powder Technology* 26: 920–927.

Xiao, J., Li, F.C., Zhong, Q.F. et al. (2015). Separation of aluminum and silica from coal gangue by elevated temperature acid leaching for the preparation of alumina and SiC. *Hydrometallurgy* 155: 118–124.

Xiao, H.M., Ma, X.Q., and Liu, K. (2010). Co-combustion kinetics of sewage sludge with coal and coal gangue under different atmospheres. *Energy Conversion and Management* 51: 1976–1980.

Xu, J.P. (2014). Water allocation modelling and policy simulation for the min river basin of China under changing climatic conditions. *Research Presentation*, Spain, Barcelona, INFORMS (16 July 2014).

Xu, J.P. and Zhou, X.Y. (2011). *Fuzzy-Like Multiple Objective Decision Making*. Heidelber: Springer-Verlag.

Yang, M., Guo, Z.X., Deng, Y.S. et al. (2012). Preparation of CaO-Al_2O_3-SiO_2 glass ceramics from coal gangue. *International Journal of Mineral Processing* 102: 112–115.

Yang, L.X., Ji, X.Y., Gao, Z.Y., and Li, K.P. (2007). Logistics distribution centers location problem and algorithm under fuzzy environment. *Journal of Computational and Applied Mathematics* 208: 303–315.

You, C.F. and Xu, X.C. (2010). Coal combustion and its pollution control in China. *Energy* 35: 4467–4472.

Yunna, W., Shuai, G., Haobo, Z., and Gao, M. (2014). Decision framework of solar thermal power plant site selection based on linguistic Choquet operator. *Applied Energy* 136: 303–311.

Zadeh, L.A. (1965). Fuzzy sets. *Information and Control* 8: 338–353.

Zhang, J.X., Gao, R., Li, M. et al. (2015a). Basic characteristics and effective control of gangue piles in mining areas: a case study. *Journal of Residuals Science and Technology* 12: 145–154.

Zhang, Y.Y., Xu, L., Seetharaman, S. et al. (2015b). Effects of chemistry and mineral on structural evolution and chemical reactivity of coal gangue during calcination: towards efficient utilization. *Materials and Structures* 48: 2779–2793.

Zhou, C., Liu, G., Yan, Z. et al. (2012). Transformation behavior of mineral composition and trace elements during coal gangue combustion. *Fuel* 97: 644–650.

Zhou, C., Liu, G., Wu, S., and Lam, P.K.S. (2014). The environmental characteristics of usage of coal gangue in bricking-making: a case study at Huainan, China. *Chemosphere* 95: 274–280.

6

Dynamic Investment Strategy Toward Emissions Reduction and Energy Conservation of Coal Mining

Coal mining activities have caused many environmental problems, such as water, soil, and air pollution produced, which further bring some indirect impact on human activities, constraining economic and social development [Tiwary, 2001, Bian and et al., 2009, Zhengfu and et al., 2010]. However, almost 30% of global primary energy demand is met using coal, making it difficult to replace in the near future. The main five coal producing countries all suffer from coal mining-related environmental problems: China, the United States, India, Australia, and Indonesia [Dudley, 2016]. China and the United States have tried to stop this kind of problem through special laws and regulations for the coal industry, indicating that ecological coal mining comes to be a dominant development trend for improving the coal industry sustainability and clean transition. In practical problems, technological and management improvements are the main directions for emissions reductions (ER) and energy conservation (EC), and they have different advantages and disadvantages.

6.1 Background Review

Technological improvements can have a significant and direct effect on the environmental pollution and traditional mining methods but require large investments in both time and money. On the other hand, management improvements are easier to implement, but the effect is auxiliary and indirect. Therefore, to combine the advantages from two sides, integration method has been seen as the most appropriate way to achieve maximum ecological benefit with minimal capital and time investment. Considering the types of pollutants produced from coal mining activities are various, this chapter adopts a combination of technological investment and management optimization to reduce overall pollution emissions and energy consumption. To find the optimal equilibrium strategy for emissions reduction and energy conservation, this chapter built a dynamic multiobjective mixed 0–1 model based on ecological coal mining, to describe the ecological and economic benefits conflict and the environmental investment time arrangement. In this model, environmental technological investment and production adjustments are integrated

Innovative Approaches towards Ecological Coal Mining and Utilization, First Edition.
Jiuping Xu, Heping Xie, and Chengwei Lv.

into a dynamic system to seek ecological and economic equilibrium, and emissions reductions and energy conservation are controlled using double dynamic transfer equations. Meanwhile, an antithetic method-based particle swarm optimization (PSO) is developed to solve this model in mathematics level. The strategy results and comparison analyses based on Chaohua Colliery showed that the dynamic collaboration promoted global ecological and economic optimization, integrated production adjustment, and environmental investment planning provided coal mines with a superior method to solve conflicts between the ecology and the economy.

Ecological coal mining requires emissions reductions and energy conservation during coal mining. Therefore, to achieve ecological coal mining, the emission system and energy system of coal mining activities should be introduced as some basic background. Meanwhile, the main problems in emissions reductions and energy conservation of coal mining are also analyzed as the background of modeling.

6.1.1 Multi-system Consideration of Emission and Energy

To achieve ecological coal mining, the complexity of environmental problem in coal mining should be recognized. During coal mining processes, coal mine water discharge in tunneling process would destruct local groundwater resources and pollute the surface water resources; coal gangue emission and accumulation in extracting process would result in heavy metal pollution of land and groundwater; a large number of dust and boiler exhaust gas from mining area would cause air pollution [Yang and et al., 2003, Khadse and et al., 2007]. Meanwhile, coal mine methane (CMM) direct emission aggravates greenhouse effect and brings energy waste; the production system consumes lost of electricity; transportation system needs oil support; and heat system supply heat to mining area through burning raw coal [Hilson and Nayee, 2002, Ghose, 2003]. The environmental influences of coal mining are multiple dimensions, so it is necessary to consider these environmental problems comprehensively and solve them systematically.

Emissions reductions (ER) and energy conservation (EC) are achieved through environmental investment and production adjustments, but different pollutants emissions and energy consumption occur in different processes or systems [Higgins and et al., 2008, Myšková et al., 2013]. When environmental investment is adopted to improve the capacities of emissions reductions and energy conservation in coal mine, the technology improvement and equipment upgrade need to be implemented in different processes or systems with different characteristics. Environmental investment in emissions reduction projects improved mining technologies to reduce pollutants at the source and treatment technology improvement projects to reduce pollutants after mining [Geller et al., 2006]. Environmental investment in energy conservation refers to equipment upgrade and technology improvement in electricity system, transportation system, and heating system [Tanaka, 2008]. Production adjustment means coal mines can directly and quickly reduce pollutants production and energy consumption through reducing coal production to fix the disadvantage of environmental investment in time period.

6.1.2 Multidimensional Consideration of Economic and Ecological Benefits

Ecological coal mining aims to transform coal mining industry from environmental damage mode to clean mode, so the first pursuit is the ecological benefit, namely emission reductions and energy conservation. However, ecological coal mining is to promote the sustainable development of coal mining industry rather than giving up development for ecology, and economic benefit is the important guarantee for normal operation and long-term development for coal mining industry, so economic benefits are also necessary factor to be considered [Nan et al., 2010, Lin and et al., 2007]. However, in this multi-system management problem, the ecological and economic benefits are influenced by too much factors and produce different effects, as shown in Figure 6.1.

Production adjustments influence coal sale income and mining costs, and environmental project operations increase colliery operating costs but may bring some economic returns. The economic benefits would be damaged when coal mines decide to increase ecological benefits through reducing coal production [Liu and Liu, 2010, Hu and Kao, 2007]. Meanwhile, technology improvement also need capital investment and increase investment cost. Coal production produces pollution and consumes energy, emissions reduction project investment reduces pollution emissions, and energy conservation project investment reduces energy consumption, all of which directly influence ecological benefit [Lundgren, 2014]. Therefore, a multiobjective equilibrium problem needs to be solved in ecological coal mining.

6.1.3 Multi-stage Consideration of Environmental Investment

Environmental investment also needs to be dynamically considered. Ecological coal mining is a gradual process and all project constructions need time to be implemented, while the fluctuations and changes of the external factors and the internal condition at the coal mines all increase the difficulty of achieving an economic and ecological equilibrium strategy. Moreover, in this problem, multiple pollutants and multiple energy are considered together, therefore, different projects will be arranged in a system [Zeng and et al., 2014]. However, different projects have different construction periods, that means the time of these projects bringing the economic and ecological influences are different even if they are begun at the same time [Yu and et al., 2015]. Moreover, due to the limited available capital, coal mines cannot guarantee to begin these projects at the same time. Therefore, in a longer decision period, it is reasonable to plan these investment activities in many subtimes period.

In addition, under the different construction periods, external environmental and inner condition will bring more significant influence on ecological and economic

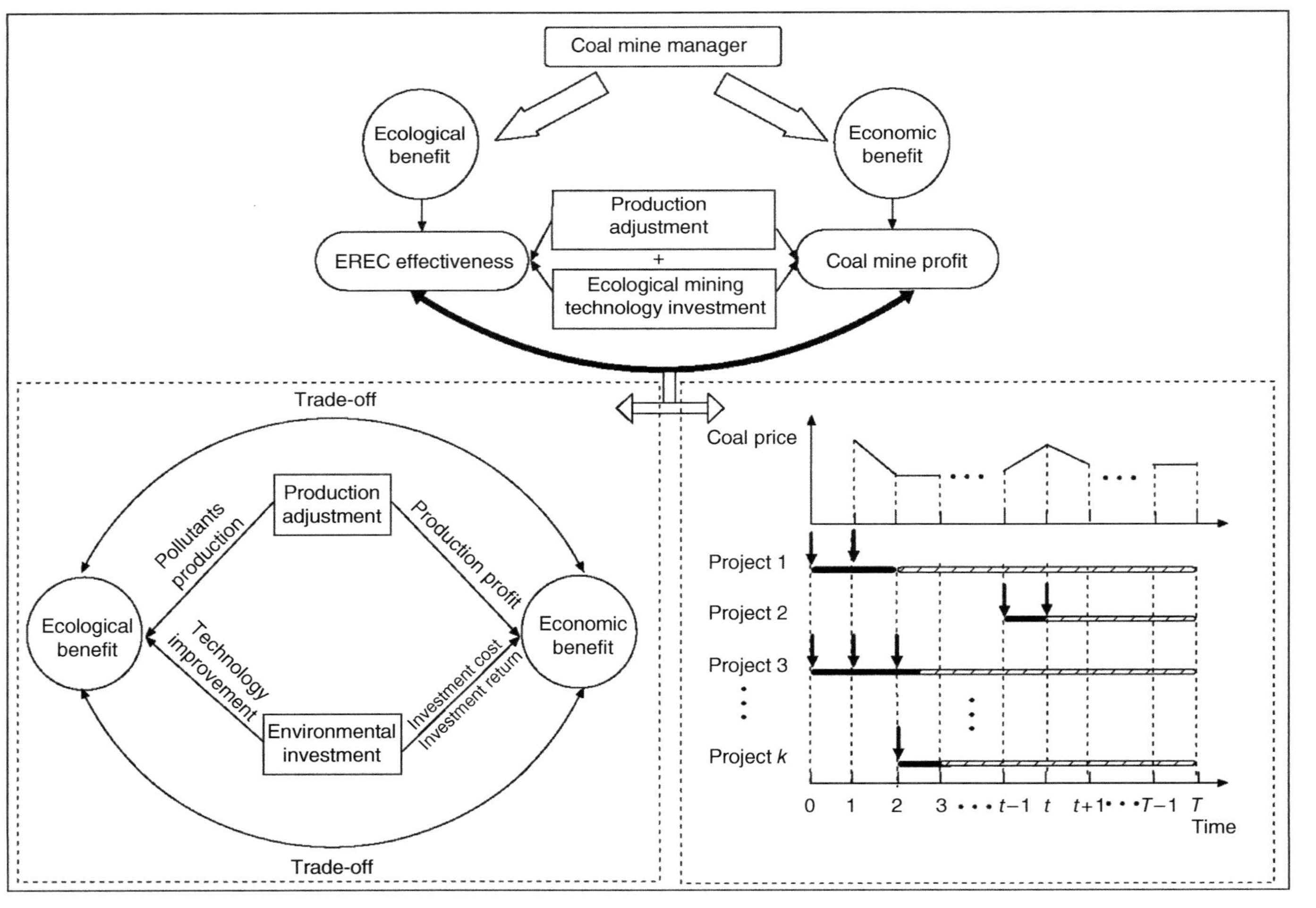

Figure 6.1 The relationship of ecological and economic benefit in production and environmental investment.

benefits of coal mines. The main external factors come from coal market, and the main internal factors influencing ecological benefit are the emissions reduction and energy conservation capacities. If the production plan cannot be made according to market changes, the overall economic benefit could suffer. Considering the characteristics of investment activities and market fluctuations, it is necessary to divide the whole period into stages to allow for a more precise determination of the emission reduction and energy conservation capacities to better control pollution emissions and energy consumption [Lin and et al., 2007, Hilson, 2000].

6.2 Modeling

To solve this ecological coal mining problem, a dynamic programming and a multi-objective programming are adopted to address the trade-offs between the ecological and economic objectives in the production and investment planning for coal mine emissions reductions and energy conservation. The model is built to describe the investment and production activities of a single coal mine, to achieve both emissions reductions and energy conservation, and pursue an objective of economic and ecological equilibrium.

6.2.1 Assumptions

(1) There is only one coal mine in the regional market.
(2) Materials prices are uncertain as they are influenced by national and international markets.
(3) Relevant policies have been determined.
(4) Coal imports and exports are not considered.

6.2.2 Notations

The following notations will be used in this chapter.

Indices

t = Decision stage index, $t \in \Phi^t = \{1, 2, \dots, T\}$;
i = Coal production pollutant index, $i \in \Phi^i = \{1, 2, \dots, I\}$;
j = Coal production energy index, $j \in \Phi^j = \{1, 2, \dots, J\}$;
k = Emission reductions project index, $k \in \Phi^{ik} = \{1, 2, \dots, K_2\}$, in which $k \in \Phi_1^{ik} = \{1, 2, \dots, K_1\}$, $(K_1^i \leq K_2^i)$ refers to the projects that can improve mining technology and reduce pollutant i emissions per unit of coal production, $k \in \Phi_2^{ik} = \{K_1^i + 1, K_1^i + 2, \dots, K_2^i\}$ refers to the projects to improve the treatment or reuse technology and reduce the amount of pollutant i after mining;
l = Energy conservation project index, $l \in \Phi^{jl} = \{1, 2, \dots, L^j\}$.

(Continued)

(Continued)

Decision variables

x_t = Coal mine exploitation amount in stage t;
y_{ikt} = If the coal mine invests in ER project k for pollution i in stage t, $y_{ikt} = 1$, otherwise, $y_{ikt} = 0$;
z_{jlt} = If the coal mine invests in EC project l for energy j in stage t, $z_{jlt} = 1$, otherwise $z_{jlt} = 0$.

Certain parameters

γ, φ = Coal mining recovery rate and raw coal loss rate;
RES = Mineable coal reserves at the coal mine;
AE, AI = Extracting and inventory capacity at the coal mine;
$I(t)$ = Coal inventory at the coal mine during stage t;
d^r_{ik}, d^c_{jl} = Construction duration for emissions reductions projects and energy conservation projects;
ERI_{ikd} = Total investment costs when the coal mine invests in pollutant i emissions reductions project k in stage d during the construction period, $d = 1, 2, \ldots, d^r_{ik}$;
ECI_{jld} = Total investment costs when the coal mine invests in energy j and energy conservation project l in stage d during the construction period, $d = 1, 2, \ldots, d^c_{jl}$;
$TC(t)$ = Total capital amount invested for all projects under construction at the beginning of stage t;
OPE_i, OEC_j = Original pollutant i emissions and energy j consumption per unit of coal output;
IR_{ik} = The reduction amount of pollutant i after investing in emission reduction project k;
NAT_{ikt} = The new added treatment capacity of project k in pollutant i at the beginning of stage t;
IS_{jl} = The energy j consumption reduction amount per unit of coal production after investing in the energy conservation project l;
NAC_{jlt} = The new added conservation capacity of project l to energy j at the beginning of stage t;
AER_{ikt} = Actual emission reduction amount of pollutant i from project k at stage t;
$CER_i(t)$ = Cumulative pollutant i emission reduction amount at the end of stage t;
$UEC_j(t)$ = Energy j consumption per unit of coal production at the end of stage t;
$M_t, AM(t)$ = Newly added and actual available capital for environmental investment at the beginning of stage t.
B_{ikt}, C_{ikt} = Unit benefit and unit operation costs of treating pollutant i in project k;

Uncertain parameters

$\widetilde{PC}_t$, $\widetilde{CE}_{jt}$ = Prices of unit coal and unit energy j in stage t;
$\widetilde{CF}_t$, $\widetilde{CI}_t$ = Production cost and inventory cost per unit of coal;
$\widetilde{D}_t^{\min}$, $\widetilde{D}_t$ = Coal market minimum demand and normal demand during stage t.
$\widetilde{r}_t$ = Capital interest rate in stage t;

6.2.3 Colliery Economic Benefit: Profit Objective

The priority for market-based collieries is profit maximizing, of which comprises coal production income, production costs, inventory costs and investment costs, operating costs, and investment income of the project.

Generally, total coal production costs can be expressed as the product of total unit costs and total coal production ($\widetilde{P}_t \times S_t$). In addition, the energy consumption per unit of coal output $UEC_j(t)$ and in part of the overall unit energy consumption costs should be adjusted after energy conservation project implementation in the process, in which energy consumption cost at each stage can be calculated by $\sum_{j\in\Phi^j} \widetilde{CE}_{jt} \times UEC_j(t) \times x_t$ and other production costs can be obtained by $\widetilde{CF}_t \times x_t$. To sum up, the total production costs at each stage are $\left(\sum_{j\in\Phi^j} \widetilde{CE}_{jt} \times UEC_j(t) + \widetilde{CF}_t\right) \times x_t$. Similarly, total inventory costs could be expressed as $\widetilde{CI}_t \times I(t)$.

Project profit is also affected by investment plan, which is mainly reflected in investment cost and project operating profit. In fact, at the beginning of each stage, capital investment is needed to construct emission reduction and energy conservation projects. However, considering different construction durations for them [Altman and et al., 1996, Yu and et al., 2015], the investment costs should be discussed in detail. For example, if project k with construction durations of one time unit or shorter ($d_{ik}^r \leq 1$), investment is completed in the current stage, which could be expressed as $ERI_{ik1} \times y_{ikt}$; if more than one time unit, the total capital investment for project k is the sum of the current investment of all unfinished projects. Therefore, when $1 < d_{ik}^r \leq t$, the total investment cost is $\sum_{d=1}^{d_{ik}^r} ERI_{ikd} \times y_{ik(t-d+1)}$; when $d_{ik}^r > t$, the current total investment cost is $\sum_{d=1}^{t} ERI_{ikd} \times y_{ik(t-d+1)}$.

The total investment cost in emission reductions projects at stage t can be obtained as follows:

$$\sum_{i\in\Phi^i} \sum_{k\in\Phi^i} \sum_{d=1}^{\min\{\max\{d_{ik}^r,1\},t\}} ERI_{ikd} \times y_{ik(t-\min\{t,d\}+1)} \tag{6.1}$$

Project operating profit is positively correlated with the total treated pollutant quantities and unit net profit. Therefore, the equation relationship is expressed by Eq. (6.2).

$$\sum_{i\in\Phi^i} \sum_{k\in\Phi^i} (B_{ikt} - C_{ikt}) \times AER_{ikt} \tag{6.2}$$

6.2.4 Colliery Ecological Benefit: Emission Reduction and Energy Conservation

Minimizing pollutant emissions and energy consumption is second goal for coal mines. However, there are different evaluation indexes for governments in different countries applied to evaluate emissions reductions, energy conservation ecological benefits, and certain emission reductions pollutants. Therefore, let $g_i(CPE_i(t))$ is the pollutant i emission reductions effectiveness and let $u_j(UEC_j(t))$ is the energy j consumption reduction performance. Meanwhile, functions $g_i(\cdot)$ and $u_j(\cdot)$ can be determined based on local laws and regulations. Therefore, making the ecological benefit of all emission reductions and EC projects maximizing could be obtained by Eq. (6.3).

$$\max f_2 = \min \{g_1(CER_1(T)), \dots, g_i(CER_i(T)), \dots, g_t(CER_m(T)), \\ u_1(UEC_1(t)), \dots u_j(UEC_i(T)), \dots, u_n(UEC_n(T))\} \tag{6.3}$$

6.2.5 Coal Production and Environmental Investment Activities

(1) *Total mining quantity restriction*: Equation (6.4) considers that the total mining volume must not exceed the mineable coal reserves RES, in which raw coal losses (γ) are considered caused by the technical and cost restrictions during mining process.

$$\frac{\sum_{t\in\Phi^t} x_t}{\gamma} \leq RES \tag{6.4}$$

(2) *Mining capacity restrictions*: Equation (6.4) restricts that the colliery's total mining quantity cannot exceed its mining capacity in each decision stage, where the total mining quantity could be expressed as the quotient of the planned mining amount x_t and the recovery rate (γ).

$$\frac{x_t}{\gamma} \leq A^E, \quad t \in \Phi^t \tag{6.5}$$

(3) *Social responsibility restriction*: Yu and Wei [2012] proposed that coal mines must meet the minimal demand, because coal, as a special status, can maintain the safety and stability of the national economy. In addition, considering the inevitable loss of coal, the total available amount of coal must meet the basic social demand, and mathematical relationship of which is expressed as follows:

$$(1-\varphi) \times x_t + I(t) \geq D_t^{\min}, \quad t \in \Phi^t \tag{6.6}$$

(4) *Inventory capacity restrictions*: Similar to mining activities, there are also capacity limitations for inventory activities, as shown below:

$$I(t) \leq AI, \quad t \in \Phi^t \tag{6.7}$$

(5) *Limited available investment capital*: Equation (6.8) ensures that investment decisions in emission reductions and energy conservation project must not more than the available investment capital.

$$TC_t \leq AM(t), \quad t \in \Phi^t \tag{6.8}$$

(6) *Non-stacking technological investment*: Equation (6.9) maintains that coal mines only invest one time for these technologies. Because, at the source, the technological effect of emission reductions or energy conservation cannot be superimposed, that is repeatedly investing in the same technology toward coal mines not only fails to achieve better emission reductions or energy conservation capacities but also causes waste.

$$\sum_{t\in\Phi^t} y_{ikt} \leq 1, \quad i \in \Phi^i, \quad k \in \Phi_1^i \tag{6.9}$$

$$\sum_{t\in\Phi^t} z_{jlt} \leq 1, \quad j \in \Phi^j, \quad l \in \Phi^{jl} \tag{6.10}$$

6.2.6 State Process Control Colliery Operations

(1) *State process inventory control*: Equation (6.11) maintains that the inventory at each stage is the sum of the previous stage's inventory and the current production minus the current coal sold to meet market demand.

$$I(t) = I(t-1) + x_t - S_t, \quad t \in \Phi^t \tag{6.11}$$

$$I(0) = 0 \tag{6.12}$$

where $S_t = \min\{\widetilde{D}_t, I(t-1) + x_t\}$ is the current coal supply for the market.

(2) *State process control of available investment capital*: Equation (6.13) represents that available investment capital at each stage is the sum of the actual value of the surplus capital from the previous stage in the current stage $((1+\widetilde{r}_{t-1}) \times (AM(t-1) - TC_{t-1}))$, and the newly added capital M_t in the current stage.

$$AM(1) = M_1, \tag{6.13}$$

$$AM(t) = M_t + (1+\widetilde{r}_{t-1}) \times (AM(t-1) - TC_{t-1}), \quad t \in \Phi^t \tag{6.14}$$

(3) *State process control of cumulative energy reduction*: Actually, cumulative emission reductions quantities and strategies of investment and production must be known and adjusted for coal mines to ensure that the emission reductions rate of the entire period compared to previous period maintains the government standards. Therefore, state transition functions are applied into this chapter to represent the dynamic characterizes in the model, where state variables are used to describe the cumulative emissions reduction quantities. For which, let $\sum_{k=1}^{K_1^i}\sum_{v=1}^{t} NAT_{ikt}$ and $\sum_{k=K_1^i+1}^{K_2^i}\sum_{v=1}^{t} NAT_{ikt}$ are the actual unit emission reductions capacity until stage t and the treatment capacities for pollutant i, respectively. Further, the cumulative emission reductions for pollutant i at the end of each stage could be calculated by summing the previous stage $CER_i(t-1)$ and the emission reductions in the current stage shown, and obviously, the original state is zero shown as Eq. (6.15), because the emission reductions plan has not been implemented in stage 1.

$$CER_i(t) = CER_i(t-1) + \left(\left(\sum_{k\in\Phi_1^i}\sum_{v=1}^{t} NAT_{ikt}\right) \times x_t + \sum_{k\in\Phi_2^i}\sum_{v=1}^{t} NAT_{ikt}\right),$$

$$i \in \Phi^i, \quad t \in \Phi^t \tag{6.15}$$

In Eq. (6.15), for emission reductions projects, a new contribution NAT_{ikt} could be achieved only when the construction is completed in stage t. Therefore, NAT_{ikt} must be 0, if $t \leq d^r_{ik}$, because the project k is not completed; if it could be completed, let $y_{ik(t-d^r_{ik})} = 1$ and the contribution is only decided by the investment decisions in stage $t - d^r_{ik}$ at now. Therefore, equation relationships are represented as follow:

$$NAT_{ikt} = \begin{cases} IR_{ik} \times 0, & \text{if } t \leq d^r_{ik} \\ IR_{ik} \times y_{ik(t-d^r_{ik})}, & \text{if } t > d^r_{ik} \end{cases}, \quad i \in \Phi^i, \quad k \in \Phi^i, \quad t \in \Phi^t \tag{6.16}$$

(4) *State process control of unit energy conservation*: Similarly, Eq. (6.17) calculates the actual unit energy conservation capacity in stage t by summing unit energy conservation in the last stage and all new added capacity.

$$UEC_j(t) = UEC_j(t-1) + \sum_{l \in \Phi^{jl}} NAC_{jlt}, \quad j \in \Phi^j, \quad t \in \Phi^t \tag{6.17}$$

$$UEC_j(0) = 0 \tag{6.18}$$

where denote NAC_{jlt} as the new contribution of the projects l for energy j in stage t, which could be obtained as follows:

$$NAC_{jlt} = \begin{cases} ES_{jl} \times 0, & \text{if } t \leq d^c_{jl} \\ ES_{jl} \times z_{jl(t-d^c_{jl})}, & \text{if } t > d^c_{jl} \end{cases}, \quad j \in \Phi^j, \quad l \in \Phi^{jl}, \quad t \in \Phi^t \tag{6.19}$$

6.2.7 Ecological Coal Mining Economic-Ecological Equilibrium Model

Based on the previous description, a multiobjective dynamic model is built to optimize the strategies for environmental investment and production adjustments to trade off the ecological and economic benefits and achieve higher emissions and energy reductions. This model has a more comprehensive and systematic structure, in which production and environmental investment plans are coordinated to seek a global optimal solution, and dynamic adjustment is used to ensure better coordination. By considering the different environmental projects, the model has more extensive applicability and can therefore assist management select the most appropriate investment projects to ensure greater precision in the economic and ecological objectives. In addition, different project durations and phased capital investment are also considered to ensure the model is more practicable, thus allowing management to have greater control over the actual emission reductions and energy conservation effectiveness.

$$\max f_1 \tag{6.20}$$

$$\max f_2 \tag{6.21}$$

$$
\text{s.t.}\begin{cases}
\frac{\sum_{t\in\Phi^t} x_t}{\gamma} \le RES, \\
(1-\varphi)\times x_t \ge \widetilde{D}_t^{\min}, \quad t\in\Phi^t \\
AM(1) = M^1, \\
AM(t) = M^t + (1+r_t)\times(AM(t-1) - TC_{t-1}), \quad t\in\Phi^t \\
TC_t = \sum_{i\in\Phi^i}\sum_{k\in\Phi^i}\sum_{d=1}^{\min\{\max\{d_{ik}^r,1\},t\}} ERI_{ikd}\times y_{ik(t-\min\{t,d\}+1)} \\
\qquad + \sum_{j\in\Phi^j}\sum_{l\in\Phi^{jl}}\sum_{d=1}^{\min\{\max\{d_{jl}^c,1\},t\}} ECI_{jld}\times z_{jl(t-\min\{t,d\}+1)} \\
TC_t \le AM(t), \quad t\in\Phi^t \\
I(t) = I(t-1) + x_t - S_t \le A^I, \quad t\in\Phi^t \\
I(0) = 0 \\
S_t = \min\{\widetilde{D}_t, I(t-1)+x_t\}, \quad t\in\Phi^t \\
NAT_{ikt} = \begin{cases} IR_{ik}\times 0, & \text{if } t\le d_{ik}^r \\ IR_{ik}\times y_{ik(t-d_{ik}^r)}, & \text{if } t> d_{ik}^r \end{cases}, \quad i\in\Phi^i, \quad k\in\Phi^{ik}, \quad t\in\Phi^t \\
CER_i(t) = CER_i(t-1) + \left(\left(\sum_{k\in\Phi_1^i}\sum_{v=1}^{t} NAT_{ikt}\right)\times x_t \right. \\
\qquad \left. + \sum_{k\in\Phi_2^i}\sum_{v=1}^{t} NAT_{ikt}\right) \\
CER_i(0) = 0, \quad i\in\Phi^i \\
NAC_{jlt} = \begin{cases} ES_{jl}\times 0, & \text{if } t\le d_{jl}^c \\ ES_{jl}\times z_{jl(t-d_{jl}^c)}, & \text{if } t> d_{jl}^c \end{cases}, \quad j\in\Phi^j, \quad l\in\Phi^{jl}, \quad t\in\Phi^t \\
UEC_j(t) = UEC_j(t-1) + \sum_{l\in\Phi^{jl}} NAC_{jlt}, \quad j\in\Phi^j, \quad t\in\Phi^t \\
UEC_j(0) = EC_j, \quad j\in\Phi^j \\
\sum_{t\in\Phi^t} y_{ikt} \le 1, \quad i\in\Phi^i, \quad k\in\Phi_1^{ik} \\
\sum_{t\in\Phi^t} z_{jlt} \le 1, \quad j\in\Phi^j, \quad l\in\Phi^{jl} \\
x^t \ge 0, \quad t\in\Phi^t \\
y_{ikt}, z_{jlt} = 0 \text{ or } 1, \quad i\in\Phi^i, \quad k\in\Phi^{ik}, \quad j\in\Phi^j, \quad l\in\Phi^{jl}, \quad t\in\Phi^t
\end{cases}
\tag{6.22}
$$

6.3 Economic-Ecological Equilibrium Model Solution Approach

6.3.1 General Parameterization

Model (6.20) is a universal model, with the practical problems being specified. Therefore, when model (6.20) is applied to a practical coal mine problem, it can be transformed into a specific model depending on the coal mine circumstances. In particular, functions $g_i(CPE_i(t))$ and $u_j(UEC_j(t))$ can be determined based on local emissions policies.

6.3.2 Fuzzy Goals for the Multiobjective Model

In this chapter, as the economic and ecological objectives have different dimensions, they cannot be directly dealt with using weighted sum scalarization. Fuzzy goal programming (FGP) [Mohamed, 1997] can transform objectives with different dimensions into fuzzy goals and the corresponding membership functions, after which the objectives can be optimized to minimize the weighted sum of the deviational variable between the actual membership level and the aspiration membership level for each fuzzy goal. With this method, decision-makers (DMs) can pursue an optimal objective value by pursuing the highest membership level for each objective, and the original objective functions are equivalent to the objective functions that express the actual membership levels. As the membership levels are in the same dimension, this method can solve the weighted sum scalarization with different dimensions in model (6.20) This method has been previously adopted to solve multi-objective decision making (MODM) problems in different areas [El-Wahed and Sang, 2006, Hu et al., 2007, Baky, 2009, 2010] and has proved to be highly effective. Therefore, this chapter adopts FGP to transform the multiobjective model into a single-objective model.

First, the optimal solution for each objective f_ξ^B is calculated and the DMs decided on the worst acceptable value f_ξ^W (ξ is the objective index), from which the fuzzy goal $f_1(x,y,z) \gtrsim f_1^B$ and $f_2(x,y,z) \gtrsim f_2^B$ can be determined, with the membership functions being as in Eq. (6.23)

$$\mu_\xi(f_\xi(x,y,z)) = \begin{cases} 1, & \text{if } f_\xi(x,y,z) \leq f_\xi^B, \\ \frac{f_\xi(x,y,z)-f_\xi^W}{f_\xi^B-f_\xi^W}, & \text{if } f_\xi^B < f_\xi(x,y,z) \leq f_\xi^W, \\ 0, & \text{if } f_\xi(x,y,z) \geq f_\xi^W \end{cases} \tag{6.23}$$

The fuzzy goals are then rewritten as:

$$\mu_\xi(f_\xi(x,y,z)) + d_\xi^- - d_\xi^+ = 1, \quad d_\xi^-,\ d_\xi^+ \geq 0, \quad \forall \xi \tag{6.24}$$

where the auxiliary variables d_ξ^-, d_ξ^+ are the negative and positive deviation variables between the actual satisfactory level and the aspiration satisfactory level for each objective.

Using Eqs. (6.23) and (6.24), the multiobjective model can be transformed into a single-objective model by weighting and summing the negative deviations between the actual satisfactory level and the aspiration satisfactory level for each objective, as shown in model (6.25):

$$\begin{aligned} &\min F = \textstyle\sum_{\xi\in\Xi} \omega_\xi \times d_\xi^- \\ &\text{s.t.} \begin{cases} \mu_\xi(f_\xi(x,y,z)) + d_\xi^- - d_\xi^+ = 1, & \xi \in \Xi \\ (x,y,z) \in S \end{cases} \end{aligned} \tag{6.25}$$

where the auxiliary variable ω_ξ is the importance degree for objective ξ and $\sum_{\xi\in\Xi} \omega_\xi = 1$; and S is the feasible region of Eq. (6.20). By minimizing the weighted sum of these deviations, DMs can determine an overall satisfactory solution close to the optimal solution.

6.3.3 Standard and AM-Based PSO for Nonlinear Dynamic Model

Model (6.20) integrates the continuous variables and 0–1 variables into the system, which increases the difficulties of finding an optimal solution. Further, as coal mine production is complex in practice, model (6.20) uses some nonlinear functions, making it difficult to solve using traditional exact solution methods. Therefore, a heuristic intelligent algorithm is adopted to solve this problem.

PSO proposed by Kenndy and Eberhart [1995] is not unduly influenced by objective function continuity as it uses primary math operators and can achieve good results, even in static, noisy, and continuously changing environments [Song and Gu, 2004]. Since PSO can be implemented easily and effectively, it has been applied to solve real-world nonlinear problems [Zhu et al., 2011, Chang and Shi, 2011, Eghbal et al., 2011]. Many studies have proven that both PSO and genetic algorithm (GA) can obtain high-quality solutions; however, the PSO has outperformed the GA in computational efficiency in nonlinear model with continuous variables [Hassan et al., 2005, Latiff et al., 2007, Duan et al., 2009, Eghbal et al., 2011]. Further, Zeng and et al. [2014] proposed an antithetic method (AM)-based PSO and proved that this AM-based PSO was able to search for the same global optima as the standard GA, was faster than the standard PSO, and had a better computational stability when solving integer programming than the standard PSO and the standard GA.

Therefore, to improve calculation efficiency, a combinational algorithm with a standard PSO and an AM-based PSO is proposed. The difference between the standard PSO and the AM-based PSO lies in the particle updating mechanism. In the standard PSO, formulas (6.26) and (6.27) are applied to update the position and velocity of each particle:

$$v_j(\tau+1) = \omega(\tau)v_j(\tau) + c_p r_p \left(\text{pbest}(j) - X_j(\tau)\right) + c_g r_g \left(\text{gbest}(\tau) - X_j(\tau)\right), \quad (6.26)$$

$$X_j(\tau+1) = v_j(\tau+1) + X_j(\tau) \quad (6.27)$$

where τ is the current iteration number. Then the inertia weight $\omega(t)$ is used to determine the influence of the previous velocity on the new velocity.

AM-based PSO updating is based on two kinds of elements: original elements (positive) and antithetic elements (non-positive). The maximum number of original elements is plus 1, and the minimum number of antithetic elements is minus 1, as shown in Figure 6.2. The new particle position can then be determined.

Therefore, in the economic-ecological equilibrium model solution approach, the standard and AM-based PSOs are integrated to solve the mixed integer dynamic model, for which the standard PSO is used to seek the optimal solution to the production plan and the AM-based PSO is used to seek the optimal solution to the investment plan. The updating process is shown in Figure 6.2.

Based on the above description, the step of the algorithm is as Figure 6.3.

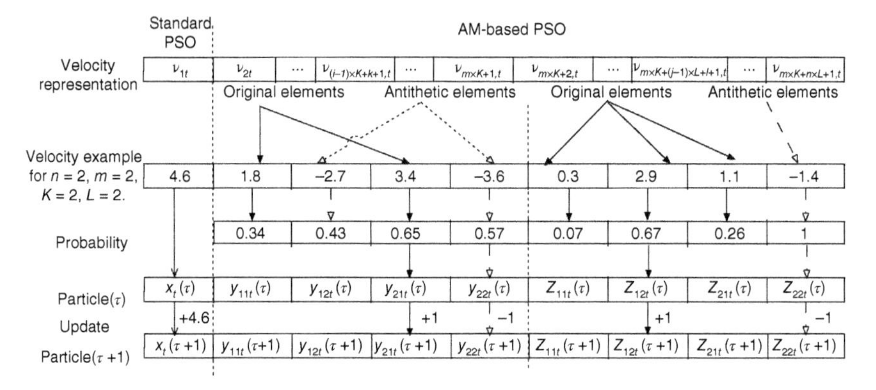

Figure 6.2 The particle updating process of standard and AM-based PSO.

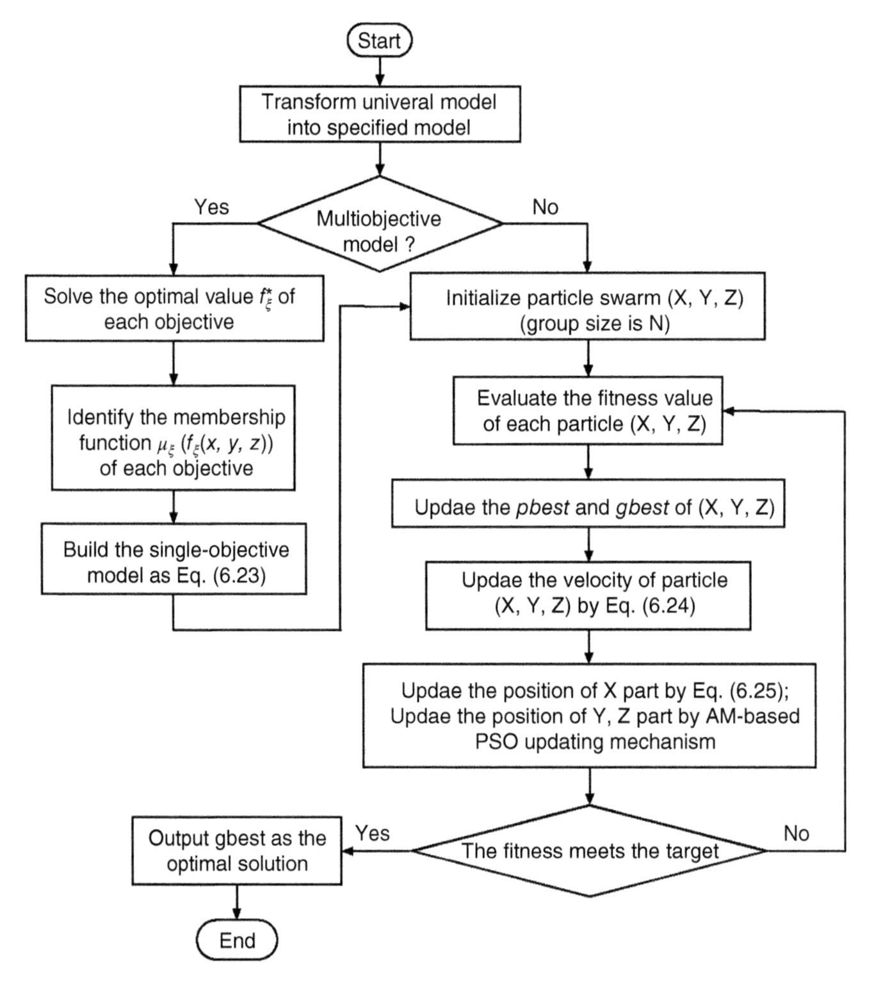

Figure 6.3 The flowchart of the algorithm solving economic-ecological equilibrium model.

6.4 Case Study

In this section, a real-world application will be given to demonstrate the effectiveness of the proposed method.

6.4.1 Case Description

Chaohua Coal Mine is Zhengzhou Coal Industry Co., Ltd.'s main mine and is one of the first hundred collieries in China. In the mining process, the main energy elements consumed are electricity, oil, and raw coal, and the main pollution emissions are gangue, waste water, CMM and fly ash, as shown in Figure 6.4. For the sustainable development of the environment, the Henan Provincial government issued the 13th Five-Year Plan in 2015 following the National 13th Five-Year Plan to reduce emission and achieve energy conservation for coal mines: total pollutant emissions should reduce by 5% compared to the previous period, and unit pollutant emissions and energy consumption should reduce by 5% compared to the beginning of the period. The new energy saving and emission reduction targets require Chaohua Coal Mine to invest in environmental projects. Since the Zhengzhou Coal Industry Co., Ltd. already has advanced technologies of CMM and sulfur and nitrogen, Chaohua Coal Mine only needs to focus on the treatment of the main coal mining pollutants: gangue, waste water, and fly ash.

6.4.2 Parametrization

In this case, as China is still a developing country and economic development is still the first objective of enterprises, the Chaohua Coal Mine makes decisions based on a profit priority, with the ecological objectives being transformed into constraints

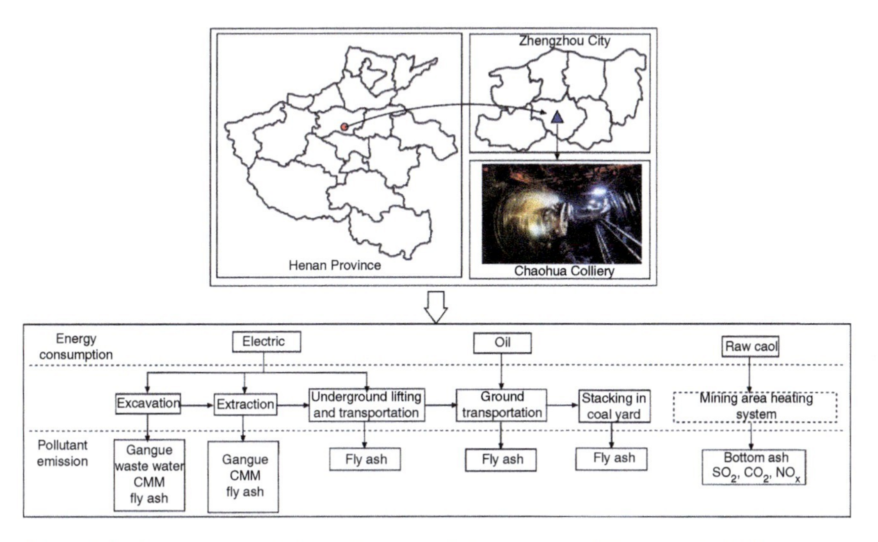

Figure 6.4 The geographical location and mining process of Chaohua Coal Mine.

to meet the local minimum policy; therefore, model (6.20) applied in this case should as shown in Eq. 6.28, where S is the flexible region of model (6.20), α and β are the lowest total pollutant and unit emission reductions rate standards released by the government; and ω is the lowest unit energy conservation rate standards. R_j is the conversion rate for converting one unit of energy j into standard coal. $ERR^t(T)$ is the actual total emission reductions rate, $ERR^u(T)$ is the actual unit emission reductions rate and $ECR^u(T)$ is the actual unit energy conservation rate. U_{it} is the pollutant i emissions during stage t in the last period.

$$\max f_1 = \sum_{t\in\Phi^t}\left(E(\widetilde{P}_t)\times S_t - E(\widetilde{C}_t^F)\times x_t - \sum_{j\in\Phi^j} E(\widetilde{C}_{jt}^E)\times UEC_j(t)\times x_t - E(\widetilde{CI}_t)\right.$$
$$\left.\times I(t) - TC_t + \sum_i^m\sum_k^{K^i}(B_{ikt} - C_{ikt})\times AER_{ikt}\right)\times (1+E(\widetilde{r}_t))^{T-t}$$

$$\text{s.t.}\begin{cases} g_{i1}(CPE_i(T)) = ERR^t(T) = 1 - \dfrac{CPE_i(T)}{\sum_{t\in\Phi^t} U_{it}} \geq \alpha, \quad i\in\Phi^i \\ g_{i2}(CPE_i(T)) = ERR^u(T) = 1 - \dfrac{CPE_i(T)}{\sum_{t\in\Phi^t} x_t\times PE_i} \geq \beta \quad i\in\Phi^i \\ u_j(UEC_j(T)) = ECR^u(T) = 1 - \dfrac{\sum_{t\in\Phi^t}\sum_{j\in\Phi^j}(UEC_j(t)\times R_j)\times x_t}{\sum_{j\in\Phi^j}(EC_j\times R_j\times\sum_{t\in\Phi^t} x_t)} \geq \omega \\ (x,y,z)\in S \end{cases} \tag{6.28}$$

6.4.3 Data Collection

The basic Chaohua Coal Mine data shown in Tables 6.1 and 6.2 were obtained from the Zhengzhou Coal Group Company annual report.

The coal mine plans to invest in five pollutant treatment projects from 2016 to 2020 in accordance with the three kinds of pollutant emissions. The capital

Table 6.1 The basic parameters of Chaohua Colliery.

RES (10^4 t)	C^{max} (10^4 t)	γ
10 000	256	0.9

Table 6.2 The last five years pollutant emissions of Chaohua Colliery.

Year	2011	2012	2013	2014	2015
Gangue (t)	226 270	244 057	242 121	248 050	231 473
Waste water (t)	551 089	594 409.9	589 694.7	604 135	563 761.1
Fly ash (t)	87.2604	86.569 64	89.882 92	87.986	82.105 96

Table 6.3 The parameters of emission reduction.

Pollutant i	PE_i (t/t)	Project k	d^r (year)	ERI_{ikd} (10^4 CNY) $d = 1$	$d = 2$	$d = 3$	IR_{ik} (t/t)	IR_{ik} (t)	B_{ikt}	C_{ikt}
Gangue	1210	Project 1	0	1135	—	—	31.6	—	0	0
		Project 2	2	680	400	—	—	3700	35	12
		Project 3	1	763	—	—	—	1500	50	31
Waste water	2947	Project 1	3	860	600	350	397.25	—	0	0
		Project 2	1	1200	—	—	—	4700	0	1.03
		Project 3	0	630	—	—	—	1340	0	0.84
		Project 4	1	830	—	—	—	2400	0	0.96
Fly ash	0.4292	Project 1	1	750	—	—	0.002 66	—	0	0
		Project 2	0	500	—	—	0.0014	—	0	0
		Project 3	0	690	—	—	—	22	0	8.5

Table 6.4 The parameters for energy conservation.

Energy j	EC_j (t/t, KHW/t, L/t)	Project l	d^c (year)	ECI_{jld} (10^4 CNY) $d = 1$	$d = 2$	ES_j (t/t, KHW/t, L/t)
Raw coal	0.0056	Project 1	0	923	—	0.000 32
		Project 2	0	545	—	0.000 22
		Project 3	1	402	—	0.000 18
Electricity	29.3682	Project 1	2	650	700	1.5199
		Project 2	0	748	—	0.6835
		Project 3	1	930	—	0.8035
		Project 4	1	1230	—	1.2024
Oil	0.2291	Project 1	0	860	—	0.0106
		Project 2	0	610	—	0.0075
		Project 3	0	928	—	0.0114

investments into the emission reductions projects and energy conservation projects at the beginning of each construction year and the completed treatment capacities for each project are, respectively, listed in Tables 6.3 and 6.4.

Besides the detailed Chaohua Coal Mine coal seam data, some further parameters based on the coal mining industry were obtained from the Chinese Coal Industry Development Plan from the 13th Five-Year Plan and the enterprise development plan, as shown in Table 6.5.

Table 6.5 The parameters for coal market.

T	1	2	3	4	5
$\widetilde{P}_t$	(510,520,528)	(487,500,509)	(502,514,530)	(481,496,505)	(496,510,523)
$\widetilde{C_t^F}$	(92,103,112)	(98,111,123)	(104,117,128)	(109,115,130)	(101,113,122)
$\widetilde{C_t^E}$	(510,520,528)	(487,500,509)	(502,514,530)	(481,496,505)	(496,510,523)
	(0.68,0.75,0.84)	(0.70,0.78,0.85)	(0.75,0.83,0.92)	(0.73,0.86,0.91)	(0.74,0.85,0.92)
	(2.7,3.3,3.8)	(2.5,3.0,3.5)	(2.8,3.03,3.5)	(2.81,3.17,3.62)	(2.9,3.37,3.72)
$\widetilde{C_t^I}$	(1.2,1.5,1.7)	(1.1,1.4,1.7)	(1.3,1.6,1.7)	(1.3,1.5,1.7)	(1.4,1.6,1.7)
$\widetilde{r}_t$	(0.23,0.27,0.3)	(0.17,0.21,0.23)	(0.17,0.24,0.26)	(0.16,0.19,0.22)	(0.18,0.23,0.25)
$\widetilde{D_t^{\min}}$ 10^4 t	(165,173,182)	(165,176,187)	(163,171,182)	(163,170,183)	(162,174,181)
$\widetilde{D}_t$ 10^4 t	(215,226,231)	(203,215,223)	(200,208,216)	(196,205,212)	(193,208,219)

6.4.4 Results and Different Scenarios

6.4.4.1 Results Analysis

Using the algorithm proposed in Section 6.3, the optimal production and investment planning solution for the coal mine can be determined, as shown in Table 6.6. The changes for each ER–EC index condition are shown in Figure 6.5.

From Table 6.6, it can be seen that the coal mines, respectively, produce 230.64, 213.50, 164.70, 169.17, and 208.92 × 10^4 t of coal from the first year to the fifth year and invest in gangue emission reductions project 3, waste water emission reductions project 1, and an electricity energy conservation project 4 in the first year; gangue emission reductions projects 1 and 4, and electricity energy conservation project 1 in the second year; gangue emission reductions project 3 and electricity energy conservation project 2 in the third year; gangue emission reductions project 3 in the fourth year; and finally the fly ash ER project 3 in the fifth year. The emission reductions and energy conservation statuses during the whole period are shown in Figure 6.5, which indicates how the strategy improves the total and unit emissions reductions rate for gangue, waste water, and fly ash, as well as the unit energy conservation rate.

6.4.4.2 Sensitivity Analysis

Based on the different scenarios, the optimal results for the EIPA-EREC under different circumstances are presented.

6.4.4.2.1 Scenario 1: Stricter Total Emission Reductions Rate Standard α

Figure 6.6 shows the optimal results when the total emission reductions standard changes. Comparing the optimal strategies from $\alpha = 0.05$ to 0.1 and 0.15, there are obvious production cuts and the investment plan requires small

Table 6.6 The optimal solution of production and investment.

T (year)			1	2	3	4	5	$ERR^t(T)$	$ERR^u(T)$
X (10^4 t)			230.64	213.50	164.70	169.17	208.92		
Y	Gangue	Project 1	0	1	0	0	0	0.0625	0.0642
		Project 2	0	0	0	0	0		
		Project 3	1	1	1	1	0		
	Waste water	Project 1	1	0	0	0	0	0.0500	0.0516
		Project 2	0	0	0	0	0		
		Project 3	0	0	0	0	0		
		Project 4	0	0	0	0	0		
	Fly ash	Project 1	0	0	0	0	0	0.0743	0.0519
		Project 2	0	0	0	0	0		
		Project 3	0	0	0	0	1		
Z	Raw coal	Project 1	0	0	0	0	0		
		Project 2	0	0	0	0	0		
		Project 3	0	0	0	0	0		
	Electricity	Project 1	0	1	0	0	0		
		Project 2	0	1	0	0	0		
		Project 3	0	0	1	0	0		
		Project 4	1	0	0	0	0		
	Oil	Project 1	0	0	0	0	0		
		Project 2	0	0	0	0	0		
		Project 3	0	0	0	0	0		
f_{opt} (million CNY)					6695.67				

adjustments, which differ depending on the increases in α. Correspondingly, the coal mine's profits decrease from 6695.67 10^6 CNY to 6493.51 10^6 CNY and 62 279.16 10^6 CNY.

6.4.4.2.2 Scenario 2: Stricter Unit Emission Reductions Rate Standard β

In this scenario, the unit emission reductions standard β was increased from 0.05 to 0.06 and 0.07 to determine the optimal results. In Figure 6.7, when comparing the optimal strategies from $\beta = 0.05$ to 0.06 and 0.07, the emission reductions project investment obviously increases and is earlier, while the energy conservation project investment is later. Production has an adjustment in time when β is high. The results

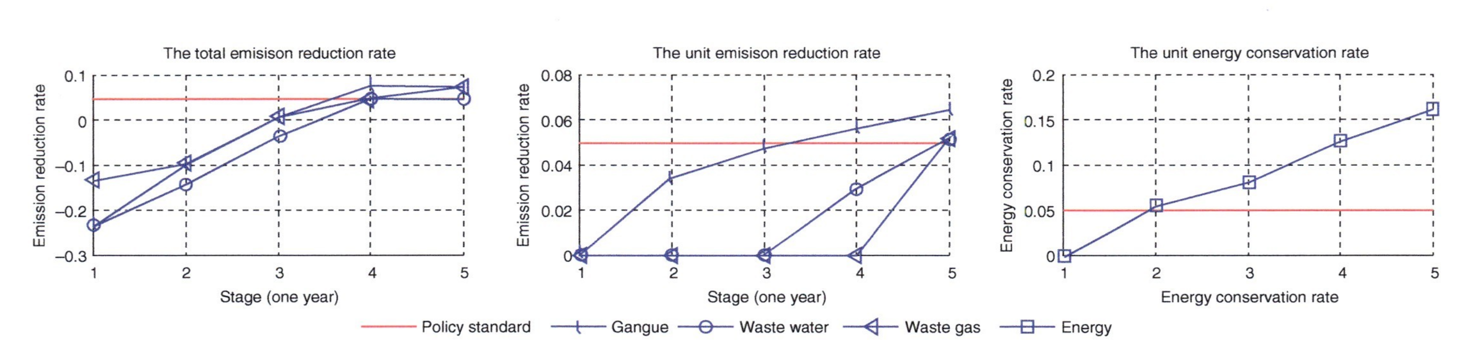

Figure 6.5 The emission reductions and energy conservation status at each stage when $\alpha = \beta = \omega = 0.05$.

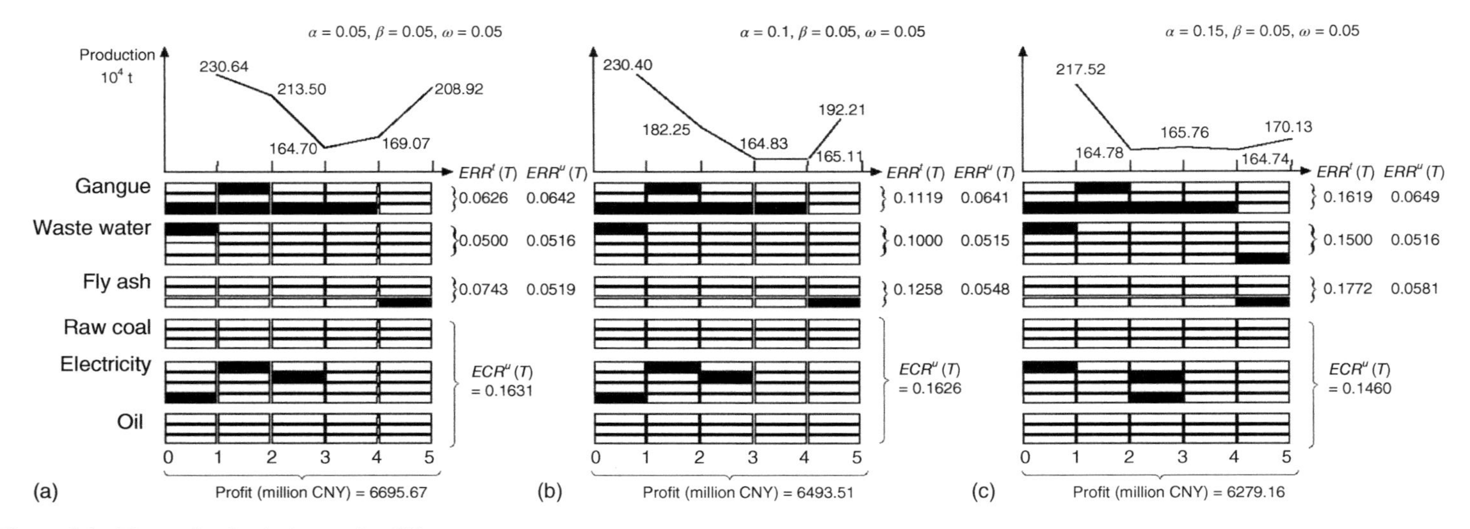

Figure 6.6 The optimal solution under different α.

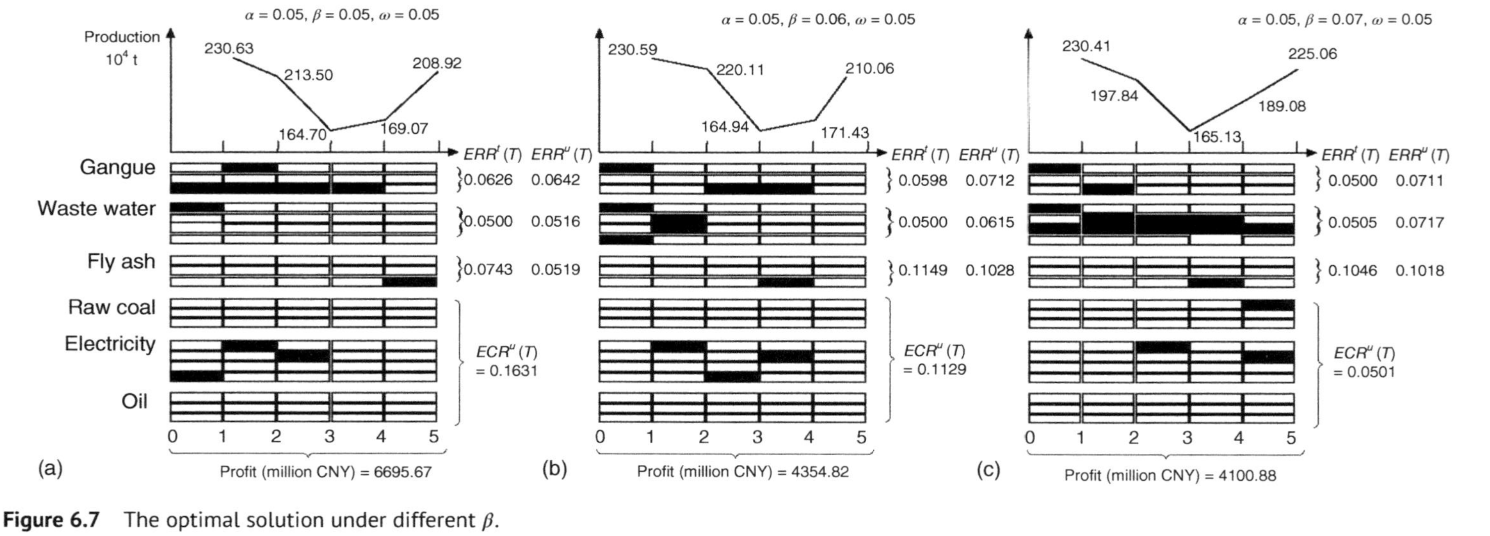

Figure 6.7 The optimal solution under different β.

show that the coal mine can achieve an actual emission reductions rate improvement to 6% with a 4354.82 10^6 CNY profit, or 7% with 4100.88 10^6 CNY profit.

6.4.4.2.3 Scenario 3: Stricter Unit Energy Conservation Standard ω

In this scenario, the unit energy conservation rate ω was increased from 0.05 to 0.15 and 0.25 to determine the optimal results. The results in Figure 6.8 show that energy conservation project investment obviously increases and production has an adjustment in time when ω is high. Therefore, the coal mine can improve the actual unit energy conservation rate to 15% with 6692.88 million CNY profit or achieve a 25% energy conservation rate with 4067.09 million CNY profit.

6.5 Discussion and Analysis

From the analyses in these scenarios, it can be seen that an integration of production adjustments and environmental investment can achieve ecological and economic equilibrium in the EIPA-EREC, and that dynamic coordination can promote this effectiveness. Therefore, in the following, a detailed discussion is given focused on these results.

6.5.1 Comprehensive Discussion for Results

First, by fully analyzing the results under different scenarios, it can be concluded that production and investment plans coordination can trade off the ecological and economic benefit. As shown in Figure 6.9, to reach the required standards, the production and investment plans have different effects but are not wholly independent of each other. The total emission reductions standard α is mainly affected by production reductions, when α increases, a reduction in production is needed to reduce pollution emissions, and the emission reductions project investments slightly increase, which can both improve emission reductions capacity and reduce the pressure of the production reduction. To reach the unit emission reductions standard β, emission reductions project investments can play a key role, as shown in Figure 6.9. When β increases, the coal mine needs to increase emission reductions project investment; on the other hand, increasing investment gives space for production increases. Under these circumstances, production needs to be adjusted to offset some of the losses resulting from large emission reductions project investment. Energy conservation project investments are the only approach to meet the unit energy conservation standard ω; however, when ω is very high, concessions need to be made in the emission reductions project investment to release capital for the energy conservation projects, which decreases the actual emission reductions effectiveness (as Figure 6.10). Energy conservation project investment can play a significant role in decreasing costs only if there is adequate coal production;

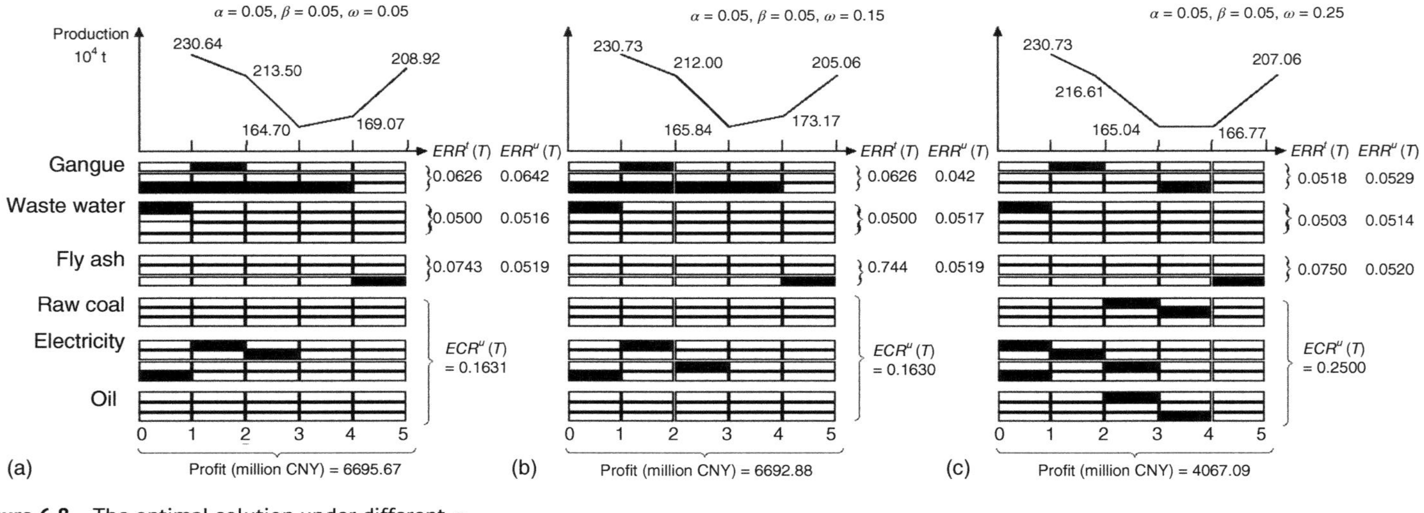

Figure 6.8 The optimal solution under different ω.

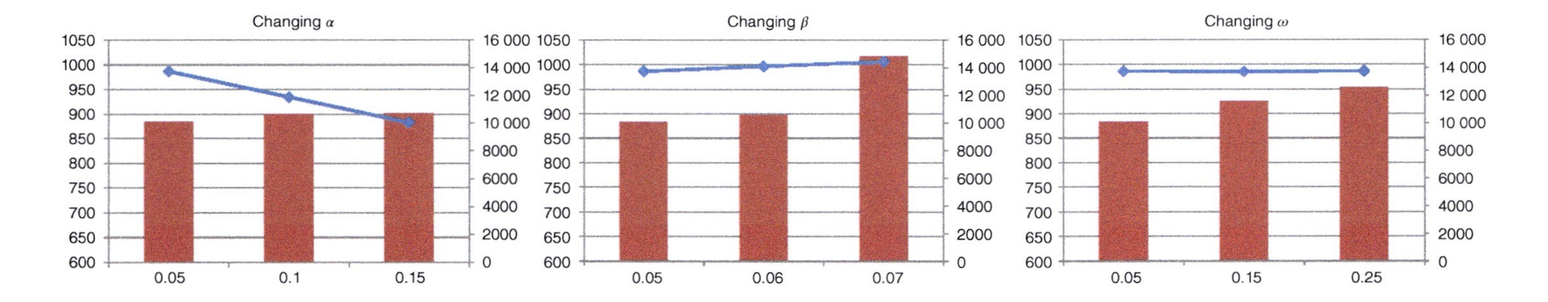

Figure 6.9 The total coal productions and investment under different conditions.

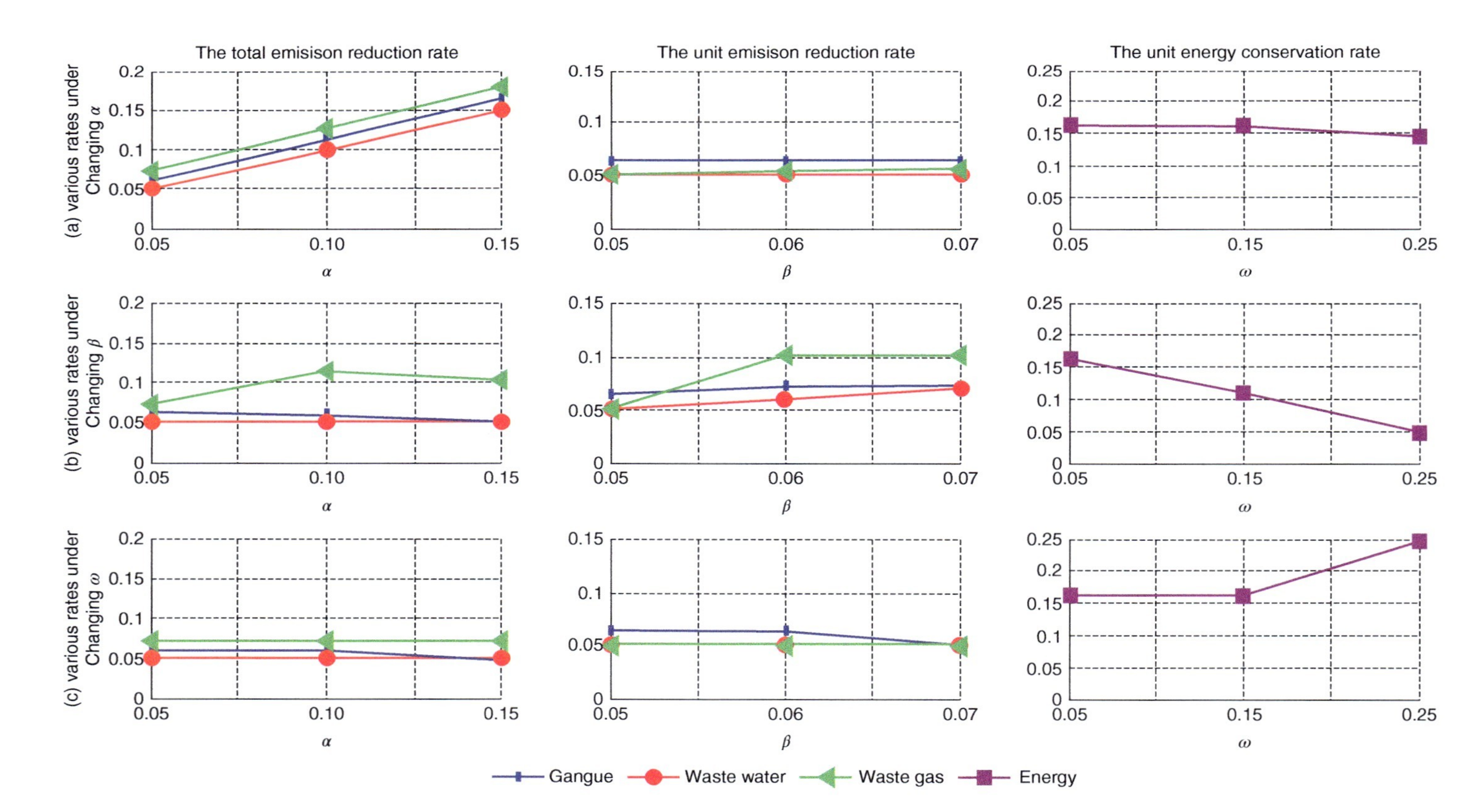

Figure 6.10 The ER–EC status of coal mines under different conditions.

however, as pointed out, the level of production is influenced by the emission reductions standards and capacity. Therefore, investment and production must be coordinated to reduce the pollution emissions and energy consumption and guarantee maximum profit.

Further, we found that the adjustments toward decision times can promote the coordination of the production and investment plans which implies that coordinating production and investment across time is necessary. As seen in Scenario 1, although there are production decreases in all stages, the different reduction levels indicate that production is adjusted from the earlier stages to the later stages when the total emission reductions standard increases from 0.05 to 0.15, which improves the total emission reductions rate (see Figure 6.10a), energy conservation project investment plan changes also need to be made to offset the losses from the reduction in total production. In Scenario 2, when the unit emission reductions standard increases from 0.06 to 0.07, the coal mines transfer production from earlier stages to later stages (see Figure 6.7) to fit with the changing emission reductions project investment plans to mitigate the greater pollutant emissions from the increase in production. Energy conservation project investments are postponed so there is capital available for the emission reductions projects. In Scenario 3, as the unit energy conservation standard increases from 0.15 to 0.25, some production adjustments are also made in line with the energy conservation capacity changes to guarantee maximum benefit. Therefore, time adjustments can resolve the ecological and economic equilibrium between production and investment plans.

In addition, it can be seen that the effect of total and unit emission reductions standard changes on production is opposite. As what was shown in Figure 6.9b), it can be seen that total production is most sensitive when the total emission reductions standard α changes (from 0.05 to 0.15), second most sensitive to the unit emission reductions standard β (from 0.05 to 0.07), and least sensitive to the unit energy conservation standard ω changes from 0.5 to 0.25. Figure 6.9 shows that a decrease in α leads to a decrease in production, while an increase in β leads to an increase in production. That is because the emission reductions rate in total amount is determined by both production and the ER capacity, while emission reductions rate per unit is only determined by emission reductions project investments. Therefore, production reduction as a strategy to achieve the α is implemented. When the β increases, larger emission reductions project investments to improve the emission reductions capacity, and then with the improvement in emissions reductions capacity, the original production plan produces less pollution so under the same α, production can increase.

Finally, we would like to draw the proposition that high-efficiency projects can effectively resolve the ecological and economic equilibrium. As Figure 6.11 shows the profits under different standards, it is obvious that profits decrease with increase in various standards; however, they are different sensitivities to these changing standards. Profit is most sensitive to changes in the unit emission reductions standard β, somewhat sensitive to changes in the total emission reductions standard α and not very sensitive to changes in the unit energy conservation standard ω. From the results shown in Figures 6.6–6.8 when α increases, coal production slightly

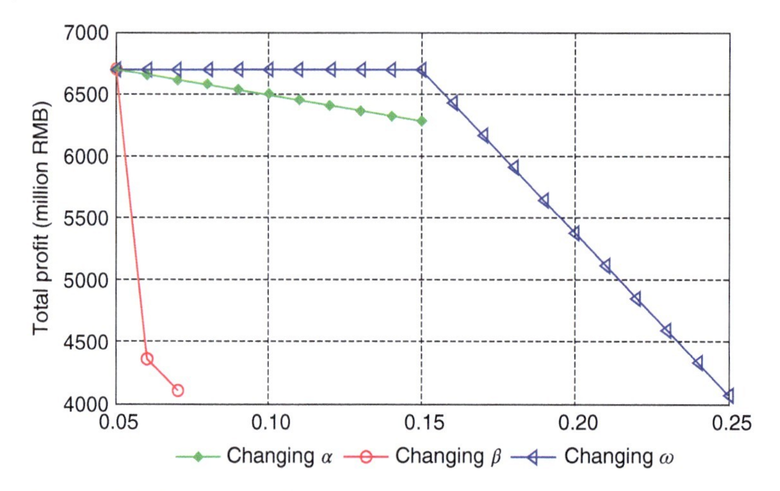

Figure 6.11 The total profit under different conditions.

decreases, investment does not change, and profit decreases moderately; when β increases, capital needs to be invested in emission reductions projects and the profitable projects need to release some money to other projects to guarantee all pollutants can reach the unit emission reductions standard; when ω increases, there is no obvious change in coal production or capital flows from the profitable emission reductions projects to the profitable energy conservation projects but there is a small reduction in profit. These results indicate that the capital used in the profitable environmental projects can solve the economic and ecological problem more effectively. However, low-efficiency projects would require a longer payback period, adversely influencing coal mine profit. Therefore, coal mine profit is related to the investment project efficiency; the higher the efficiency of the projects, the greater the feedback and the higher the profit.

6.5.2 Management Implications

Based on the above analysis and comparisons, in the following some suggestions regarding production and environmental investment at coal mines are given.

First of all, based on what was discussed above, the environmental projects need capital investment, but not all projects can bring returns. As all projects have capital recovery periods, when the capital recovery period is over, a project may not seem beneficial in the short term. However, environmental projects improve coal mine sustainability; emission reductions projects can improve pollution treatment capacity and allow the coal mine to produce more coal under the same policy standard and energy conservation projects can improve efficiency, enlarging the long-term colliery profit space. From a long-term development perspective, these projects also improve the overall economic benefit. Therefore, for sustainable colliery development, it is necessary for coal mines to invest in technological emission reductions and EC.

Further, the dynamic coordination between production and investment plans should be highly paid attention to. The coal mine's two goals of environmental protection and economic benefit can be achieved through both production and investment plans. Production and project investment are affected by different standards; however, considering only one aspect to solve problems can result in substantial losses. If one side is given the main role, the other side can have an auxiliary effect by achieving the standard or decreasing the losses. For example, to reduce pollutant emissions under the same production levels, it is better to increase production in the stages after the completion of emission reductions investment projects, or complete the emission reductions projects before the stages in which production is expected to be very high. To minimize economic losses, emission reductions project investment could be arranged in the later stages; correspondingly, as a reduction in production is necessary to maintain the emission reductions rate, more coal should be produced in those stages when the coal price is high and more investment is being put into emission reductions projects. Therefore, by coordinating production and environmental investment, collieries can achieve ecological and economic balance.

At the same time, it is necessary for the government to conduct and promote the policy-driven incentives. As coal mines are profit-driven organizations, they are only concerned with economic benefits if there are no policy limits. Therefore, the government acting on behalf of the public should develop policies and regulations to control pollution emissions and promote emission reductions in enterprises [Hilson, 2000, Jenkins and Yakovleva, 2006], especially in heavily polluting industries. Further, environmental investment can result in an increased economic burden on the coal mines. To relieve the conflict between the ecological aim and coal mine profit, governments should provide some policy or financial support to coal mines to develop environmental technologies [Zhang and Wen, 2008], especially waste reuse technologies, which can result in increased economic benefits for the coal mines.

References

Altman, A., et al. (1996). Cost-effective sulphur emission reduction under uncertainty. *European Journal of Operational Research* 90: 395–412.

Baky, I.A. (2009). Fuzzy goal programming algorithm for solving decentralized bi-level multi-objective programming problems. *Fuzzy Sets and Systems* 160: 2701–2713.

Baky, I.A. (2010). Solving multi-level multi-objective linear programming problems through fuzzy goal programming approach. *Applied Mathematical Modelling* 34: 2377–2387.

Bian, Z. et al. (2009). The impact of disposal and treatment of coal mining wastes on environment and farmland. *Environmental Geology* 58: 625–634.

Chang, J.F. and Shi, P. (2011). Using investment satisfaction capability index based particle swarm optimization to construct a stock portfolio. *Information Sciences* 181: 2989–2999.

Duan, Y., Harley, R.G., and Habetler, T.G. (2009). Comparison of particle swarm optimization and genetic algorithm in the design of permanent magnet motors. *Power Electronics and Motion Control Conference, 2009. IPEMC '09. IEEE International*, pp. 822–825.

Dudley, B. (2016). BP Statistical Review of World Energy. *Technical Report*. BP Technical Report.

Eghbal, M., Saha, T.K., and Hasan, K.N. (2011). Transmission expansion planning by meta-heuristic techniques: a comparison of shuffled frog leaping algorithm, PSO and GA. *IEEE Power and Energy Society General Meeting*, pp. 1–8.

El-Wahed, W.F.A. and Sang, M.L. (2006). Interactive fuzzy goal programming for multi-objective transportation problems. *Omega* 34: 158–166.

Geller, H., Harrington, P., Rosenfeld, A.H. et al. (2006). Polices for increasing energy efficiency: thirty years of experience in OECD countries. *Energy Policy* 34: 556–573.

Ghose, M. (2003). Promoting cleaner production in the Indian small-scale mining industry. *Journal of Cleaner Production* 11: 167–174.

Hassan, R., Cohanim, B., De Weck, O., and Venter, G. (2005). A comparison of particle swarm optimization and the genetic algorithm. *Proceedings of the 1st AIAA Multidisciplinary Design Optimization Specialist Conference*, pp. 18–21.

Higgins, A.J. et al. (2008). A multi-objective model for environmental investment decision making. *Computers & Operations Research* 35: 253–266.

Hilson, G. (2000). Sustainable development policies in Canada's mining sector: an overview of government and industry efforts. *Environmental Science & Policy* 3: 201–211.

Hilson, G. and Nayee, V. (2002). Environmental management system implementation in the mining industry: a key to achieving cleaner production. *International Journal of Mineral Processing* 64: 19–41.

Hu, J.L. and Kao, C.H. (2007). Efficient energy-saving targets for APEC economies. *Energy Policy* 35: 373–382.

Hu, C.F., Teng, C.J., and Li, S.Y. (2007). A fuzzy goal programming approach to multi-objective optimization problem with priorities. *European Journal of Operational Research* 176: 1319–1333.

Jenkins, H. and Yakovleva, N. (2006). Corporate social responsibility in the mining industry: exploring trends in social and environmental disclosure. *Journal of Cleaner Production* 14: 271–284.

Kenndy, J. and Eberhart, R. (1995). Particle swarm optimization. *Proceedings of IEEE International Conference on Neural Networks*, pp. 1942–1948.

Khadse, A. et al. (2007). Underground coal gasification: a new clean coal utilization technique for India. *Energy* 32: 2061–2071.

Latiff, N.M.A., Tsimenidis, C.C., and Sharif, B.S. (2007). Performance comparison of optimization algorithms for clustering in wireless sensor networks. *IEEE International Conference on Mobile Ad Hoc & Sensor Systems Conference*, pp. 1–4.

Lin, T.T. et al. (2007). Applying real options in investment decisions relating to environmental pollution. *Energy Policy* 35: 2426–2432.

Liu, H. and Liu, Z. (2010). Recycling utilization patterns of coal mining waste in China. *Resources, Conservation and Recycling* 54: 1331–1340.

Lundgren, T. (2014). The effects of climate policy on environmental expenditure and investment: evidence from Sweden. *Journal of Environmental Economics & Policy* 3: 148–166.

Mohamed, R.H. (1997). The relationship between goal programming and fuzzy programming. *Fuzzy Sets and Systems* 89: 215–222.

Myšková, R., Ilona, O., Petr, C., and Karel, S. (2013). Assessment of environmental and economic effects of environmental investment as a decisions problem. *Wseas Transactions on Environment & Development* 9: 268–277.

Nan, Z., Levine, M.D., and Price, L. (2010). Overview of current energy-efficiency policies in China. *Energy Policy* 38: 6439–6452.

Song, M.P. and Gu, G.C. (2004). Research on particle swarm optimization: a review. *Proceedings of 2004 International Conference on Machine Learning and Cybernetics, 2004*. IEEE. pp. 2236–2241.

Tanaka, K. (2008). Assessment of energy efficiency performance measures in industry and their application for policy. *Energy Policy* 36: 2887–2902.

Tiwary, R. (2001). Environmental impact of coal mining on water regime and its management. *Water, Air, & Soil Pollution* 132: 185–199.

Yang, L. et al. (2003). Clean coal technology-study on the pilot project experiment of underground coal gasification. *Energy* 28: 1445–1460.

Yu, S. and Wei, Y. (2012). Prediction of China's coal production-environmental pollution based on a hybrid genetic algorithm-system dynamics model. *Energy Policy* 42: 521–529.

Yu, S. et al. (2015). A dynamic programming model for environmental investment decision-making in coal mining. *Applied Energy*. 166.

Zeng, Z. et al. (2014). Antithetic method-based particle swarm optimization for a queuing network problem with fuzzy data in concrete transportation systems. *Computer-Aided Civil and Infrastructure Engineering* 29: 771–800.

Zhang, K. and Wen, Z. (2008). Review and challenges of policies of environmental protection and sustainable development in China. *Journal of Environmental Management* 88: 1249–1261.

Zhengfu, B. et al. (2010). Environmental issues from coal mining and their solutions. *Mining Science and Technology (China)* 20: 215–223.

Zhu, H., Wang, Y., Wang, K., and Chen, Y. (2011). Particle swarm optimization (PSO) for the constrained portfolio optimization problem. *Expert Systems with Applications* 38: 10161–10169.

7

Carbon Dioxide Emissions Reduction-Oriented Integrated Coal-Fired Power Operation Method

Climate change caused by greenhouse gas (GHG) emissions has become the most challenging issue in recent years. More and more countries are striving to find ways to reduce energy consumption and carbon dioxide emissions [Francey et al., 2013]. Compared with the period before the Industrial Revolution, with the current energy infrastructure, the cumulative CO_2 emissions from the burning of fossil fuels between 2010 and 2060 may cause the temperature to rise by 1.3 °C. Climate change caused by the increase in CO_2 concentration has also caused a series of dangers, including irreversible atmospheric warming, reduced rainfall during dry seasons, and unstoppable sea level rise [Solomon et al., 2009]. For the global anthropogenic CO_2 emissions, the power sector is an important source, accounting for nearly 40% of energy CO_2 emissions and 24% of total CO_2 emissions [Stern, 2006].

7.1 Background Review

In 2014, about 20% of the world's electricity was provided by coal-fired power plants (CPPs), and the proportion was even higher in some developing countries and, this proportion in China exceeded 65% [BP Global, 2015, CEC, 2017]. With the development of the global economy, fossil fuel power generation may increase accordingly, which will further aggravate CO_2 emissions in the atmosphere. Although it has been proven that wind energy, solar energy, and innovative fossil fuel power plants such as NET Power generate zero CO_2 emissions while generating electricity, there are still two-thirds of electricity generated using fossil fuels; therefore, better ones must be sought ways to reduce CO_2 emissions to alleviate global warming [Service, 2017a,b].

At present, the current important research on reducing CO_2 emissions can be summarized from two perspectives of hard-path and soft-path. From a hard-path perspective, sustainable energy technology (SET) can save energy and reduce CO_2 emissions [Lonsdale et al., 2012, Wang et al., 2017]. To promote the economic development of fossil fuels, carbon capture and storage (CCS) is an effective technology to reduce CO_2 emissions, which has been successfully applied to unit 3 of the Boundary Dam Power Station in Saskatchewan, Canada Oelkers and Cole [2008] and Schrag [2009]. This is the first time that most of the CO_2 emissions

Innovative Approaches towards Ecological Coal Mining and Utilization, First Edition.
Jiuping Xu, Heping Xie, and Chengwei Lv.

have been used and stored in a commercial-scale power plant [Van Noorden, 2014]. Dang et al. studied the performance of a new type of co-firing bio-oil and sequestering biochar technology in agricultural land to reduce CO_2 emissions from CPPs and concluded that co-firing of bio-oil co-firing fuel (BCF) is more attractive approach can reduce the CO_2 emissions in the existing CPPs [Dang et al., 2015]. There are also some other studies on hard-path methods to reduce CPPs emissions can be found in Mukherjee and Borthakur [2001], Liu et al. [2010], and Pavlish et al. [2003]. In general, the high efficiency of the hard-path method is regarded as the most powerful tool to reduce CO_2 emissions. However, commercial-scale applications are extremely expensive for coal-fired power generation companies [Porter et al., 2015]. In addition, given that negative emission technologies are still in the early stages of development, it is impossible to limit global warming below 2° with negative emissions alone [Fuss et al., 2014, Scott et al., 2013, Lomax et al., 2015, Van Der Meer et al., 2014, Friedlingstein et al., 2014]. Therefore, the soft-path solution may be more effective for coal-fired power generation companies to reduce CO_2 emissions [Goto et al., 2013, Li, 2012].

The soft-path uses policy control or operational management methods to reduce CO_2 emissions. Wang and Chen used two computable general equilibrium (CGE) models to simulate carbon taxes to assess the impact of carbon taxes on China's economy and environment and ultimately proved that carbon taxes can improve energy use, save and reduce CO_2 emissions [Wang and Chen, 2015]. Shif and Frey employed an uncertain multiobjective opportunity-constrained optimization model to optimize emission performance and reduce the total cost by adjusting the coal blending ratio [Shih and Frey, 1995]. Taking into account the trade-off between economic development and emission reduction, Lv et al. established a coal blending method based on an equilibrium strategy in an uncertain environment to reduce CO_2 and PM_{10} emissions [Lv et al., 2016]. Although these soft-path innovations may alleviate climate change caused by anthropogenic CO_2 emissions, the current situation is still not optimistic, and more in-depth research is needed.

In the past, CO_2 emission reduction studies mainly considered CPPs coal blending. Few people consider blending coal in a coal storage and distribution center (CSDC). For large coal-fired power companies with multiple subordinate CPPs, various coal types are purchased at CSDC and then adjusted and distributed according to the coal blending ratio of each subordinate CPP to reduce CO_2 emissions and save energy consumption. In the past, CO_2 emissions research only think out single-cycle situations, but in reality, companies making production and operation plans to reduce CO_2 emissions and maximize profits usually involve long-term development plans, which are divided into multiple interrelated stages. The constant changes in the internal and external environment necessitate decision-making at every stage. This situation requires the introduction of multi-stage decision-making. Currently, multi-stage decision-making is usually achieved through dynamic programming. In addition, there is a conflict between reducing CO_2 emissions and economic development, so multiple goals need to be used to balance income and the environment. The dynamic multiobjective optimization model (DMOM) has been proven to perform well in dealing with such complex situations and has been used in many fields. Ganguly et al. proposed a multiobjective dynamic programming method to determine the optimal line path and branch conductor size of the power distribution system by simultaneously optimizing cost and reliability [Ganguly et al., 2013]. Huang et al. presented a DMOM based on membrane calculation to get the best control for time-varying unstable plants, and this method is very effective when the rise time is short and the overshoot is small [Huang et al., 2011]. Further studies on the application of DMOM can be found in Nwulu and Xia [2015], Zhang et al. [2013], and Dubey et al. [2015]. In this chapter, DMOM optimized for CSDC coal blending is applied to the strategy of carbon purchase, blending and distribution, so as to maximize economic benefits while reducing CO_2 emissions.

7.2 Key Problem Statement

To combine the DMOM and integrated coal purchasing, blending, and distribution (ICPBD) issues to balance the reduction of CO_2 emissions and economic development, the relevant background and instructions are introduced below.

In large coal-fired power generation companies with multiple subordinate CPPs, coal procurement and coal blending are done by each subordinate CPP itself. However, the market power demand is uncertain, and the coal supply fluctuates [Giri and Chakraborty, 2016]. Therefore, the coal required for CPP mixing at each subordinate power plant is not necessarily feasible. The coal sent to power generation may not meet the requirements of coal-based boilers that will seriously damage the boiler. In addition, each subordinate CPP mixes the coal they need, so they also need to bear the cost of purchasing, ordering, transportation, and processing.

Compared with low-quality coal, high-quality coal with high carbon content can improve emission performance during power generation. That is, the fewer incombustible impurities and the less air pollutants are produced [Smoot and Smith, 2013, Tola and Pettinau, 2014, Hu et al., 2014]. Under the environmental protection policy, CPPs prefer to use high-quality coal. However, the reality is that the higher the quality of coal, the higher the price [Yang et al., 2012, Li, 2012, Heinberg and Fridley, 2010]. At this time, CPPs have a difficult choice. More use of scarce high-quality coal, which cannot often meet demand, will reduce emissions of pollutants, but purchase costs will greatly increase. Therefore, to solve the problem of cost and demand conflicts, the method of coal blending can be employed, allowing CPPs to blend several different coal qualities [Dai et al., 2014, Shih and Frey, 1995]. Under such circumstances, for large coal-fired power companies with multiple subordinate CPPs, CSDC's ICPBD strategy can reduce CO_2 emissions, reduce the harm caused by uncertainty, and reduce overall costs.

The goal of ICPBD is to determine the best coal procurement strategy for large coal-fired power companies and implement deployment and distribution plans to balance the contradiction between environmental protection and economic development. Multi-stage decision-making methods and dynamic ICPBD strategies are used to maximize total profits throughout the planning cycle. However, procurement, coal blending adjustment, and distribution control will directly affect the two goals of CO_2 emissions and economic benefits. Therefore, while balancing economic and environmental goals, policymakers also need to consider the equality of all subordinate CPPs. Decision-makers need to predict coal prices based on market demand, supplier supply capacity, and capacity constraints to determine the demand for different coal types. After the coal is transported to CSDC, it needs to be stored, but the storage quantity cannot exceed the total storage capacity of the warehouse. The safety stock level needs to be set to eliminate the risk of coal demand and supply uncertainty. Finally, decision-makers need to make a distribution plan based on the appropriate coal blending quality and transportation conditions to meet the needs of each subordinate CPP. The logical framework diagram of ICPBD is shown in Figure 7.1.

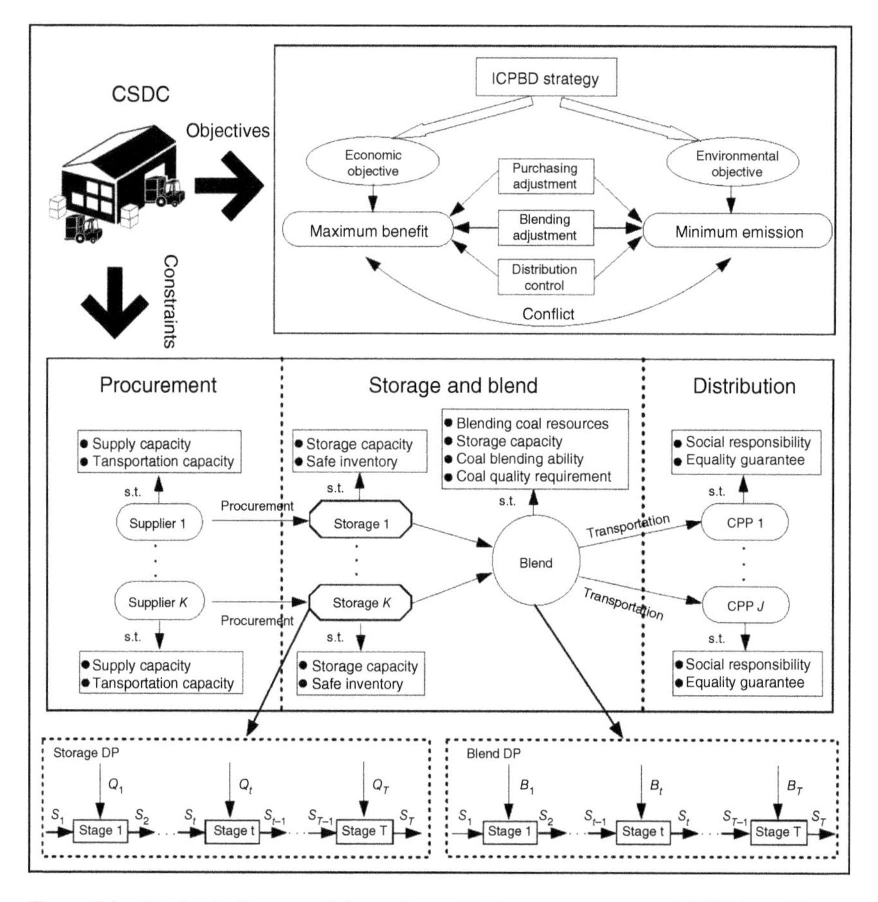

Figure 7.1 The logic diagram of dynamic equilibrium strategy toward ICPBD problem.

7.3 Modeling

The DMOM for balancing the conflict between profits and CO_2 emissions can be mathematically formulated as follows.

7.3.1 Assumptions

The various assumptions involved in this chapter are described as follows:

1. The inventory carrying cost per coal type per stage for the raw coal and the blended coal remains the same throughout the study period.
2. The transportation costs per product from the CSDC to each CPP remains fixed for all given periods.
3. The loading, unloading, and other handling costs per product from the supplier to the CSDC are included in the transportation costs.

7.3.2 ICPBD Strategy Intentions

In order to ensure the sustainable development of the corporation, the decision-maker needs to make decisions on purchase, storage, blending, and distribution. Meanwhile, as a modern state-owned enterprise, environmental protection must be taken into account. The specific objectives are given as following.

7.3.2.1 Maximizing Economic Benefit

Firstly, for a market-oriented power group, the first priority is to pursue the highest profit, which is the difference between the power sales revenue and expenditure, including tax, coal procurement, freight, and inventory cost; therefore, the profit objective is considered from two aspects.

Indices

k = Index for coal supplier, $k \in \Psi = \{1, 2, \dots, K\}$;
j = Index for coal power plant (CPP), $j \in \Phi = \{1, 2, \dots, J\}$;
t = Index for decision stage, $t \in \Omega = \{1, 2, \dots, T\}$;
w = Index for coal quality, $w \in \Pi = \{1, 2, \dots, W\}$.

Decision variables

Q_k^t = Quantity transported from supplier k to the CSDC in stage t;
Q_j^t = Quantity transported from the CSDC to CPP j in stage t;
B_{kj}^t = Quantity of coal component k used to blend coal for CPP sj in stage t.

Crisp parameters

σ = Price of a unit of electric power;

(Continued)

(Continued)

δ = Taxes that the corporation pays for each power generation unit;
h = Inventory carrying cost per unit per stage for each raw coal and each blended coal;
RS_k^t = Storage quantity of raw coal component k at the CSDC at the end of stage t;
$RCS^{\max}$ = Maximum CSDC raw coal storage capacity;
BS_j^t = Storage quantity of the blended coal for CPP j at the CSDC at the end of stage t;
$BCC^{\max}$ = Maximum CSDC coal blending capacity;
$BCS^{\max}$ = Maximum CSDC blended coal storage capacity.

Uncertain parameters

$\widetilde{SS}_k^{\max}$ = Maximum supply capacity of supplier k;
$\widetilde{T}_{kj}$ = Conversion parameters for a unit of coal to a unit of power for coal component k at CPP j;
$\widetilde{P}_k^t$ = Purchase price of coal component k at the stage t;
$\widetilde{C}_k^t$ = Unit transportation cost from supplier k to the CSDC in stage t;
$\widetilde{C}_j^t$ = Unit transportation cost from the CSDC to CPP j in stage t;
$\widetilde{V}_{kj}$ = CO_2 emissions caused by burning a unit of coal component k at CPP j;
$\widetilde{RSI}_k^{\min}$ = Coal component k safety stock quantity at the CSDC in stage t;
$\widetilde{LB}_{jw}$ = Lower bounds for coal quality w to meet CPP j operations;
$\widetilde{R}_{kw}$ = The wth quality of coal component k;
$\widetilde{UB}_{jw}$ = Upper bounds for coal quality w to meet CPP j operations;
$\widetilde{D}_j^{t,\min}$ = Coal that CPP j needs to meet minimum electricity demand in stage t in the region;
$\widetilde{D}_j^{t,\max}$ = Maximum amount of coal CPP j needs in stage t in the region.

7.3.2.1.1 *Electricity Sales Income*

Let $\widetilde{T}_{kj}$ be the conversion parameters from a unit of coal to a unit of power for coal component k at CPP j. The parameters $\widetilde{T}_{kj}$ are difficult to determine because of many factors such as coal quality, whose value can therefore be estimated within a certain range based on historical data; therefore, as for the historical data characteristics, some uncertain situations should be considered fuzzy and can be described using trapezoidal fuzzy numbers, which is written as $T_{kj} = (r_{kj}^1, r_{kj}^2, r_{kj}^3, r_{kj}^4)$, where $r_{kj}^1 \le r_{kj}^2 \le r_{kj}^3 \le r_{kj}^4$ [Puri, 1986, Liu and Liu, 2002, Ferrero and Salicone, 2004]. As the fuzzy parameters cannot be directly calculated, the expected value operator method is employed to transform the trapezoidal fuzzy number into its corresponding expected value [Xu and Zhou, 2011]. Let σ denote the price of a unit electric power and B_{kj}^t be the quantity used to blend coal for CPP j for coal component k in stage t. Therefore, the total electricity sales income can be calculated as $\sigma \sum_{t=1}^{T} \sum_{k=1}^{K} \sum_{j=1}^{J} E\left[\widetilde{T_{kj}}\right] B_{kj}^{t}$.

7.3.2.1.2 Total Expenditure

There are four main expenditures at the electric power company, including taxes, coal purchases, freight, and inventory costs. First, let δ be the taxes the corporation need to pay for unit power generation; therefore, the tax is $\delta \sum_{t=1}^{T} \sum_{k=1}^{K} \sum_{j=1}^{J} E\left[\widetilde{T_{kj}}\right] B_{kj}^{t}$. Then, let $\widetilde{P}_{k}^{t}$ denote the unit price of coal component k in stage t and Q_{k}^{t} denote the quantity transported from supplier k to CSDC in stage t; therefore, the total procurement cost is $\sum_{t=1}^{T} \sum_{k=1}^{K} E\left[\widetilde{P}_{k}^{t}\right] Q_{k}^{t}$. There are two parts of corporation's potential transportation costs: from the suppliers to the CSDC, and from the CSDC to all CPPs. Let $\widetilde{C}_{k}^{t}$ denote the unit transportation cost from supplier k to the CSDC in stage t and let Q_{k}^{t} denote the quantity transported from supplier k to the CSDC in stage t; then, the transportation costs from the supplier to the CSDC is $\sum_{t=1}^{T} \sum_{k=1}^{K} E\left[\widetilde{C}_{k}^{t}\right] Q_{k}^{t}$. Similarly, let $\widetilde{C}_{j}^{t}$ be unit transportation cost from the CSDC to CPP j in stage t and let Q_{j}^{t} be the quantity transported from the CSDC to CPP j in stage t; then, the transportation cost from the CSDC to all CPPs is $E\left[\widetilde{C}_{j}^{t}\right] Q_{j}^{t}$, and the total transportation costs are $\sum_{t=1}^{T} \sum_{k=1}^{K} E\left[\widetilde{C}_{k}^{t}\right] Q_{k}^{t} + \sum_{t=1}^{T} \sum_{j=1}^{J} E\left[\widetilde{C}_{j}^{t}\right] Q_{j}^{t}$. The potential inventory costs of corporation are generally divided into two parts: the raw coal from all suppliers and the blended coal at the CSDC. Let h denote the inventory carrying cost per unit per stage for each raw coal and each blended coal and let RS_{k}^{t} denote the storage quantity of raw coal k at the CSDC at the end of stage t; then, the total raw coal inventory cost at the CSDC is $\sum_{t=1}^{T} \sum_{k=1}^{K} h(RS_{k}^{t} + RS_{k}^{t-1})/2$. Similarly, let BS_{j}^{t} denote the storage quantity of blended coal for CPP j at the CSDC at the end of stage t; then, the blended coal inventory cost at the CSDC is $\sum_{t=1}^{T} \sum_{j=1}^{J} h(BS_{j}^{t} + BS_{j}^{t-1})/2$. Combining the two parts, the total inventory cost is $\sum_{t=1}^{T} \sum_{k=1}^{K} h(RS_{k}^{t} + RS_{k}^{t-1})/2 + \sum_{t=1}^{T} \sum_{j=1}^{J} h(BS_{j}^{t} + BS_{j}^{t-1})/2$; therefore the total expenditure is $\delta \sum_{t=1}^{T} \sum_{k=1}^{K} \sum_{j=1}^{J} E\left[\widetilde{T_{kj}}\right] B_{kj}^{t} + \sum_{t=1}^{T} \sum_{k=1}^{K} E\left[\widetilde{P}_{k}^{t}\right] Q_{k}^{t} + \sum_{t=1}^{T} \sum_{k=1}^{K} E\left[\widetilde{C}_{k}^{t}\right] Q_{k}^{t} + \sum_{t=1}^{T} \sum_{j=1}^{J} E\left[\widetilde{C}_{j}^{t}\right] Q_{j}^{t} + \sum_{t=1}^{T} \sum_{k=1}^{K} h(RS_{k}^{t} + RS_{k}^{t-1})/2 + \sum_{t=1}^{T} \sum_{j=1}^{J} h(BS_{j}^{t} + BS_{j}^{t-1})/2$.

Based on above analysis, the total profit for the corporation can be described as following:

$$
\begin{aligned}
\max TR = (\sigma - \delta) \sum_{t=1}^{T} \sum_{k=1}^{K} \sum_{j=1}^{J} E\left[\widetilde{T_{kj}}\right] B_{kj}^{t} - \sum_{t=1}^{T} \sum_{k=1}^{K} E\left[\widetilde{P}_{k}^{t}\right] Q_{k}^{t} - \sum_{t=1}^{T} \sum_{k=1}^{K} E\left[\widetilde{C}_{k}^{t}\right] Q_{k}^{t} \\
- \sum_{t=1}^{T} \sum_{j=1}^{J} E\left[\widetilde{C}_{j}^{t}\right] Q_{j}^{t} - \sum_{t=1}^{T} \sum_{k=1}^{K} h(RS_{k}^{t} + RS_{k}^{t-1}) \Big/ 2 \\
- \sum_{t=1}^{T} \sum_{j=1}^{J} h(BS_{j}^{t} + BS_{j}^{t-1}) \Big/ 2
\end{aligned}
\tag{7.1}
$$

7.3.2.2 Minimizing CO_2 Emissions

With the rapid development of the global society, the demand for electricity is increasing, which means that more coal-fired power will be needed to meet demand. However, a lot of CO_2 emissions from the CPP can cause irreversible climate change effects [Solomon et al., 2009], from which, enterprises not only consider their own steady development, but also have an unshirkable responsibility for environmental protection. Let $\widetilde{V}_k^j$ denote the CO_2 emissions from burning a unit of coal component k at CPP j and B_{kj}^t be the quantity used to blend the coal for CPP j for coal component k in stage t; therefore, therefore, the total CO_2 emissions can be written as:

$$\min \mathrm{TCO_2} = \sum_{t=1}^{T} \sum_{k=1}^{K} \sum_{j=1}^{J} E[\widetilde{V}_{kj}] B_{kj}^t \tag{7.2}$$

7.3.3 ICPBD Strategy Limitations

To pursue the largest possible profit and ensure steady development, coal purchase, storage, blending, and distribution phase constraints must be met in addition to the nonnegative constraints, the details of which are explained in the following.

7.3.3.1 Coal Purchase Phase Restriction

In order to pursue maximum profits and ensure stable development, in addition to nonnegative constraints, it is also necessary to meet the constraints of coal purchase, storage, blending, and distribution phase stages, which are as following.

$$Q_k^t \le E\left[\widetilde{SS}_k^{\max}\right], \quad \forall k \in \Psi, \ \forall t \in \Omega \tag{7.3}$$

7.3.3.2 Coal Storage Phase Restrictions

7.3.3.2.1 Raw Coal State Transition Equation

The period in this chapter is divided into T stages for analysis, and the condition in each stage is influenced by the previous stage. Let RS_k^t be the storage quantity of the raw coal component k at the CSDC in the end of stage t, while Q_k^t be the quantity transported from supplier k to the CSDC, and B_{kj^t} be the quantity of coal component k used to blend coal for CPP j in stage t. Therefore, the state transition equation for the raw coal component can be expressed as:

$$RS_k^t = RS_k^{t-1} + Q_k^t - \sum_{j=1}^{J} B_{kj}^t, \quad \forall k \in \Psi, \ \forall t \in \Omega \tag{7.4}$$

7.3.3.2.2 Raw Coal Resources Storage Capacity limitation

The raw coal delivered cannot exceed the maximum storage capacity due to its limited raw coal storage capacity. Therefore, the coal storage at the end of the stage t is the sum of the initial coal inventory and the raw coal procurement. Let RS_k^t represents the storage quantity of the raw coal component k at the CSDC at the end of stage t, Q_k^t be the quantity transported from supplier k to the CSDC in stage t, and

$RCS^{\max}$ be the maximum storage capacity at CSDC. Therefore, this constraint can be expressed as:

$$\sum_{k=1}^{K} (RS_k^{t-1} + Q_k^t) \le RCS^{\max}, \quad \forall t \in \Omega \tag{7.5}$$

7.3.3.2.3 Raw Coal Resources Safety Stock Limitation

Electricity is very significant to society development and coal is essential for CPPs; thus, the coal supply for each CPP should be adequate. However, due to unforeseen events such earthquakes, storms, and floods, the coal supply requirements sometimes will not met. Therefore, to reduce the adverse such effects, the CSDC must establish a safety stock. Demand variations also influence inventory management and the organization's ability to meet its customer service needs [Fisher and Raman, 1996, Giri and Chakraborty, 2016]. Then, let $RSI_k^{\min}$ denote the coal k safety stock quantity at the CSDC in stage k. Therefore, the CSDC raw coal storage at the end of the current stage should not be less than the safety stock quantity at the CSDC, which is described as:

$$RS_k^t \ge E\left[\widetilde{RSI}_k^{\min}\right], \quad \forall k \in \Psi, \ \forall t \in \Omega \tag{7.6}$$

7.3.3.3 Coal Blending Phase Restrictions

7.3.3.3.1 Limited Available Coal Blending Resources

CSDC coal blending is limited by the current available coal resources of CSDC, which is generally the sum of the CSDC storage at the beginning of the stage and the coal purchased during that period. Let RS_k^{t-1} denote the raw coal k storage quantity at the end of stage $t-1$ and B_{kjt} be the quantity of coal component k used to blend coal for CPP j in stage t; then the available raw coal quantity is written as $RS_k^{t-1} + Q_k^t$, and the blended raw coal required from supplier k at the CSDC in stage t is subject to this restriction:

$$\sum_{j=1}^{J} B_{kj}^t \le RS_k^{t-1} + Q_k^t, \quad \forall k \in \Psi, \ \forall t \in \Omega \tag{7.7}$$

7.3.3.3.2 Blended Coal State Transition Equation

The period in this chapter is divided into T stages for analysis, in which, the condition in each stage is affected by the decisions in previous stage. The blended coal has uncertain losses because of environmental factors. Let BS_j^t be the storage quantity of the blended coal for CPP j in the CSDC at the end of stage t and B_{kjt} be the quantity of coal component k used to blend coal for CPP j during the period; therefore, the state transition equation for the blended coal is expressed as:

$$BS_j^t = BS_j^{t-1} + \sum_{k=1}^{K} B_{kj}^t - Q_j^t, \quad \forall j \in \Phi, \ \forall t \in \Omega \tag{7.8}$$

7.3.3.3.3 Limited Available Blended Coal for Transportation

The transportation quantity from CSDC to each CPP is limited by the current available mixed coal resources of CSDC, which is generally the sum of the CSDC blended coal storage at the beginning of the current stage and the blended coal in this stage. Let BS_j^{t-1} represent the blended coal storage quantity at the end of stage $t-1$; then, the available blended coal quantity in the stage t is written as $BS_j^{t-1} + \sum_{k=1}^{K} B_{kj}^t$. Therefore, this constraint can be given as:

$$Q_j^t \leq BS_j^{t-1} + \sum_{k=1}^{K} B_{kj}^t, \quad \forall j \in \Phi, \ \forall t \in \Omega \tag{7.9}$$

7.3.3.3.4 Coal Blending Capacity Limitation

The quantity of coal blending should not exceed the coal blending capacity of the CSDC due to existing equipment conditions. Let $BC^{\max}$ denote the maximum coal blending capacity at the CSDC. Therefore, this constraint can be expressed as:

$$\sum_{k=1}^{K} \sum_{j=1}^{J} B_{kj}^t \leq BC^{\max}, \quad \forall t \in \Omega \tag{7.10}$$

7.3.3.3.5 Blended Coal Storage Capacity Limitation

Due to the limited storage capacity of blended coal, in which, the blended coal cannot exceed the maximum storage capacity. Therefore, the decision-maker must ensure that the blended coal storage at the CSDC does not exceed the maximum storage capacity when deciding on the coal blending plan. The coal storage at the end of the stage t is the sum of the blended coal storage at the beginning of the stage and the coal blending; therefore, this constraint can be expressed as following:

$$\sum_{j=1}^{J} BS_j^{t-1} + \sum_{k=1}^{K} \sum_{j=1}^{J} \mathrm{B}_{kj}^t \leq BCS^{\max}, \quad \forall t \in \Omega \tag{7.11}$$

7.3.3.3.6 Coal Quality Requirement

Each electricity generation boiler has special coal quality requirements to do with the five coal properties: volatile matter content, heat rate, ash content, moisture content, and sulfur content [Dai et al., 2014, Samimi and Zarinabadi, 2012]. If the coal properties do not meet these conditions, either the boiler must be redesigned to adapt to the available coal properties or the coal properties must be adjusted through coal blending to meet the existing boiler requirements. Since the power generation boilers of each CPP are predetermined, it is difficult to change the specific coal requirements; therefore, in this chapter, the coal blending method is used to limit the properties of the above five kinds of coal within the specified range to ensure the normal boiler operations; therefore, this constraint can be expressed as follows:

$$E[\widetilde{LB}_{jw}] \leq \frac{\sum_{k=1}^{K} E[\widetilde{R}_{kw}] B_{kj}^t}{\sum_{k=1}^{K} B_{kj}^t} \leq E[\widetilde{UB}_{jw}], \quad \forall j \in \Phi, \ \forall w \in \Theta, \ \forall t \in \Omega \tag{7.12}$$

where $w = 1$ denotes volatile matter content, $w = 2$ denotes heat rate, $w = 3$ denotes ash content, $w = 4$ denotes moisture content and $w = 5$ denotes sulfur content, $\widetilde{R}_{kw}$ denotes the wth quality of the coal component k, and LB_{jw} and UB_{jw} denote the lower and the upper bounds of the wth quality at each CPP j to meet CPP j operations.

7.3.3.4 Coal Distribution Phase Restrictions

7.3.3.4.1 Social Responsibility Limitation

Demand variations have a significant impact on inventory management and an organization's ability to meet customer service needs [Giri and Chakraborty, 2016]. For electric power companies, at any given time, demand variations can effect the requirements of CPPs to meet the needs of customers. Modern enterprises also need to bear social responsibility, in which, providing basic power is the basic social responsibility of enterprises, since electricity is critical for social development. Let $\widetilde{D}_j^{t,\min}$ denote the amount of coal CPP j needs to meet minimum electricity demand in the region; therefore, this constraint can be expressed as:

$$Q_j^t \geq E\left[\widetilde{D}_j^{t,\min}\right], \quad \forall j \in \Phi, \ \forall t \in \Omega \tag{7.13}$$

7.3.3.4.2 Equality Guarantee

For sustainable development, equality must be considered. In this chapter, the satisfactory degree method proposed by Xu et al. Xu et al. [2015, 2016] is used to measure the equality of all subordinate CPPs; therefore, the satisfactory degree of each CPP can be defined as follows:

$$\theta_j = \begin{cases} 0, & \sum_{t=1}^{T} Q_j^t \leq \sum_{t=1}^{T} E\left[\widetilde{D}_j^{t,\min}\right] \\ \dfrac{\sum_{t=1}^{T} Q_j^t - \sum_{t=1}^{T} E[\widetilde{D}_j^{t,\min}]}{- \sum_{t=1}^{T} E[\widetilde{D}_j^{t,\min}]}, & \sum_{t=1}^{T} E\left[\widetilde{D}_j^{t,\min}\right] \leq \sum_{t=1}^{T} Q_j^t \leq \sum_{t=1}^{T} D_j^{t,\max} \\ 1, & \sum_{t=1}^{T} Q_j^t \geq \sum_{t=1}^{T} E\left[\widetilde{D}_j^{t,\min}\right] \end{cases} \tag{7.14}$$

Let α be the minimal satisfaction of the operator's choice. The satisfactory degree should be greater than or equal to the minimum satisfactory degree set by the decision-maker; therefore, the equality guarantee constraint can be expressed as follows:

$$\theta_j \geq \alpha, \quad \forall j \in \Phi \tag{7.15}$$

7.3.4 Global Model

Based on the previous analyses, a DMOM with CSDC coal blending is proposed to optimize spatiotemporal conflict strategy in the ICPBD problem. In this chapter, dynamic programming is employed to address the fluctuant environmental time conflicts, while equilibrium strategy is used to deal with the supply and coal demand space conflicts. The model process can be summarized as follows. First of all, in the coal purchasing phase, the decision-maker purchases various coal under the limitation of supply and transportation capacity according to the predicted market price of coal and demand of each subplant. Secondly, in the storage and blending phase, decisions are made based on the quantities purchased of the various kinds of coal, demand of each CPP and the coal property boiler requirements at each subordinate CPP after the CSDC coal blending. Thirdly, in the distribution and transportation phases, depending on coal combustion efficiency and CO_2 emissions performance at each subordinate CPP, the blended coal supply quantity can be decided aiming at improving overall emissions performances and guaranteeing steady development to meet the social responsibilities. Finally, the global DMOM is formulated as shown in Eq. (7.16) by integrating Eqs. (7.1)–(7.15).

$$
\begin{aligned}
\max TR = {} & (\sigma - \delta)\sum_{t=1}^{T}\sum_{k=1}^{K}\sum_{j=1}^{J} E\left[\widetilde{T_{kj}}\right] B_{kj}^{t} - \sum_{t=1}^{T}\sum_{k=1}^{K} E\left[\widetilde{P_{k}^{t}}\right] Q_{k}^{t} - \sum_{t=1}^{T}\sum_{k=1}^{K} E\left[\widetilde{C_{k}^{t}}\right] Q_{k}^{t} \\
& - \sum_{t=1}^{T}\sum_{j=1}^{J} E\left[\widetilde{C_{j}^{t}}\right] Q_{j}^{t} - \sum_{t=1}^{T}\sum_{k=1}^{K} h\left(RS_{k}^{t} + RS_{k}^{t-1}\right)/2 \\
& - \sum_{t=1}^{T}\sum_{j=1}^{J} h\left(BS_{j}^{t} + BS_{j}^{t-1}\right)/2 \\
\min \mathrm{TCO}_2 = {} & \sum_{t=1}^{T}\sum_{k=1}^{K}\sum_{j=1}^{J} E[\widetilde{V}_{kj}]B_{kj}^{t}
\end{aligned}
$$

$$
\text{s.t.}\begin{cases}
Q_{k}^{t} \le E\left[\widetilde{SS_{k}^{\max}}\right], & \forall k \in \Psi,\ \forall t \in \Omega \\
RS_{k}^{t} = RS_{k}^{t-1} + Q_{k}^{t} - \sum_{j=1}^{J} B_{kj}^{t}, & \forall k \in \Psi,\ \forall t \in \Omega \\
\sum_{j=1}^{J} B_{kj}^{t} \le RS_{k}^{t-1} + Q_{k}^{t}, & \forall k \in \Psi,\ \forall t \in \Omega \\
\sum_{k=1}^{K} \left(RS_{k}^{t-1} + Q_{k}^{t}\right) \le RCS^{\max}, & \forall t \in \Omega \\
RS_{k}^{t} \ge E\left[\widetilde{RSI_{k}^{\min}}\right], & \forall k \in \Psi,\ \forall t \in \Omega \\
RS_{k}^{0} = 3, & \forall k \in \Psi \\
BS_{j}^{t} = BS_{j}^{t-1} + \sum_{k=1}^{K} B_{kj}^{t} - Q_{j}^{t}, & \forall j \in \Phi,\ \forall t \in \Omega \\
BS_{j}^{0} = 0, & \forall j \in \Phi \\
BS_{j}^{T} = 0, & \forall j \in \Phi \\
Q_{j}^{t} \le BS_{j}^{t-1} + \sum_{k=1}^{K} B_{kj}^{t}, & \forall j \in \Phi,\ \forall t \in \Omega \\
\sum_{k=1}^{K}\sum_{j=1}^{J} B_{kj}^{t} \le BCC^{\max}, & \forall t \in \Omega \\
\sum_{j=1}^{J} BS_{j}^{t-1} + \sum_{k=1}^{K}\sum_{j=1}^{J} B_{kj}^{t} \le BCS^{\max}, & \forall t \in \Omega \\
E[\widetilde{LB}_{jw}] \le \dfrac{\sum_{k=1}^{K} E[\widetilde{R}_{kw}]B_{kj}^{t}}{\sum_{k=1}^{K} B_{kj}^{t}} \le E[\widetilde{UB}_{jw}], & \forall j \in \Phi,\ \forall w \in \Theta,\ \forall t \in \Omega \\
Q_{j}^{t} \le E[\widetilde{D}_{j}^{t,\max}], & \forall j \in \Phi,\ \forall t \in \Omega \\
Q_{j}^{t} \ge E[\widetilde{D}_{j}^{t,\min}], & \forall j \in \Phi,\ \forall t \in \Omega \\
\theta_{j} \ge \alpha, & \forall j \in \Phi
\end{cases}
\tag{7.16}
$$

The proposed dynamic multiobjective programming model was established to determine the ICPBD strategy to get balance between economic development and CO_2 emissions reduction conflicts. In this model, parameter α is the minimum value of coal blending satisfaction at each CPPs, set by the decision-maker. Employing the parameter setting method means that there will be more decision-making flexibility, which can better adapt to the market environment and promote sustainable development.

7.4 Case Study

In this section, the effectiveness and practicality of the proposed model are demonstrated through a real-world case study.

7.4.1 Presentation of Case Region

Shaanxi, Shanxi, and the northwestern regions of Inner Mongolia are the main source of coal resources in China. However, the highest consumption of coal resource is in the industrially developed southeast region and the central coastal provinces, such as Jiangsu Province and Guangdong province. Jiangsu Province is in southeast China and is one of the most economically developed provinces in China, ranking second in China in 2019. With the economic development, Jiangsu Province is also a region of carbon emissions due to the electricity and is primarily supplied by CPPs. To protect the atmospheric environment, the relevant departments of Jiangsu Province have implemented a series of CO_2 emission trading projects to reduce CO_2 emissions. Under the CO_2 emissions trading scheme, corporate will be assigned an emissions quota and can profit by selling excess permits to other firms if they have a surplus of carbon emissions. Under the CO_2 emissions reduction pressure from the Jiangsu Province authority, and to guarantee to the stability and safety of the coal energy supply chain, Binhai port CSDC was established by the State Power Investment Corporation in 2017.

Based on the above situation, the State Power Investment Corporation in Jiangsu Province was chosen as the case study to test the practicability and efficiency of the proposed optimization method. Besides, considering the feasibility of computer calculation and data acquisition, three of the main power plants in Jiangsu Province were chosen for this case, which are the Kanshan CPP, the Wangting CPP, and the Xiexin CPP.

7.4.2 Model Transformation

As a market-based State Power Investment Corporation, economic development is still the main objective of the enterprises; therefore, the decision-maker needs to guarantee continued economic development. However, as a responsible large-scale state-owned enterprise, the CO_2 emissions also need to be considered. Therefore, the minimal CO_2 emissions can be transformed into an acceptable range. Based on research by Zeng et al. Zeng et al. [2014], in this chapter, let T denote the decision-maker's attitude toward the CO_2 emissions reduction; therefore, the proposed model can be transformed into its equivalent single-objective form, as shown in Eq. (7.17).

$$\max TR = (\sigma - \delta)\sum_{t=1}^{T}\sum_{k=1}^{K}\sum_{j=1}^{J} E\left[\widetilde{T_{kj}}\right] B_{kj}^{t} - \sum_{t=1}^{T}\sum_{k=1}^{K} E\left[\widetilde{P_k^t}\right] Q_k^t - \sum_{t=1}^{T}\sum_{k=1}^{K} E\left[\widetilde{C_k^t}\right] Q_k^t$$
$$- \sum_{t=1}^{T}\sum_{j=1}^{J} E\left[\widetilde{C_j^t}\right] Q_j^t - \sum_{t=1}^{T}\sum_{k=1}^{K} h(RS_k^t + RS_k^{t-1})/2$$
$$- \sum_{t=1}^{T}\sum_{j=1}^{J} h(BS_j^t + BS_j^{t-1})/2$$

$$\text{s.t.}\begin{cases}
\sum_{t=1}^{T}\sum_{k=1}^{K}\sum_{j=1}^{J} E[\widetilde{V}_{kj}]B_{kj}^t \le \mathrm{TCO}_2^* \\
Q_k^t \le E\left[\widetilde{SS_k^{\max}}\right], \quad \forall k \in \Psi,\ \forall t \in \Omega \\
RS_k^t = RS_k^{t-1} + Q_k^t - \sum_{j=1}^{J} B_{kj}^t, \quad \forall k \in \Psi,\ \forall t \in \Omega \\
\sum_{j=1}^{J} B_{kj}^t \le RS_k^{t-1} + Q_k^t, \quad \forall k \in \Psi,\ \forall t \in \Omega \\
\sum_{k=1}^{K} (RS_k^{t-1} + Q_k^t) \le RCS^{\max}, \quad \forall t \in \Omega \\
RS_k^t \ge E\left[\widetilde{RSI_k^{\min}}\right], \quad \forall k \in \Psi,\ \forall t \in \Omega \\
RS_k^0 = 3, \quad \forall k \in \Psi \\
BS_j^t = BS_j^{t-1} + \sum_{k=1}^{K} B_{kj}^t - Q_j^t, \quad \forall j \in \Phi,\ \forall t \in \Omega \\
BS_j^0 = 0, \quad \forall j \in \Phi \\
BS_j^T = 0, \quad \forall j \in \Phi \\
Q_j^t \le BS_j^{t-1} + \sum_{k=1}^{K} B_{kj}^t, \quad \forall j \in \Phi,\ \forall t \in \Omega \\
\sum_{k=1}^{K}\sum_{j=1}^{J} B_{kj}^t \le BCC^{\max}, \quad \forall t \in \Omega \\
\sum_{j=1}^{J} BS_j^{t-1} + \sum_{k=1}^{K}\sum_{j=1}^{J} \mathrm{B}_{kj}^t \le BCS^{\max}, \quad \forall t \in \Omega \\
E[\widetilde{LB}_{jw}] \le \frac{\sum_{k=1}^{K} E[\widetilde{R}_{kw}]B_{kj}^t}{\sum_{k=1}^{K} B_{kj}^t} \le E[\widetilde{UB}_{jw}], \quad \forall j \in \Phi,\ \forall w \in \Theta,\ \forall t \in \Omega \\
Q_j^t \le E\left[\widetilde{D}_j^{t,\max}\right], \quad \forall j \in \Phi,\ \forall t \in \Omega \\
Q_j^t \ge E\left[\widetilde{D}_j^{t,\min}\right], \quad \forall j \in \Phi,\ \forall t \in \Omega \\
\theta_j \ge \alpha, \quad \forall j \in \Phi,
\end{cases} \tag{7.17}$$

where, λ is the decision-maker's attitude toward CO_2 emissions reductions, and TCO_2^* is the maximum CO_2 emissions calculated by plugging the optimal solution without considering the total CO_2 emissions objective. Therefore, the model Eq. (7.16) is already transformed into a single target model as shown in Eq. (7.17).

Table 7.1 Certain parameters of Binhai port CSDC.

σ (RMB/kwh)	δ (RMB/kwh)	h (RMB/10^4t)	RCS (10^4t)	BCS^{max} (10^4t)	BCC^{max} (10^4t)
0.45	0.01	0.2	430	360	350

7.4.3 Data Collection

Practical parameter values were collected before the proposed model calculations. The input data and parameters for the proposed model can be divided roughly into two categories: uncertain data and crisp data.

The crisp data for input into the proposed model were obtained from the State Power Investment Corporation in China, as shown in Table 7.1. Fuzzy set theory was employed to describe the uncertainty in the real-world situations when using dynamic equilibrium strategy toward ICPBD problem. Therefore, the uncertain data (i.e. coal properties, coal prices, and so on) were collected in their fuzzy form according to fuzzy set theory; each parameter was within a certain range with the minimum value set as the lower bound (e.g. r^1_{kj} of T_{kj}, $T_{kj} = (r^1_{kj}, r^2_{kj}, r^3_{kj}, r^4_{kj})$) and the maximum value set as the upper bound (e.g. r^4_{kj} of T_{kj}). The range between the two values with the highest probability was set as the most possible range for the uncertain parameter (e.g. from r^2_{kj} to r^3_{kj} of T_{kj}). Therefore, in this article, all uncertain parameters were collected in their fuzzy form using this method, as shown in Tables 7.2–7.4.

7.5 Results and Discussion

This chapter used Lingo 13 to run the solution by inputting the collected data into the proposed optimization model (i.e. Eq. (7.17)) without considering the total CO_2 emissions objective and then calculating the maximum CO_2 emission by plugging this solution. By inputting the collected data and the maximum CO_2 emissions (TCO_2^*) into the model Eq. (7.17) and running the solution approach on Lingo 13, the results of model Eq. (7.17) were calculated.

7.5.1 Results for Different Scenarios

In this section, the results under different decision-maker attitudes (i.e. different control parameter: λ, α) are given.

Scenario 1: $\alpha = 0$, λ is changing. In this scenario, the decision-maker's satisfactory degree was set at 0 (i.e. $\alpha = 0$), which indicated that the decision-maker preferred that the total blended coal demand meet the minimum amount at each CPP. The changing λ represents the decision-maker's attitude toward the CO_2 emissions target, the results for which are shown from Tables 7.5–7.8. It can be seen that when $\lambda = 1$, the total profit was 6.388×10^9 RMB. Here, the parameter analysis step length

Table 7.2 Uncertain parameters of each CPP.

Bi-month	Xiexin CPP	Kanshan CPP	Wangting CPP
Minimum demand: $\widetilde{D}_j^{t,\min}$ (10^4 tonnes)			
1	(64.8,65.5,65.9,66.2)	(111.7,114.9,116.8,118.2)	(78.2,80.3,81.2,82.7)
1	(56.3,57.2,57.4,58.7)	(97.1,98.3,101.2,101.8)	(69.8,71.9,73.1,74.8)
3	(41.4,42.2,43.5,43.7)	(70.8,71.3,74.5,76.2)	(54.6,55.2,55.5,56.3)
4	(56.9,58.2,60.3,62.2)	(98.5,101.2,102.1,102.6)	(70.7,73.4,75.2,79.1)
5	(40.9,45.2,47.2,49.9)	(85.1,85.3,85.7,86.3)	(57.2,58.8,60.3,60.5)
6	(55.2,57.7,58.1,58.6)	(93.3,94.9,96.9,98.1)	(71.9,72.8,73.9,75.8)
Maximum demand: $\widetilde{D}_j^{t,\max}$ (10^4 tonnes)			
1	(79.2,80.1,81.2,81.9)	(145.2,146.8,147.8,149.8)	(117.5,118.1,118.9,119.1)
2	(79.2,80.1,81.2,81.9)	(145.2,146.8,147.8,149.8)	(117.5,118.1,118.9,119.1)
3	(79.2,80.1,81.2,81.9)	(145.2,146.8,147.8,149.8)	(117.5,118.1,118.9,119.1)
4	(79.2,80.1,81.2,81.9)	(145.2,146.8,147.8,149.8)	(117.5,118.1,118.9,119.1)
5	(79.2,80.1,81.2,81.9)	(145.2,146.8,147.8,149.8)	(117.5,118.1,118.9,119.1)
6	(79.2,80.1,81.2,81.9)	(145.2,146.8,147.8,149.8)	(117.5,118.1,118.9,119.1)
Transportation price: (RMB/tonne) $\widetilde{C}_j^t$			
1	(0.9,1.1,1.3,1.5)	(103.6,104.8,106.2,107.4)	(129.8,130.7,132.4,134.7)
2	(0.9,1.1,1.3,1.5)	(103.6,104.8,106.2,107.4)	(129.8,130.7,132.4,134.7)
3	(0.9,1.1,1.3,1.5)	(103.6,104.8,106.2,107.4)	(129.8,130.7,132.4,134.7)
4	(0.9,1.1,1.3,1.5)	(103.6,104.8,106.2,107.4)	(129.8,130.7,132.4,134.7)
5	(0.9,1.1,1.3,1.5)	(103.6,104.8,106.2,107.4)	(129.8,130.7,132.4,134.7)
6	(0.9,1.1,1.3,1.5)	(103.6,104.8,106.2,107.4)	(129.8,130.7,132.4,134.7)
Coal to electric (kwh/tonne)			
Datong	(2585,2595,2603,2617)	(2510,2520,2540,2550)	(2330,2345,2360,2365)
Shenghua	(2496,2515,2528,2541)	(2410,2425,2435,2450)	(2275,2285,2295,2305)
Yitai	(2462,2477,2486,2495)	(2410,2435,2450,2465)	(2325,2330,2345,2360)
Zhongmei	(2430,2446,2458,2466)	(2355,2370,2385,2410)	(2266,2276,2283,2295)
Carbon emission factor (kg/tonne)			
Datong	(2095,2099,2109,2121)	(2065,2080,2090,2109)	(2025,2035,2050,2070)
Shenghua	(2086,2095,2106,2109)	(2045,2055,2070,2078)	(2018,2030,2045,2059)
Yitai	(2072,2082,2096,2106)	(2020,2030,2050,2072)	(1990,1999,2013,2022)
Zhongmei	(2049,2056,2068,2079)	(1995,2010,2025,2030)	(1975,1980,1990,1995)

Table 7.3 Uncertain parameters of each coal component.

Bi-month	Datong	Shenghua	Yitai	Zhongmei
Procurement price: $\widetilde{P}_k^t$ (RMB/tonne)				
1	(572,575,576,577)	(541,544,545,550)	(519,521,528,532)	(471,472,477,480)
2	(570,573,578,579)	(529,538,540,545)	(515,516,520,521)	(472,474,477,481)
3	(571,573,580,588)	(531,533,536,540)	(508,514,517,521)	(470,471,474,477)
4	(572,580,585,587)	(546,548,550,552)	(522,526,531,533)	(478,483,489,494)
5	(575,576,580,585)	(538,540,551,559)	(521,524,526,533)	(481,483,485,487)
6	(576,579,582,583)	(521,524,530,537)	(521,523,529,531)	(462,464,469,473)
Transportation price: $\widetilde{C}_k^t$ (RMB/tonne)				
1	(126.2,127.3,130.1,131.6)	(124.1,124.8,130.9,131.4)	(168.1,170.3,170.7,172.5)	(90.1,90.3,91.5,92.5)
2	(136.6,139.2,141.7,146.1)	(136.8,137.1,137.9,138.2)	(179.2,181.3,182.9,186.6)	(99.9,100.1,101.1,101.7)
3	(132.5,133.8,134.8,135.3)	(127.6,132.8,133.7,134.3)	(173.2,175.7,175.8,178.1)	(93.8,95.2,95.9,96.3)
4	(139.1,139.7,143.5,144.9)	(137.1,137.6,137.9,140.2)	(181.8,183.4,183.9,184.5)	(99.9,100.8,102.2,102.7)
5	(146.3,147.3,147.7,149.1)	(138.9,141.2,142.9,148.2)	(188.3,189.1,189.5,189.9)	(104.2,105.7,106.8,107.7)
6	(137.2,137.9,138.6,138.7)	(131.9,134.1,137.2,137.6)	(178.6,179.2,179.9,181.1)	(97.5,98.2,98.6,99.3)
Maximum supply capacity: $\widetilde{SS}_k^{max}$ (10^4 tonnes)				
	(106,110,114,118)	(145,154,161,172)	(53,57,62,64)	(118,124,127,131)
Safety stock: RSI^{min} (10^4 tonnes)				
	(2.7,2.9,3.1,3.3)	(2.7,2.9,3.1,3.3)	(2.7,2.9,3.1,3.3)	(2.7,2.9,3.1,3.3)

Table 7.4 Parameters of coal properties and quality requirement in fuzzy form.

	Volatile matter (% weight)	Heating value (Gj/tonne)	Ash content (% weight)	Moisture content (% weight)	Sulfur content (% weight)
Coal properties: $\widetilde{R}_{kw}$					
Tongmei	(23.8,24.7,25.4,26.1)	(22.3,22.7,23.1,23.9)	(19.2,19.7,20.2,20.9)	(14.2,14.7,15.3,15.8)	(0.63,0.66,0.69,0.74)
Shenhua	(24.7,25.2,25.6,26.5)	(21.2,21.7,22.1,22.2)	(15.4,15.9,16.3,16.4)	(14.8,15.3,15.7,16.2)	(0.57,0.59,0.62,0.66)
Yitai	(28.4,28.9,29.3,29.4)	(20.1,20.7,21.2,21.6)	(11.1,11.8,12.2,12.9)	(15.7,16.9,17.3,18.1)	(0.41,0.42,0.46,0.51)
Zhongmei	(30.9,31.3,32.4,33.4)	(18.8,19.1,19.4,19.9)	(16.6,17.4,18.5,19.5)	(13.5,13.9,14.2,14.4)	(0.72,0.74,0.76,0.78)
CPP lower bound: $\widetilde{LB}_{jw}$					
Kanshan CPP	(5.3,5.7,6.2,6.4)	(21.6,22.2,22.6,23.2)	—	—	—
Wangting CPP	(6.4,6.8,7.1,7.3)	(21.5,21.7,21.9,22.1)	—	—	—
Xiexin CPP	(8.5,8.9,9.3,9.7)	(20.3,20.7,21.1,21.5)	—	—	—
CPP upper bound: $\widetilde{UB}_{jw}$					
Kanshan CPP	(24.9,25.6,26.3,27.6)	—	(19.7,20.1,20.8,21.4)	(14.6,15.6,16.6,17.2)	(0.58,0.61,0.63,0.66)
Wangting CPP	(28.6,29.3,29.8,30.3)	—	(20.6,21.1,21.6,22.3)	(16.3,16.8,17.2,17.7)	(0.61,0.66,0.69,0.72)
Xiexin CPP	(31.1,32.1,33.1,33.3)	—	(21.7,22.4,23.4,23.7)	(17.2,17.8,18.3,18.7)	(0.68,0.69,0.72,0.75)

was set at 0.1, and after effective testing, in this scenario, the lowest λ was 0.7, which means that when $\alpha = 0$, total CO_2 emissions can be controlled at 70% of maximum emissions, at which point, the revenue was 4.754×10^9 RMB.

Scenario 2: $\alpha = 0.5$, λ is changing. In this scenario, the decision-maker's satisfactory degree was set at 0.5 (i.e. $\alpha = 0.5$). For the changing λ, which represents the decision-maker's attitude toward the CO_2 emissions target, when $\alpha = 0.5$ and $\lambda = 1$ it was found to be equal to $\alpha = 0$ and $\lambda = 1$ (i.e. Table 7.5), as shown in Tables 7.5,

Table 7.5 Results of the proposed model when $\lambda = 1$ and $\alpha = 0$.

Bi-month	1	2	3	4	5	6
Transportation volume: Q_k^t (10^4 tonnes)						
Datong	112.0	112.0	112.0	112.0	112.0	112.0
Shenghua	150.1	152.7	158.0	130.2	147.8	147.8
Yitai	30.9	5.1	53.9	0.0	0.0	18.0
Zhongmei	125.0	12.2	94.1	107.8	4.2	68.7
Transportation volume: Q_j^t (10^4 tonnes)						
Xiexin CPP	80.6	80.6	80.6	80.6	80.6	80.6
Kanshan CPP	147.4	147.4	147.4	147.4	147.4	147.4
Wangting CPP	118.4	118.4	118.4	118.4	118.4	118.4
Xiexin CPP: B_{k1}^t (10^4 tonnes)						
Datong	52.6	52.6	52.6	52.6	52.6	52.6
Shenghua	10.0	10.0	10.0	10.0	10.0	10.0
Yitai	18.0	18.0	18.0	18.0	18.0	18.0
Zhongmei	0.0	0.0	0.0	0.0	0.0	0.0
Kanshan CPP: B_{k2}^t (10^4 tonnes)						
Datong	59.4	59.4	59.4	59.4	59.4	59.4
Shenghua	60.2	67.5	52.9	60.2	60.2	60.2
Yitai	0.0	0.0	0.0	0.0	0.0	0.0
Zhongmei	27.8	27.7	27.9	27.8	27.8	27.8
Wangting CPP: B_{k3}^t (10^4 tonnes)						
Datong	0.0	0.0	0.0	0.0	0.0	0.0
Shenghua	79.9	75.2	84.6	70.4	77.5	77.5
Yitai	0.0	0.0	0.0	0.0	0.0	0.0
Zhongmei	42.1	39.6	44.6	37.2	40.9	40.9

In this situation (i.e. $\lambda = 1, \alpha = 0$), the total profit is 6.388×10^9 RMB and total car emission amount is 4.275×10^7 tonnes.

Table 7.6 Results of the proposed model when $\lambda = 0.9$ and $\alpha = 0$.

Bi-month	1	2	3	4	5	6
Transportation volume: Q_k^t (10^4 tonnes)						
Datong	112.0	112.0	112.0	112.0	112.0	112.0
Shenghua	150.1	158.0	158.0	134.1	0.0	147.8
Yitai	30.9	5.1	53.9	0.0	0.0	18.0
Zhongmei	125.0	6.9	94.1	44.2	0.0	68.7
Transportation volume: Q_j^t (10^4 tonnes)						
Xiexin CPP	80.6	80.6	80.6	80.6	80.6	80.6
Kanshan CPP	147.4	147.4	147.4	147.4	147.4	147.4
Wangting CPP	118.4	72.7	55.4	74.6	59.2	118.4
Xiexin CPP: B_{k1}^t (10^4 tonnes)						
Datong	52.6	52.6	58.3	46.9	52.6	52.6
Shenghua	10.0	10.0	11.1	8.9	10.0	10.0
Yitai	18.0	18.0	19.9	16.0	18.0	18.0
Zhongmei	0.0	0.0	0.0	0.0	0.0	0.0
Kanshan CPP: B_{k2}^t (10^4 tonnes)						
Datong	59.4	59.4	53.7	65.1	59.4	59.4
Shenghua	60.2	82.5	46.5	51.7	60.2	60.2
Yitai	0.0	0.0	0.0	0.0	0.0	0.0
Zhongmei	27.8	27.5	25.2	30.6	27.8	27.8
Wangting CPP: B_{k3}^t (10^4 tonnes)						
Datong	0.0	0.0	0.0	0.0	0.0	0.0
Shenghua	79.9	65.5	49.5	15.4	38.7	77.5
Yitai	0.0	0.0	0.0	0.0	0.0	0.0
Zhongmei	42.1	34.5	26.1	8.1	20.4	40.9

In this situation (i.e. $\lambda = 0.9, \alpha = 0$), the total profit is 5.823×10^9 RMB and total car emission amount is 3.848×10^7 tonnes.

7.9–7.11. It can be seen that when $\lambda = 0.95$, the revenue was 6.157×10^9 RMB. Here, the step length for the parameter analysis was set at 0.05, and after effective testing, it can be seen that in this scenario, the lowest λ was 0.85, which means when $\alpha = 0.5$, total CO_2 emissions can be controlled at 85% of the maximum emissions amount, at which time, the total revenue was 5.595×10^9 RMB.

Table 7.7 Results of the proposed model when $\lambda = 0.8$ and $\alpha = 0$.

Bi-month	1	2	3	4	5	6
Transportation volume: Q_k^t (10^4 tonnes)						
Datong	112.0	112.0	112.0	112.0	112.0	112.0
Shenghua	139.8	133.4	158.0	16.7	0.0	118.4
Yitai	41.2	0.0	48.7	0.0	0.0	18.0
Zhongmei	125.0	5.2	99.3	29.1	0.0	53.2
Transportation volume: Q_j^t (10^4 tonnes)						
Xiexin CPP	80.6	80.6	80.6	80.6	80.6	80.6
Kanshan CPP	147.4	99.6	147.4	131.2	85.6	147.4
Wangting CPP	80.6	72.4	55.4	74.6	59.2	73.6
Xiexin CPP: B_{k1}^t (10^4 tonnes)						
Datong	52.6	68.0	37.2	52.6	52.6	52.6
Shenghua	10.0	13.0	7.1	10.0	10.0	10.0
Yitai	18.0	23.2	12.7	18.0	18.0	18.0
Zhongmei	0.0	0.0	0.0	0.0	0.0	0.0
Kanshan CPP: B_{k2}^t (10^4 tonnes)						
Datong	59.4	44.0	65.1	58.0	37.8	59.4
Shenghua	60.2	34.9	51.7	46.0	30.0	60.2
Yitai	0.0	0.0	0.0	0.0	0.0	0.0
Zhongmei	27.8	20.7	30.6	27.3	17.8	27.8
Wangting CPP: B_{k3}^t (10^4 tonnes)						
Datong	0.0	0.0	0.0	11.2	21.6	0.0
Shenghua	59.3	95.7	0.0	13.5	6.3	48.2
Yitai	0.0	0.0	0.0	0.0	0.0	0.0
Zhongmei	31.3	50.5	0.0	21.5	31.2	25.4

In this situation (i.e. $\lambda = 0.8, \alpha = 0$), the total profit is 5.377×10^9 RMB and total car emission amount is 3.420×10^7 tonnes.

7.5.2 Propositions and Analysis

A comprehensive discussion based on the results under different scenarios will be given and some propositions will be extracted in this section.

First of all, it can concluded that dynamic equilibrium strategy toward ICPBD problem has potential to promote environmental-friendly electricity generation.

Table 7.8 Results of the proposed model when $\lambda = 0.7$ and $\alpha = 0$.

Bi-month	1	2	3	4	5	6
Transportation volume: Q_k^t (10^4 tonnes)						
Datong	112.0	112.0	112.0	112.0	112.0	112.0
Shenghua	149.1	29.3	124.0	0.0	0.0	51.1
Yitai	31.9	0.0	57.5	0.0	0.0	18.0
Zhongmei	125.0	10.7	124.4	0.0	0.0	58.9
Transportation volume: Q_j^t (10^4 tonnes)						
Xiexin CPP	80.6	62.4	80.6	80.6	80.6	80.6
Kanshan CPP	115.4	99.6	73.2	101.1	85.6	95.8
Wangting CPP	80.6	72.4	55.4	74.6	59.2	73.6
Xiexin CPP: B_{k1}^t (10^4 tonnes)						
Datong	52.6	40.7	52.6	52.6	52.6	52.6
Shenghua	10.0	7.8	10.0	10.0	10.0	10.0
Yitai	18.0	13.9	18.0	18.0	18.0	18.0
Zhongmei	0.0	0.0	0.0	0.0	0.0	0.0
Kanshan CPP: B_{k2}^t (10^4 tonnes)						
Datong	51.0	44.0	32.3	44.6	37.8	42.3
Shenghua	40.5	34.9	25.7	35.4	30.0	33.6
Yitai	0.0	0.0	0.0	0.0	0.0	0.0
Zhongmei	24.0	20.7	15.2	21.0	17.8	19.9
Wangting CPP: B_{k3}^t (10^4 tonnes)						
Datong	8.4	27.3	0.0	41.8	11.5	27.2
Shenghua	77.7	7.5	0.0	0.0	2.8	7.4
Yitai	0.0	0.0	0.0	3.5	0.1	0.0
Zhongmei	51.9	39.1	0.0	54.1	16.4	39.0

In this situation (i.e. $\lambda = 0.7, \alpha = 0$), the total profit is 4.754×10^9 RMB and total car emission amount is 2.993×10^7 tonnes.

By in-depth analysis on CO_2 emissions in a large-scale coal-fired power enterprise, a dynamic equilibrium strategy toward ICPBD problem was established to reduce CO_2 emissions, as shown in the proposed mathematical model. From the effectiveness test on control parameters λ and α, which represented the decision-maker's attitude toward the CO_2 emissions reduction target and the satisfactory degree for each respective CPP, the lowest possible CO_2 emissions level were determined, as shown in Table 7.12. It was observed that regardless of the change in the satisfactory

Table 7.9 Results of the proposed model when $\lambda = 0.95$ and $\alpha = 0.5$.

Bi-month	1	2	3	4	5	6
Transportation volume: Q_k^t (10^4 tonnes)						
Datong	112.0	112.0	112.0	112.0	112.0	112.0
Shenghua	150.1	154.6	158.0	158.0	48.8	147.8
Yitai	30.9	5.1	53.9	0.0	0.0	18.0
Zhongmei	125.0	10.3	94.1	77.4	0.0	68.7
Transportation volume: Q_j^t (10^4 tonnes)						
Xiexin CPP	80.6	80.6	80.6	80.6	80.6	80.6
Kanshan CPP	147.4	147.4	147.4	147.4	147.4	147.4
Wangting CPP	118.4	113.0	118.4	77.2	59.2	118.4
Xiexin CPP: B_{k1}^t (10^4 tonnes)						
Datong	52.6	52.6	52.6	52.6	52.6	52.6
Shenghua	10.0	10.0	10.0	10.0	10.0	10.0
Yitai	18.0	18.0	18.0	18.0	18.0	18.0
Zhongmei	0.0	0.0	0.0	0.0	0.0	0.0
Kanshan CPP: B_{k2}^t (10^4 tonnes)						
Datong	59.4	59.4	59.4	59.4	59.4	59.4
Shenghua	60.2	73.0	47.4	60.2	60.2	60.2
Yitai	0.0	0.0	0.0	0.0	0.0	0.0
Zhongmei	27.8	27.6	27.9	27.8	27.8	27.8
Wangting CPP: B_{k3}^t (10^4 tonnes)						
Datong	0.0	0.0	0.0	0.0	0.0	0.0
Shenghua	79.9	71.6	86.4	41.6	38.7	77.5
Yitai	0.0	0.0	0.0	0.0	0.0	0.0
Zhongmei	42.1	37.8	45.6	22.0	20.4	40.9

In this situation (i.e. $\lambda = 0.95$, $\alpha = 0.5$), the total profit is 6.157×10^9 RMB and total car emission amount is 4.062×10^7 tonnes.

degree, the proposed dynamic equilibrium strategy toward ICPBD problem had the potential to improve the environment. Taking scenario 1 (i.e. $\alpha = 0$, λ is changing) as an example, the lowest possible CO_2 emissions level was 0.65, and when $\lambda = 0.7$ and $\alpha = 0$, as shown in Table 7.8, using the dynamic equilibrium strategy, the total CO_2 emissions could be reduced by about 30% compared with the maximum CO_2 emissions (i.e. TCO_2^* in model Eq. (7.17)). A similar situation was also seen in the other scenarios. A similar situation was also seen in the other scenarios.

Table 7.10 Results of the proposed model when $\lambda = 0.9$ and $\alpha = 0.5$.

Bi-month	1	2	3	4	5	6
Transportation volume: Q_k^t (10^4 tonnes)						
Datong	112.0	112.0	112.0	112.0	112.0	112.0
Shenghua	147.8	145.5	158.0	127.2	0.0	147.8
Yitai	33.2	2.8	53.9	0.0	0.0	18.0
Zhongmei	125.0	21.7	94.1	52.5	0.0	68.7
Transportation volume: Q_j^t (10^4 tonnes)						
Xiexin CPP	80.6	80.6	80.6	80.6	80.6	80.6
Kanshan CPP	141.0	134.5	134.5	134.5	129.5	147.4
Wangting CPP	118.4	118.4	74.1	74.6	59.2	118.4
Xiexin CPP: B_{k1}^t (10^4 tonnes)						
Datong	52.6	52.6	52.6	52.6	52.6	52.6
Shenghua	10.0	10.0	10.0	10.0	10.0	10.0
Yitai	18.0	18.0	18.0	18.0	18.0	18.0
Zhongmei	0.0	0.0	0.0	0.0	0.0	0.0
Kanshan CPP: B_{k2}^t (10^4 tonnes)						
Datong	59.4	59.4	59.4	59.4	57.2	59.4
Shenghua	53.7	47.2	47.2	47.2	45.4	60.2
Yitai	0.0	0.0	0.0	0.0	0.0	0.0
Zhongmei	27.9	27.9	27.9	27.9	26.9	27.8
Wangting CPP: B_{k3}^t (10^4 tonnes)						
Datong	0.0	0.0	0.0	0.0	2.2	0.0
Shenghua	84.1	88.3	77.5	2.5	35.4	77.5
Yitai	0.0	0.0	0.0	0.0	0.0	0.0
Zhongmei	44.3	46.6	40.9	1.3	21.5	40.9

In this situation (i.e. $\lambda = 0.90$, $\alpha = 0.5$), the total profit is 5.892×10^9 RMB and total car emission amount is 3.848×10^7 tonnes.

When the dynamic equilibrium strategy toward ICPBD problem was employed, the total CO_2 emissions were the main factor used by the decision-maker to balance total profits and carbon emissions. Indeed, the purpose of the modeling for the decision-maker is to guarantee the quality requirements for the steam coal in CPPs and scientifically reduce CO_2 emissions. To maximize total revenue, the decision-maker attempts to adjust the coal blending ratio for each CPP to improve

Table 7.11 Results of the proposed model when $\lambda = 0.85$ and $\alpha = 0.5$.

Bi-month	**1**	**2**	**3**	**4**	**5**	**6**
Transportation volume: Q_k^t (10^4 tonnes)						
Datong	112.0	112.0	112.0	112.0	112.0	112.0
Shenghua	147.3	145.7	158.0	22.4	0.0	147.8
Yitai	33.7	0.2	53.9	0.0	0.0	18.0
Zhongmei	125.0	24.1	94.1	54.3	0.0	68.7
Transportation volume: Q_j^t (10^4 tonnes)						
Xiexin CPP	74.2	77.8	80.6	80.6	80.6	80.6
Kanshan CPP	147.4	138.6	107.4	101.1	85.6	147.4
Wangting CPP	118.4	118.4	74.1	74.6	59.2	118.4
Xiexin CPP: B_{k1}^t (10^4 tonnes)						
Datong	48.4	50.8	52.6	52.6	52.6	52.6
Shenghua	9.2	9.7	10.0	10.0	10.0	10.0
Yitai	16.5	17.4	18.0	18.0	18.0	18.0
Zhongmei	0.0	0.0	0.0	0.0	0.0	0.0
Kanshan CPP: B_{k2}^t (10^4 tonnes)						
Datong	63.6	61.2	47.4	44.6	37.8	59.4
Shenghua	54.0	48.6	37.7	35.4	30.0	60.2
Yitai	0.0	0.0	0.0	0.0	0.0	0.0
Zhongmei	29.9	28.8	22.3	21.0	17.8	27.8
Wangting CPP: B_{k3}^t (10^4 tonnes)						
Datong	0.0	0.0	0.0	26.7	21.6	0.0
Shenghua	84.1	87.4	32.0	8.8	6.3	77.5
Yitai	0.0	0.0	0.0	0.0	0.0	0.0
Zhongmei	44.3	46.1	16.9	39.1	31.2	40.9

In this situation (i.e. $\lambda = 0.85$, $\alpha = 0.5$), the total profit is 5.595×10^9 RMB and total car emission amount is 3.634×10^7 tonnes.

CO_2 emissions performance even though this could lead to an increase in unit production costs.

Further, we found that the marginal CO_2 emissions reduction rate is larger than the profit when the environmental protection constraint are tightened. From Table 7.12, it can be seen that with a tightening of the CO_2 emission constraint (i.e. changing the CO_2 emissions control parameter λ from 1 to the possible lowest level) in Eq. (7.17), both total CO_2 emissions and total profits decrease; however, the profit decrease is smaller than the total CO_2 emissions decrease. Take scenario 2 as an example, the results for which are shown in Tables 7.5, 7.9–7.11; the possible CO_2 emissions control level is from 1 to 0.85 (i.e. $\lambda \in (0.85, 1)$), and when the CO_2 emissions control level is most relaxed (i.e. $\lambda = 1$), the total profit is 6.388×10^9 RMB and the total CO_2 emissions are 4.275×10^7 tonnes. When the decision-maker tightens the attitude toward total CO_2 emissions constraints (e.g. sets $\lambda = 0.95$), the total profit is 5.892×10^9 RMB and the total CO_2 emissions are 4.062×10^7 tonnes; that is, the ratio decreases by 3.6% and 5% for the two factors, respectively, with the reduction in the CO_2 emissions being greater than the reduction in profits. When the decision-maker continues to tighten the CO_2 emissions attitude (i.e. sets $\lambda = 0.9$), a similar situation is seen, with the ratio decrease being 4.3% and 5.3%, respectively, with the reduction in the CO_2 emissions being greater than the reduction in profits. Similar conclusions can also be seen when the decision-maker continues to tighten up the attitude toward total CO_2 emissions (i.e. $\lambda = 0.85$). Figure 7.2 shows the changing trends for the total CO_2 emissions and total profits under various control levels λ when $\alpha = 0$. As can be seen in Figure 7.2, the slope for the changing total CO_2 emissions is larger than that for total profit, which indicates that the CO_2 emissions marginal reduction rate is larger than that of profit when the CO_2 emissions constraints are tightened. Based on the above analysis, it can be concluded that tightening up the CO_2 emissions constraints to control GHGs leads to a reduction in profits.

At the same time, through the sensitive analysis on the satisfactory degree, it can be seen that this indicator has a significant impact on the total economic profits. To ensure equality for each subordinate power plant in the proposed model, satisfactory measures are employed to guarantee that each CPP receives a relatively satisfactory amount of blended coal. Parameter α represents the lowest satisfactory degree for each CPP, which was subjectively determined by the decision-maker. As shown in Table 7.12, a parameter analysis was conducted by changing α under various CO_2 emission control levels λ. Taking the situation of $\lambda = 0.95$ as an example, as shown in

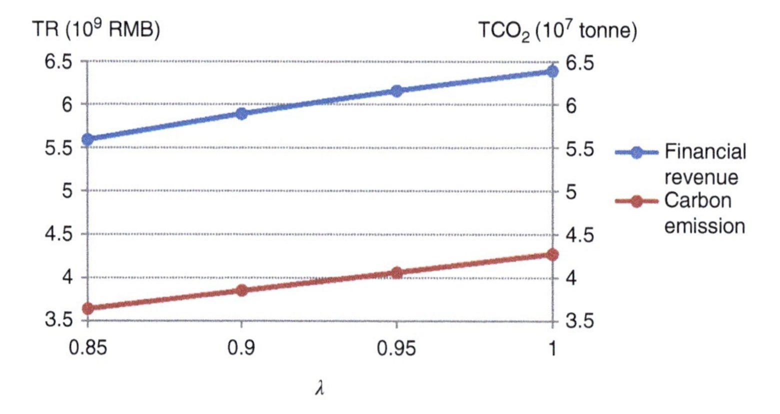

Figure 7.2 Changing trends for total CO_2 emissions and the total benefits under different levels λ and $\beta = 0.5$.

Table 7.12 The total profits and CO_2 emissions under different situations.

λ	$\alpha = 0$	$\alpha = 0.1$	$\alpha = 0.2$	$\alpha = 0.3$	$\alpha = 0.4$	$\alpha = 0.5$	$\alpha = 0.6$	$\alpha = 0.7$	$\alpha = 0.8$	$\alpha = 0.9$	$\alpha = 1.0$	CO_2 emissions
1	6.388	6.388	6.388	6.388	6.388	6.388	6.388	6.388	6.388	6.388	6.388	4.275
0.95	6.157	6.157	6.157	6.157	6.157	6.157	6.157	6.148	6.134			4.062
0.9	5.923	5.923	5.923	5.921	5.907	5.892	5.877	5.836				3.848
0.85	5.664	5.657	5.647	5.634	5.619	5.595						3.634
0.8	5.377	5.368	5.356	5.341	5.273							3.420
0.75	5.080	5.068	5.020	4.943								3.207
0.7	4.754	4.685										2.993
0.65	4.341											2.779

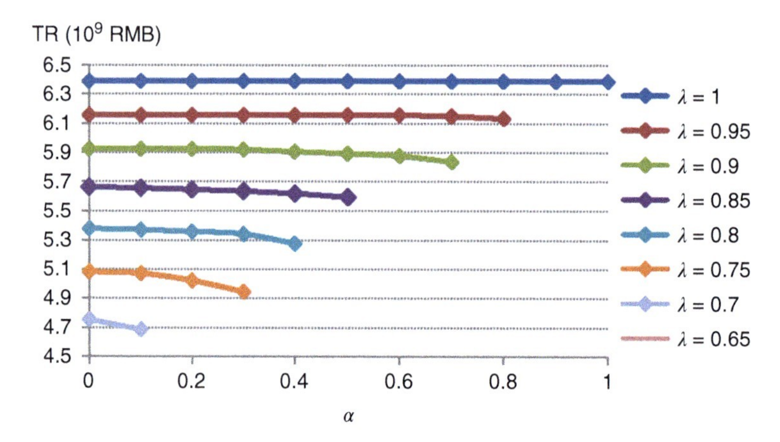

Figure 7.3 Changing trends for total financial revenue at different satisfaction degree α.

Table 7.12; when the decision-maker sets $\alpha = 0.5$, the total profit is 6.157×10^9 RMB and as α change to 0.6, the profit is the same as when $\alpha = 0.5$. When α is set at 0.7, the profit is 6.148×10^9 RMB, a decrease of 0.14%, and when α is set at 0.8, the profit is 6.134×10^9 RMB, a decrease of 0.23%. A similar result can also be found when the decision-maker decides on various CO_2 emissions control level λ, as shown in Figure 7.3. From Figure 7.3, it can be seen that an increase in the satisfactory degree α influences total profit under a certain emissions control level λ, which leads to either invariant or decreasing. When the satisfactory degree increases under a certain λ, more coal for the high-carbon CPP (e.g. Wangting CPP in this chapter) is provided, while the supply of coal for the low-carbon CPP (e.g. Xiexin CPP in this chapter) is reduced, leading to a decrease in power generation and a commensurate decrease in total profit.

In addition, the results also demonstrate that CPPs have different sensitivities toward changing CO_2 emissions constraints. As shown in Table 7.13, it can be seen that the various CPPs have different sensitivities when the decision-maker changes their attitude toward the total CO_2 emission constraint. According to Table 7.13, the satisfactory degree for each CPP is 100% (i.e. satisfying the maximum demand) when the decision-maker sets the loosest CO_2 emissions constraint (i.e. $\lambda = 1$). However, when the decision-maker tightens up their attitude toward CO_2 emissions (e.g. sets $\lambda = 0.9$), the satisfactory degree for the Wangting CPP' coal decreases to 28.1% while the others remain unchanged. When the decision-maker continues to tighten

Table 7.13 Satisfaction degrees of different CPPs at various λ, $\alpha = 0$.

	$\lambda = 1$	$\lambda = 0.95$	$\lambda = 0.9$	$\lambda = 0.85$	$\lambda = 0.8$	$\lambda = 0.75$	$\lambda = 0.7$	$\lambda = 0.65$
Xiexin CPP	1	1	1	1	1	1	0.883	0.230
Kanshan CPP	1	1	1	0.928	0.599	0.270	0	0
Wangting CPP	1	0.641	0.281	0	0	0	0	0

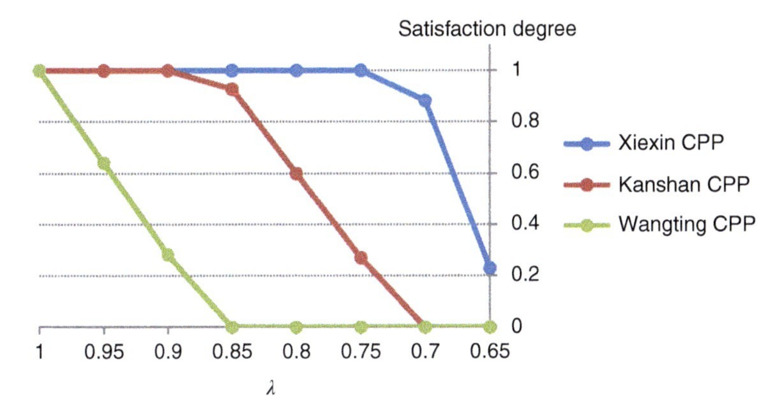

Figure 7.4 Changing trends for satisfaction degree at various CO_2 emissions control parameter λ and $\alpha = 0$.

up their attitude toward total CO_2 emissions constraints (e.g. sets $\lambda = 0.8$), a similar situation occurs, with the satisfactory degree for the three CPPs being 1, 60%, and 0. From Figure 7.4, it can be seen that under increasingly strict CO_2 emissions restrictions, the coal supply at the high-carbon CPP (e.g. Wangting CPP) decreases preferentially; that is, the coal priority is given to the CPPs with low-carbon emissions under the premise of meeting the minimum coal demand for all CPPs.

Finally, we found that the coal blending ratio at each CPP has relatively small changes toward the changing of the parameters in the proposed method. From Tables 7.5–7.11, it can be seen that the coal blending ratio for each CPP fluctuates at different stage under a certain CO_2 emissions control level λ and satisfactory degree α. However, the fluctuation range is small. Take Table 7.6 as an example, when the decision-maker sets $\lambda = 0.9$ and $\alpha = 0.5$; the Xiexin CPP always mainly depends on coal from the Datong colliery which is blended with coal from the Shenghua colliery and Yitai colliery, but not from Zhongmei colliery, with the blending ratio being 65.3%, 12.4%, and 22.3%. The Kanshan CPP always gets coal from the Datong, Shenghua, and Zhongmei collieries but not from the Yitai colliery, with the blending ratio fluctuating marginally from 40.3% to 44.2%, 35.1% to 40.9%, and 18.8% to 20.8%, respectively. The Wangting CPP depends on coal from the Datong, Shenghua, and Zhongmei collieries with the blending ratio fluctuating marginally from 0% to 3.8%, 59.8% to 65.5% and 34.5% to 36.4%, respectively. Similar results can also be seen in other situations (as shown in Figure 7.5). The above analyses show that the proposed method in this chapter is robust and has a satisfactory performance.

7.5.3 Management Recommendations

Based on the analysis and discussion above, several management recommendations are proposed to improve production operations and CO_2 emissions performance at large-scale coal-fired power enterprises.

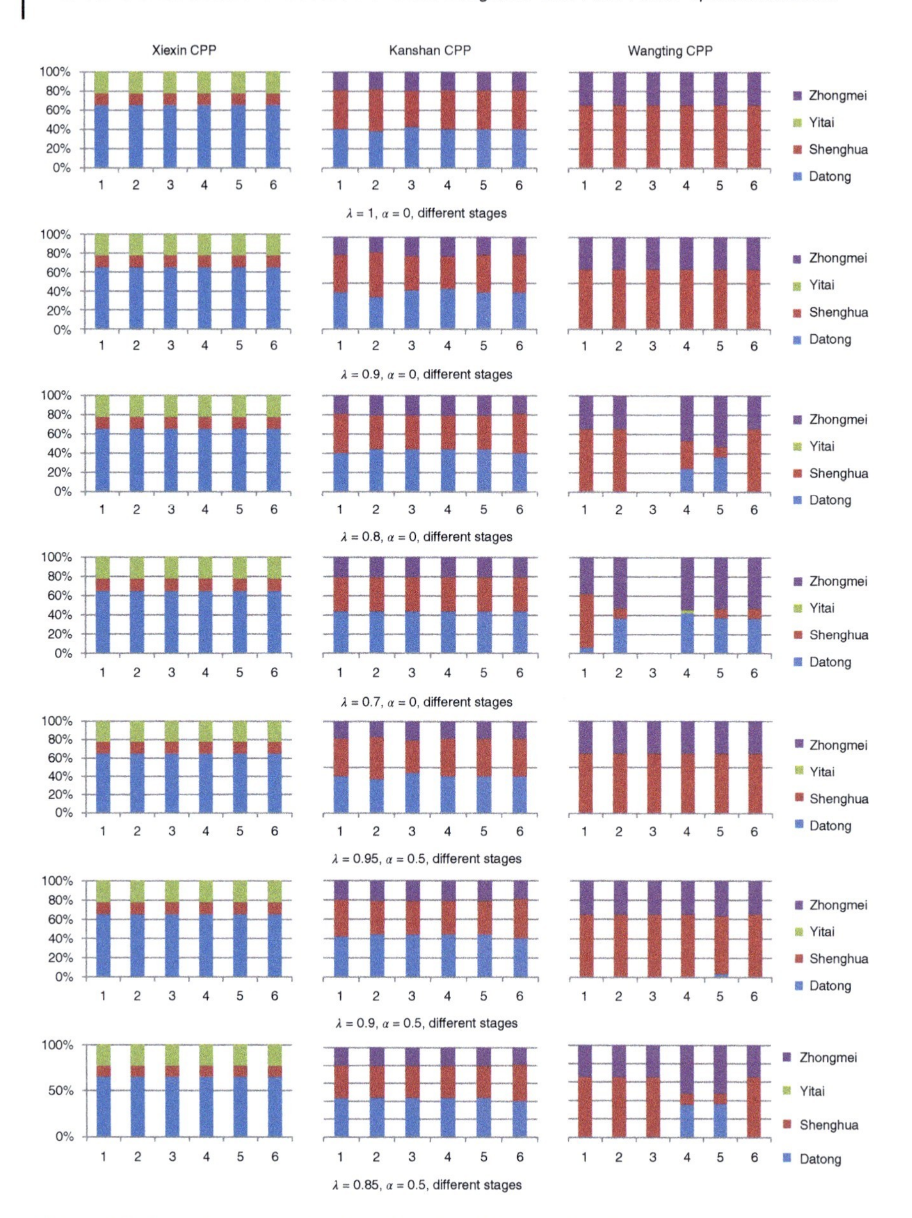

Figure 7.5 Blending ratio of each CPP under different situations.

First of all, to reduce CO_2 emissions, in large-scale coal-fired power enterprises with multiple subordinate CPPs, a dynamic equilibrium strategy toward ICPBD problem-based should be adopted to replace existing methods where coal purchasing and blending is done at each subordinate CPP. Currently, each subsidiary CPP purchases the coal they need to meet their own requirements, which means that because of coal supply uncertainty and market environment volatility, some of the purchased coal may not meet the CPP boiler combustion requirements, which can result in irreparable damage to the boiler and high CO_2 emissions due to insufficient combustion. Further, each subordinate CPP at present needs to store large quantities of coal to deal with the risks posed by volatile markets and uncertain demand, which takes up much of their current capital and substantially increases inventory costs. As discussed before, the ICPBD strategy proposed in this chapter not only promotes low-carbon power generation and guarantees supply chain reliability but also reduces the cost of the coal-to-power energy chain because of the scale advantages.

Second, to comply with the equality principle for every sub-coal power plant, the satisfactory degree at all subordinate CPPs needs to be considered. In this chapter, Eqs. (7.14) and (7.15) were used to measure the satisfactory degree. As discussed before, in Section 7.5, revenue does not increase as the satisfactory degree increases. Therefore, it is recommended that the satisfactory degree for each subordinate CPP should be emphasized when decision-makers employ a dynamic equilibrium strategy toward ICPBD problem.

References

BP Global (2015). BP Technology Outlook.

CEC (2017). National electric power industry statistical bulletin.

Dai, C., Cai, X.H., Cai, Y.P., and Huang, G.H. (2014). A simulation-based fuzzy possibilistic programming model for coal blending management with consideration of human health risk under uncertainty. *Applied Energy* 133 (10): 1–13.

Dang, Q., Wright, M.Mba., and Brown, R.C. (2015). Ultra-low carbon emissions from coal-fired power plants through bio-oil co-firing and bio-char sequestration. *Environmental Science and Technology* 49 (24): 14688.

Dubey, H.M., Pandit, M., and Panigrahi, B.K. (2015). Hybrid flower pollination algorithm with time-varying fuzzy selection mechanism for wind integrated multi-objective dynamic economic dispatch. *Renewable Energy* 83: 188–202.

Ferrero, A. and Salicone, S. (2004). The random-fuzzy variables: a new approach to the expression of uncertainty in measurement. *IEEE Transactions on Instrumentation and Measurement* 53 (5): 1370–1377.

Fisher, M. and Raman, A. (1996). Reducing the cost of demand uncertainty through accurate response to early sales. *Operations Research* 44 (1): 87–99.

Francey, R.J., Trudinger, C.M., van der Schoot, M. et al. (2013). Atmospheric verification of anthropogenic CO_2 emission trends. *Nature Climate Change* 3 (8): 520–524.

Friedlingstein, P., Andrew, R.M., Rogelj, J. et al. (2014). Persistent growth of CO_2 emissions and implications for reaching climate targets. *Nature Geoscience* 7 (10): 709–715.

Fuss, S., Canadell, J.G., Peters, G.P. et al. (2014). Betting on negative emissions. *Nature Climate Change* 4 (10): 850–853.

Ganguly, S., Sahoo, N.C., and Das, D. (2013). Multi-objective planning of electrical distribution systems using dynamic programming. *International Journal of Electrical Power & Energy Systems* 46 (46): 65–78.

Giri, B.C. and Chakraborty, A. (2016). Coordinating a vendor–buyer supply chain with stochastic demand and uncertain yield. *International Journal of Management Science & Engineering Management* 12 (2): 96–103.

Goto, K., Yogo, K., and Higashii, T. (2013). A review of efficiency penalty in a coal-fired power plant with post-combustion CO_2 capture. *Applied Energy* 111 (11): 710–720.

Heinberg, R. and Fridley, D. (2010). The end of cheap coal. *Nature* 468 (7322): 367–369.

Hu, Y., Li, H., and Yan, J. (2014). Numerical investigation of heat transfer characteristics in utility boilers of oxy-coal combustion. *Applied Energy* 130 (5): 543–551.

Huang, L., Suh, I.H., and Abraham, A. (2011). Dynamic multi-objective optimization based on membrane computing for control of time-varying unstable plants. *Information Sciences* 181 (11): 2370–2391.

Li, Y. (2012). Dynamics of clean coal-fired power generation development in China. *Energy Policy* 51 (6): 138–142.

Liu, B. and Liu, Y.K. (2002). Expected value of fuzzy variable and fuzzy expected value models. *IEEE Transactions on Fuzzy Systems* 10 (4): 445–450.

Liu, Y., Zhang, J., Sheng, C. et al. (2010). Simultaneous removal of NO and SO_2 from coal-fired flue gas by UV/H_2O_2 advanced oxidation process. *Chemical Engineering Journal* 162 (3): 1006–1011.

Lomax, G., Lenton, T.M., Adeosun, A., and Workman, M. (2015). Investing in negative emissions. *Nature Climate Change* 5 (6): 498–500.

Lonsdale, C.R., Stevens, R.G., Brock, C.A. et al. (2012). The effect of coal-fired power-plant SO_2 and NO_x control technologies on aerosol nucleation in the source plumes. *Atmospheric Chemistry and Physics* 12 (23): 11519–11531.

Lv, C., Xu, J., Xie, H. et al. (2016). Equilibrium strategy based coal blending method for combined carbon and PM10 emissions reductions. *Applied Energy* 183: 1035–1052.

Mukherjee, S. and Borthakur, P.C. (2001). Chemical demineralization/desulphurization of high sulphur coal using sodium hydroxide and acid solutions. *Fuel* 80 (14): 2037–2040.

Nwulu, N.I. and Xia, X. (2015). Multi-objective dynamic economic emission dispatch of electric power generation integrated with game theory based demand response programs. *Energy Conversion and Management* 89: 963–974.

Oelkers, E.H. and Cole, D.R. (2008). Carbon dioxide sequestration: a solution to a global problem. *Elements* 4 (5): 305–310.

Pavlish, J.H., Sondreal, E.A., Mann, M.D. et al. (2003). Status review of mercury control options for coal-fired power plants. *Fuel Processing Technology* 82 (2): 89–165.

Porter, R.T.J., Fairweather, M., Pourkashanian, M., and Woolley, R.M. (2015). The range and level of impurities in CO_2 streams from different carbon capture sources. *International Journal of Greenhouse Gas Control* 36: 161–174.

Puri, M.L. (1986). Fuzzy random variables. *Journal of Mathematical Analysis & Applications* 114 (2): 409–422.

Samimi, A. and Zarinabadi, S. (2012). Reduction of greenhouse gases emission and effect on environment. *Australian Journal of Basic and Applied Sciences* 8 (12): 1011–1015.

Schrag, D.P. (2009). Storage of carbon dioxide in offshore sediments. *Science* 325 (5948): 1658.

Scott, V., Gilfillan, S., Markusson, N. et al. (2013). Last chance for carbon capture and storage. *Nature Climate Change* 3 (2): 105–111.

Service, R.F. (2017a). Fossil power, guilt free. *Science* 356 (6340): 796.

Service, R.F. (2017b). Cleaning up coal—cost-effectively. *Science* 356. 798.

Shih, J.-S. and Frey, H.C. (1995). Coal blending optimization under uncertainty. *European Journal of Operational Research* 83 (3): 452–465.

Smoot, L.D. and Smith, P.J. (2013). *Coal Combustion and Gasification*. Springer Science & Bussiness Media.

Solomon, S., Plattner, G.K., Knutti, R., and Friedlingstein, P. (2009). Irreversible climate change due to carbon dioxide emissions. *Proc. Natl. Acad. Sci. U.S.A.* 106 (6): 1704–1709.

Stern, N. (2006). Stern review: the economics of climate change.

Tola, V. and Pettinau, A. (2014). Power generation plants with carbon capture and storage: a techno-economic comparison between coal combustion and gasification technologies. *Applied Energy* 113 (1): 1461–1474.

Van Der Meer, J.W.M., Huppert, H., and Holmes, J. (2014). Carbon: no silver bullet. *Science* 345 (6201): 1130.

Van Noorden, R. (2014). Two plants to put clean coal to test. *Nature* 509 (7498): 20.

Wang, Q. and Chen, X. (2015). Energy policies for managing China's carbon emission. *Renewable and Sustainable Energy Reviews* 50: 470–479.

Wang, C.H., Zhao, D., Tsutsumi, A., and You, S. (2017). Sustainable energy technologies for energy saving and carbon emission reduction. *Applied Energy*. 194: 223–224.

Xu, J. and Zhou, X. (2011). *Fuzzy-Like Multiple Objective Decision Making*. Berlin, Heidelberg: Springer-Veralg.

Xu, J., Qiu, R., and Lv, C. (2016). Carbon emission allowance allocation with cap and trade mechanism in air passenger transport. *Journal of Cleaner Production* 131: 308–320.

Xu, J., Yang, X., and Tao, Z. (2015). A tripartite equilibrium for carbon emission allowance allocation in the power-supply industry. *Energy Policy* 82: 62–80.

Yang, C.J., Xuan, X., and Jackson, R.B. (2012). China's coal price disturbances: observations, explanations, and implications for global energy economies. *Energy Policy* 51: 720–727.

Zeng, Z., Xu, J., Wu, S., and Shen, M. (2014). Antithetic method-based particle swarm optimization for a queuing network problem with fuzzy data in concrete transportation systems. *Computer-Aided Civil and Infrastructure Engineering* 29 (10): 771–800.

Zhang, L., Gao, L., and Li, X. (2013). A hybrid genetic algorithm and tabu search for a multi-objective dynamic job shop scheduling problem. *International Journal of Production Research* 51 (12): 3516–3531.

8

Equilibrium Coal Blending Method Toward Multiple Air Pollution Reduction

The total coal consumption in 2014 is 3881.8 million tonnes, accounting for 30% of global primary energy consumption and is ranked second of all known energy sources [BP Global, 2015]. As we know, one of the most important end uses of coal is the generation of electric power. In 2014, almost 20% of electricity worldwide is supplied by coal-fired power plants (CPPs), and some developing regions even provide a higher percentage of electricity [BP Global, 2015]. This high coal consumption has resulted in severe environmental pollution. Particularly, CPPs significantly contribute to the deterioration of the atmosphere with multiple air pollution emissions [Dai et al., 2014, Wolde-Rufael, 2010, Nowak, 2003]. Increasing global power demand means that there is an increasingly larger quantity of coal. The full use of alternative energy sources is still decades away, thus it is necessary to control or partially reduce emissions of CPP to alleviate the destructive impacts on atmosphere [Mao et al., 2014, Goto et al., 2013]. Therefore, reducing CPP emissions is critical for the future atmospheric environment.

8.1 Background Presentation

In coal combustion process, there are many hazardous air pollutants being emitted, such as sulfur dioxide, nitrogen oxides, and particulate matter that are less than 10 μm in diameter (PM_{10}) [Li et al., 2009, Xu et al., 2015]. These air pollutants threaten human health and the environment and are scientifically proved as the main causes for many diseases such as respiratory illness, chronic bronchitis, and premature death [Dai et al., 2014]. Therefore, there has been increasing concerns about reducing such air pollutants and many works had already been conducted for both the soft-path perspective which focused on improving the clean technology and the hard-path perspective which mainly pay attention to the development of effective optimization or policy control methods [Lonsdale et al., 2012]. For the hard-path perspective, Mukherjee and Borthakur investigated the demineralization and desulfurization of high sulfur coal from Macum Coal fields in India by treating successively with aqueous solution of sodium hydroxide and hydrochloric acid, and the method can remove 43–50% of the ash and total inorganic sulfur and around 10% organic sulfur [Mukherjee and Borthakur, 2001]. Simultaneous removal of NO

Innovative Approaches towards Ecological Coal Mining and Utilization, First Edition.
Jiuping Xu, Heping Xie, and Chengwei Lv.

and SO_2 from coal-fired flue gas by UV/H_2O_2 advanced oxidation process (AOP) was also studied in an ultraviolet (UV)-bubble column reactor [Liu et al., 2010]. Besides, some research has been conducted from the hard-path perspective to reduce CPP emissions can be found in Pavlish et al. [2003], Xu et al. [2000], and Franco and Diaz [2009]. In general, the hard-path solution is regarded as the most effective tool because the high efficiency can be achieved by it, while it is extremely expensive for developing countries, which are the main CPPs users. Therefore, soft-path solutions are more appropriate for many regions in the world [Goto et al., 2013, Tian et al., 2012, Yue, 2012a].

Several soft-path solutions have been proposed such as Pigovian tax levies, coal blending, and carbon emissions trading. Coal blending is one of several available methods for reducing pollutant emissions by CPPs. Coal blending before combustion involves the blending of several kinds of coal to ensure that the volatile matter content, heat rate, ash content, moisture content, and sulfur content meet requirements of the power generation utility [Dai et al., 2014]. It has proved to be an effective clean coal combustion technology for adjusting coal features, alleviating atmospheric pollutant emissions, and reducing the associated health risk and has therefore been successfully implemented [Dai et al., 2014, Baek et al., 2014, Shih and Frey, 1995]. However, to make coal blending successful, two major obstacles must be overcome. First, controlling the fuel characteristics of the blended coal has proved to be a formidable challenge. Many studies have focused on this issue. For example, to find an effective way to control the coal blending process, Wu et al. proposed an expert control strategy based on the combination of neural networks, mathematical models, and rule models to compute the target percentages accurately and ensure that the blended coal characteristics were within the target range [Wu et al., 1999]. Xia et al. investigated a case-based reasoning online decision-making method for the optimization of coal-blend combustion and successfully applied the proposed method to a 600-MW power plant [Xia et al., 2014]. These previous research indicated that controlling the blended coal qualities is possible. Second, it is difficult to determine the optimal blend ratio from economic and environmental perspectives because the trade-offs need to be fully considered. Shih and Frey developed a multiobjective chance-constrained model to determine the coal blending ratio that optimizes the performance of emissions and reduces the cost of CPPs [Shih and Frey, 1995]. Li et al. proposed an improved coal blending method based on the fuzzy programming to ensure the cleaner CPP production [Li et al., 2013]. Dai et al. recently studied the impact of CPP emissions on human health and proposed a simulation-based fuzzy possibilistic programming method that was then applied to a real-world case and the results of the case study demonstrated that coal blending was an effective way to reduce pollutant emissions and alleviate consequent effects on human health [Dai et al., 2014]. Although some progress has been made in addressing the atmospheric environmental degradation caused by CPPs, the realistic situation is still unsatisfactory and further improvement is needed.

Existing coal blending methods only consider single decision-maker; that is, the CPP was the only decision-maker considered. However, in practice, there are many parties involved when developing plans to reduce the rate of atmospheric environmental deterioration, so the viewpoints of all interested parties must be properly incorporated into the decision-making process. As many parties need to be considered and they usually have conflicting objectives, conflicts are inevitable, such as those between the local authority and the CPPs. For example, local governments are most concerned with protecting the atmospheric environment, while CPPs want to achieve the highest possible benefit [Xu et al., 2015, Tu et al., 2015]. Furthermore, improving the atmospheric environment in a region requires the joint efforts of all stakeholders, and this cannot be achieved by a single or even several CPPs or authorities alone. Therefore, in this case, the efficiency of traditional coal blending methods is reduced because they neither represent the interests of all stakeholders nor resolve the inherent stakeholder conflicts. Equilibrium strategy has been widely used in many fields and has become a powerful tool to solve the problems such situations. Tu et al. solved the conflicts among different sub-areas in an irrigation district with limited water resources by means of equilibrium strategy, and they concluded that it is more scientific to make a global optimization of the whole irrigation area than to optimize each sub-area [Tu et al., 2015]. For the expansion of a highway network, Angulo et al. introduced an equilibrium optimization method (i.e. a new bilevel continuous location model) to solve the social, economic, and environmental conflicts among stakeholders and drew some useful conclusions [Angulo et al., 2014]. The successful application of the equilibrium strategy in other fields has prompted us to use this method to determine the best coal blending method from a regional perspective, so as to realize the atmospheric environment-friendly production of CPPs. However, equilibrium strategy, as an abstract theory, needs to be transformed into a specific quantitative research method. Bilevel programming, which has the ability to describe the interests of multiple decision-makers, has been shown to be one of the most efficient tools for representing equilibrium strategy [Sherali et al., 1983, Vicente and Calamai, 1994]. Therefore, in this chapter, bilevel programming is integrated into a coal blending method to improve the atmospheric environment and to propose this bilevel model, the following key points must be analyzed.

8.1.1 Relationship Among All the Stakeholders

Both local authorities and CPPs plays critical roles in formulating plans to control CPP emissions in the region and thus need to be involved in the decision-making progress simultaneously. The authority always has a higher priority because they represents the interest of the public and has the power to set thresholds on regional carbon emissions, which directly affects the decision of each CPP production plan [Xu et al., 2015]. Therefore, the authority acts as a leader when drawing up such plans. On the other hand, CPPs do not just imply the authority's decision. Instead, they can also influence the authority to reconsider and adjust its initial decision by developing their own production plans. Based on such a kind of leader–follower relationship, the decision-making is similar to a Stackelberg game and thus can be abstractly represented as a corresponding mathematical form [Sinha et al., 2014,

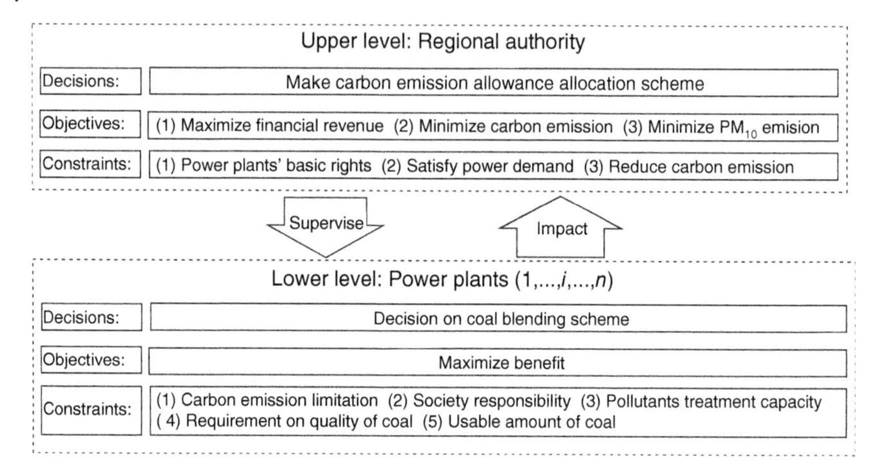

Figure 8.1 Flowchart of the bilevel structure.

Abou-Kandil and Bertrand, 1987]. As a special form of hierarchal optimization, bilevel programming can be directly mapped to this decision-making process. Thus, in this case, the authority on behalf of the public is the decision-maker in the upper level, and CPPs in the region is the subordinate decision-maker in the lower level. The relationship between the local authority and the CPPs is illustrated in Figure 8.1.

8.1.2 Decision Carrier Between All the Stakeholders

How to build the decision carrier between all the stakeholders is another concern. In detail, how an authority control the CPPs by making its own decisions is critical and has been the focus of significant research [Howell, 2012, Tang et al., 2015]. A commonly suggested method is to enforce the allocation of carbon allowances. In this method that acts as a bridge between the authority and the CPPs, the authority can allocate the allowance for carbon emissions to each CPP on the basis of each production plan.

One of the most commonly suggested methods has been the implementation of carbon emissions allowance allocations which can work as a bridge between the authority and the CPPs; the authority can allocate carbon emissions allowances to each CPP according to each individual production plan. For the purpose of control of the total carbon emission, the authority has the obligation to make sure that the allocation amount should meet its emission reduction target and, to encourage the CPPs to produce more cleaner, the CPP which has a better emission performance should be allocated more carbon emission allowance. From the CPPs' perspective, higher carbon allowances means higher production and higher possible economic benefits, so each CPP will attempt to gain as high a carbon allowance as possible. Since the authority's focus is on emissions control, CPPs with better emissions situations are allocated greater allowances. Under such an allowance allocation mechanism, the CPPs try their best to improve their individual emissions performance to compete for as great a percentage of the allowances as possible. Using high-quality coal which includes high carbon content to generate electric power will improve the emissions performance of CPPs due to the higher carbon content, the lower impurity content which cannot be combusted, and the uncombusted content will emit large quantity of air pollutant [Smoot and Smith, 2013, Tola and Pettinau, 2014, Hu et al., 2014]. Thus, the high-quality coal is the preferred of CPPs, however, generally speaking, with the increasing of the quality, the price of coal will also increase and the prices between high- and low-quality coal being often several times different [Yang et al., 2012, Yue, 2012b, Heinberg and Fridley, 2010]. This leads to a dilemma that if the CPPs depend only on high-quality coal, their emissions performance improves and they can receive a higher carbon emissions allowance; however, the net benefit may decrease due to the high procurement costs for high-quality coal. Further, high-quality coal is a limited resource, so not all CPPs are able to purchase a high enough quantity to satisfy their demand [Heinberg and Fridley, 2010]. Coal blending is an effective way to overcome this dilemma as it allows CPPs to blend several kinds of coal [Dai et al., 2014, Shih and Frey, 1995]. PM_{10} emissions performance is another concern when the CPPs make their own coal blending decisions. Generally speaking, when combusting a specific kind of coal under a certain combustion condition, the PM_{10} emissions are mainly proportional to the ash content [Dai et al., 2014, Kauppinen and Pakkanen, 1990, Meij and Te Winkel, 2004, Wang et al., 2008]. The ash content is the main attribute used when evaluating coal types [Sun et al., 2010], as the higher the ash content, the lower the heat rate of the coal, they are in the inverse relationship. Therefore, when CPPs make their own decisions as to the coal blending ratio, the ash content proportional to the PM_{10} emissions must be considered due to the authority focuses not only on the carbon emission but also on the PM_{10} emission.

From this discussion, it can be inferred that for a coal blending method to achieve a reduction in both carbon and PM_{10} emissions, a triple equilibria situation needs to be considered. First, equilibrium is sought between the authority's objective (i.e. atmospheric environmental-friendly production) and the CPPs' objective (i.e. pursuing highest economic benefit). Second, equilibrium is sought between emissions control and cost. In addition, due to the high-quality coal may result in the decrease in the net benefit, or cannot get enough quantity to satisfy demands, when the CPPs blending some different quality coal both the carbon emissions and the PM_{10} emissions should be taken into consideration simultaneously and try their best to find the balance between them.

The mathematical form for this triple equilibria situation is given in Section 8.1.1.

Previous coal blending methodological research has tended to focus on SO_2 emissions reductions or NO_x reductions; however, PM_{10} emissions control and carbon emissions reductions have rarely been mentioned, even though CPPs are one of the major PM_{10} emissions sources in the world. The Ministry of Environmental Protection of the People's Republic of China reported in 2014 that the smoke dust emitted by the 3288 CPPs was 2.36 million tonnes, accounting for nearly 14% of total emissions and ranking first of all emissions sources. As PM_{10} is one of the most important components in smoke dust, it can be inferred that CPPs contribute significantly to total PM_{10} emissions. According to European Union emissions inventory report, CPP PM_{10} emissions accounted for 4% of total emissions and were considered one of the top 10 most serious PM_{10} emissions sources [European Environment Agency, 2010]. In the United States, the automobile is the largest emission sources of the PM_{10} emissions; however, the US Environmental Protection Agency (EPA) has also claimed that the CPPs are also critical in controlling the total PM_{10} emissions amount and should be highlighted [U.S. Environmental Protection Agency, 2014]. On another hand, according to the US EPA report, during 2002–2013, the total SO_2 emissions reduced by 66% and the total NO_x emissions reduced by 46%; however, the total PM_{10} emissions reduced by only about 4% [U.S. Environmental Protection Agency, 2014]. Based on the above discussion, it can be concluded that due to energy structure differences, degree of development, and environmental policy in different regions of the world, PM_{10} emissions sources are not the same. Nonetheless, CPPs have been considered to be one of the major PM_{10} emissions sources. CPP carbon emissions are another issue that should be highlighted. The US Energy Information Administration reported that the United States was the second largest carbon emitter in the world, with CPP emissions accounting for nearly half of all emissions [International Energy Agency, 2015]. China is the largest carbon emitter in the world, accounting for 23% of the world carbon emissions in 2013, with almost 48% of carbon emissions coming from CPPs [International Energy Agency, 2015]. Under such objective fact, the combined reduction of PM_{10} and carbon emissions is an urgent mission around the world. Therefore, reducing PM_{10} and carbon emissions is an imperative for a cleaner environment. In this chapter, the proposed bilevel programming model has multiple objectives; PM_{10} emissions reduction, carbon emissions reduction, and economic benefit as the objectives, and the SO_2 and NO_x emissions as the constraints. Unlike previous coal blending methods, the proposed modeling allows for all four CPP air pollutants to be concurrently considered.

Based on the above discussion, for the purpose of combined reduction of PM_{10} and carbon emissions to achieve more environmental-friendly CPPs production, this chapter proposes an equilibrium strategy-based bilevel multiobjective coal blending method. This method is developed under the following assumptions: this is a single production period decision problem, so at the beginning of the next production period, the decision progress is reset. Further, a carbon emissions allowances bank is not considered here; that is to say, in this chapter, a carbon emissions trading problem is not focused on and a kind of static optimization problem is assumed.

8.1.3 Modeling

In this section, the mathematical form for the above-described regional coal blending method for combined carbon-PM_{10} reduction is given.

8.1.3.1 Notations

The following notations are used in this chapter.

8.1.3.2 Objectives of the Authority

As the authority acts on behalf of the public, it has priority decision-making power, so is the upper-level decision-maker. To ensure a combined reduction in both carbon emissions and PM_{10}, minimizing carbon emissions and PM_{10} emissions in the region are the main objectives. In addition, to guarantee steady local economic development, increasing financial revenue is a further objective for the authority. Details for the achievement of these objectives are shown in the following.

Indices

i = Power plant index, $i \in \Psi = 1, 2, \dots, n$.
j = Component coal index, $j \in \Omega = 1, 2, \dots, m$.
k = Pollutant index, $k \in \Upsilon = 1, 2, \dots, t$.
w = Coal quality index, $w \in \Theta = 1, 2, \dots, l$.

Decision variables

y_i = Carbon emissions quota for power plant i allocated by the authority.
x_{ij} = Amount of component coal i burned by power plant j.

Crisp parameters

φ = Price of the unit carbon emissions quota.
λ = Tax that CPP should pay for each unit of power generation.
μ = Price of a unit of electric power.
H = Actual carbon emissions in the last decision cycle.
Q = Amount of power needed to maintain regional development.
N_{ik} = Unit operating costs at power plant i to reduce pollutant k.
U_i = Amount of power that power plant i has the responsibility to produce to meet basic demand in the region.
M_i^L, M_i^U = Lower and upper bounds for the carbon emissions quota power plant i can carry.
O_{ij}^U = Total amount of component coal j that can be procured by power plant i.
PE = Actual PM_{10} emissions in the last production period.
CE = Actual carbon emissions in the last production period.

Uncertain parameters

$\widetilde{C_j}$ = Unit procurement costs for component coal j.
$\widetilde{T_{ij}}$ = Conversion parameters from a unit of coal to power for component coal j at power plant i.
$\widetilde{P_i}$ = PM_{10} emissions when power plant i discharges a tonne of carbon emissions based on historical data.

(Continued)

(Continued)

$\widetilde{E_i}$ = Conversion parameters for a unit of carbon emissions to power for power plant i.

$\widetilde{S_{jk}}$ = Amount of pollutant emission k caused by burning a unit of component coal j.

$\widetilde{V_{ij}}$ = Amount of carbon emission caused by burning a unit of component coal j at power plant i.

$\widetilde{R_{jw}}$ = The wth quality of component coal j.

$\widetilde{LR_{iw}}, \widetilde{UR_{iw}}$ = Lower and upper bounds for coal quality w to meet operations at power plant i.

Policy control parameters

π = Attitude of the authority toward the historical data and the forecast data.

γ = Attitude of the authority toward the risk of power supply demand.

α = Attitude of the authority toward PM_{10} emissions reduction.

β = Attitude of the authority toward carbon emissions reduction.

8.1.3.2.1 Objective 1: Minimizing Carbon Emissions Amount

Let y_i represents the carbon emissions allowance allocated to the CPP i, which means that the total carbon emissions produced by CPP i cannot exceed y_i. Summarize all the CPPs' carbon emissions allowance, then the total carbon emission in the region can be get and the authority would like to try their best to reduce carbon emission amount. So this objective function can be written as:

$$\min F_1 = \sum_{i=1}^{n} y_i \tag{8.1}$$

8.1.3.2.2 Objective 2: Minimizing PM_{10} Emissions Amount

Let $\widetilde{P}_i$ be the carbon emissions to PM_{10} emissions conversion factor for CPP i, which is the PM_{10} emissions power plant i discharges for a tonne of carbon emissions based on historical data. Influenced by some objective factors, the parameter $\widetilde{P}_i$ cannot be valued exactly. However, through field research, it can be found that this value can be estimated to be within a certain range, with the most likely value being in a relatively smaller range. Parameter $\widetilde{P}_i$, therefore, is a typical trapezoidal fuzzy number which can be written as $\widetilde{P}_i = (r_{i1}, r_{i2}, r_{i3}, r_{i4})$, where $r_{i1} \leq r_{i2} \leq r_{i3} \leq r_{i4}$. This indicates that the maximum value for $\widetilde{P}_i$ can be r_{i4} and the minimum value is r_{i1} and, the most likely value is between r_{i2} and r_{i3}. This situation makes this parameter cannot be calculated directly. In this chapter, we use the expected value operator method of trapezoidal fuzzy numbers proposed by Xu and Zhou to deal with this situation [Xu and Zhou, 2011]. In Xu's method, a trapezoidal fuzzy number can be represented by its corresponding expected value using the following operator:

$$\widetilde{P}_i \to E[\widetilde{P}_i] = \frac{1-\lambda}{2}(r_{i1} + r_{i2}) + \frac{\lambda}{2}(r_{i3} + r_{i4}) \tag{8.2}$$

By using this operator, we can transform the $\widetilde{P}_i$ into its equal form, which can then be directly calculated. Therefore, the total PM_{10} emissions in the region should be $\sum_{i=1}^{n} E[\widetilde{P}_i]y_i$. However, as PM_{10} is very harmful to the atmospheric environment as well as to human health, the emissions need to be estimated as exactly as possible, so just basing this figure on historical data alone may lead to some errors, especially when CPPs may intend changing their production plans. From this point of view, actual PM_{10} emissions forecasts also need to be considered. Let x_{ij} be the amount of the jth component coal that CPP i uses and $\widetilde{S_{j3}}$ be the PM_{10} emissions factor when combusting a unit of coal j. As parameter $\widetilde{S_{j3}}$ valued either, here Xu's method is also used to transform it. From this calculation, the actual forecast PM_{10} emissions should be $\sum_{i=1}^{n} \sum_{j=1}^{m} E[\widetilde{S_{j3}}]x_{ij}$. A harmonic parameter is then used to summarize the two parts for the total emissions, so this objective can finally be written as:

$$\min F_2 = \pi \sum_{i=1}^{n} E[\widetilde{P}_i]y_i + (1-\pi) \sum_{i=1}^{n} \sum_{j=1}^{m} E[\widetilde{S_{j3}}]x_{ij} \tag{8.3}$$

where $0 \leq \pi \leq 1$. This representation form allows the decision-maker to comprehensively consider both historical data and forecast values, making the decision more scientific. With an increase in π, the decision-maker has more confidence in the historical data; conversely, the decision-maker has more confidence in the forecast data.

8.1.3.2.3 Objective 3: Pursuing Potential Financial Revenue

To ensure steady regional economic and social development, the authority has an obligation to guarantee financial revenue. The authority's potential financial revenue from the CPPs can generally be divided into two; tax and financial revenue gained from the carbon emissions allowances allocations. In this chapter, let $\widetilde{T_{ij}}$ be the conversion parameter from a unit of coal to a unit of power for component coal j at CPP i and let θ be the tax rate. The tax the CPPs need to pay is written as $\theta \sum_{i=1}^{n} \sum_{j=1}^{m} E[\widetilde{T_{ij}}]x_{ij}$, where $E[\widetilde{T_{ij}}]$ is the expected value of the trapezoidal fuzzy number $\widetilde{T_{ij}}$. Let φ be the price of the unit carbon emissions allowance, so then the financial revenue from the allocation of the paid carbon emissions allowance can be written as $\varphi \sum_{i=1}^{n} y_i$. Combining the two parts, the mathematical form for this objective is

$$\max F = \varphi \sum_{i=1}^{n} y_i + \theta \sum_{i=1}^{n} \sum_{j=1}^{m} E[\widetilde{T_{ij}}]x_{ij} \tag{8.4}$$

8.1.3.3 Constrains of the Authority

8.1.3.3.1 Limitations on the CPPs' Rights

As the authority imposes a fee for the carbon emissions allowances allocated to each CPP, to protect taxpayers, the authority cannot allocate a carbon emissions quota that a CPP cannot carry. In this chapter, let M_i^U be the upper bound of carbon emissions that CPP i can carry, which is the maximum carbon emissions emitted when CPP i in under full-load production, so this constraint can be written as: $y_i \leq M_i^U$. Since all CPPs are also taxpayers, the authority has an obligation to ensure their basic rights. In this chapter, we define the basic right of the CPPs as the right to receive a carbon emissions allowance that maintains basic operations. Let M_i^L be the carbon emissions that are required to maintain the basic operations at the CPP i, so this constraint can be written as: $M_i^L \leq y_i$. To ensure a comprehensive consideration of both sides, the limitation of the CPPs' rights when the authority makes decisions is as follows:

$$M_i^L \leq y_i \leq M_i^U \tag{8.5}$$

8.1.3.3.2 Power Supply Demand Risk Control

As electric power is required to ensure economic and social development, the authority has a responsibility to ensure adequate supply. However, due to the inherent complexity and uncertainty in power generation and transmission as well as the fluctuating demand, to fully guarantee a certain power supply level is difficult to determine, as there are always some risks. The fuzzy chance constraint method which was proposed by Xu and Zhou can be a powerful tool in coping with such problems [Xu and Zhou, 2011]. In this chapter, this method is used to describe power supply demand risk control:

$$Pos\left\{\sum_{i=1}^{n} \widetilde{E}_i y_i \geq Q\right\} \geq \gamma \tag{8.6}$$

where *Pos* is the possibility measure proposed by Dubois and Prade Dubois and Prade [1983], and the parameter $\widetilde{E}_i$ is the conversion parameter from a unit of carbon emissions to power at the CPP i, Q is the power demand in the region. This constraint can be described as: the authority would like to ensure the possibility of power supply to meet the demand in the region is larger than γ, where γ is risk control policy parameter that was chosen by the authority. This constraints cannot be calculated directly, to cope with this issue, in this chapter, based on the work by Xu and Zhou, it can be transformed into its crisp form as shown follows [Xu and Zhou, 2011];

$$(1-\gamma)\sum_{i=1}^{n} r_i^1 Y_i + \gamma \sum_{i=1}^{n} r_i^2 Y_i - Q \geq 0 \tag{8.7}$$

where r_i^1 and $r_{\underset{\sim}{i}}^1$ are the parameters in the membership function of the trapezoidal fuzzy number $\widetilde{E}_i$.

8.1.3.4 Objectives of the CPPs

The CPPs are independent decision-makers and individually pursue the largest possible profit under the carbon emissions allowances constraints imposed by the authority. For market-based CPPs, the pursuit of the highest profit, which comes mainly from power generation and sales, is a priority when the managers make decisions. Generally speaking, sales revenue minus tax and costs is equal to profit. Let μ be the price of a unit of electric power, so the sales revenue at each power plant can be written as $\mu \sum_{j=1}^{m} E[\widetilde{T_{ij}}]x_{ij}$. There are three main costs at a power plant: component coal procurement, pollutant treatment, and the carbon emissions allowances purchase. Let C_{ij} be the unit procurement cost of power plant i for component coal j, so the total procurement cost at power plant i is $\sum_{j=1}^{m} C_{ij}x_{ij}$. Let $\widetilde{S_{jk}}$ be the amount of pollutant emissions k resulting from burning a unit of component coal j and N_{ik} be the unit operating costs at power plant i for treating pollutant k, then the total cost for pollutants treatment can be described as: $\sum_{j=1}^{m} \sum_{k=1}^{t} E[\widetilde{S_{jk}}]N_{ik}x_{ij}$. The cost of purchasing the carbon emissions quota is φy_i. Therefore, the profit at each power plant can be written as follows:

$$\max G_i = \mu \sum_{j=1}^{m} E[\widetilde{T_{ij}}]x_{ij} - \sum_{j=1}^{m} C_{ij}x_{ij} - \sum_{j=1}^{m} \sum_{k=1}^{t} E[\widetilde{S_{jk}}]N_{ik}x_{ij} - \varphi y_i \tag{8.8}$$

8.1.3.5 Constraints of the CPPs

To pursue the largest possible profit, there are some constraints on each CPP which include basic supply security, emissions allowances limitations, amount of usable coal component, and the coal quality requested for the operation. Nonnegative constraints also need to be satisfied.

8.1.3.5.1 Carbon Emissions Quota Restrictions

In this decision process, the CPPs are in a subordinate position, so when making their respective production plans, the authority's decisions have to be considered. The total carbon emissions at CPP i should not exceed the carbon emissions allowance that the authority allocates otherwise the CPP is punished or even deprived of the right to produce. Let $\widetilde{V_{ij}}$ be the carbon emissions factor when CPP i combusts a unit of the component coal j; here, we also use the trapezoidal fuzzy number to describe the uncertainty in measuring this parameter; let $E[\widetilde{V_{ij}}]$ be the expected value of $\widetilde{V_{ij}}$, so this constraint can be written as:

$$\sum_{j=1}^{m} E[\widetilde{V_{ij}}]x_{ij} \le y_i \tag{8.9}$$

8.1.3.5.2 Social Responsibility Limitation

The modern enterprises should not only focus on the economic benefit, they also have some social responsibility. Therefore, as electricity is critical for social development, the supply of basic electric power is the CPPs' fundamental social responsibility. Let $\widetilde{T_{ij}}$ be power generation factor when the CPP i combusts a unit of the component coal j and, let U_i be the basic electrical power supply amount that the CPP i should guarantee. Then, this constraint can be represented as follows:

$$\sum_{j=1}^{m} E[\widetilde{T_{ij}}]x_{ij} \ge U_i \tag{8.10}$$

where $E[\widetilde{T_{ij}}]$ is the expected value of $\widetilde{T_{ij}}$.

8.1.3.5.3 Component Coal Purchase Quantity Limitations

For each CPP, the quantity of each component coal that can be used is limited. That is to say, the CPPs cannot purchase any component coal without limitations. In this chapter, let O_{ij}^{U} be the upper quantity of component coal j that can be procured by the CPP i, so this constraint can be written as:

$$0 \leq x_{ij} \leq O_{ij}^{U} \tag{8.11}$$

8.1.3.5.4 Coal Quality Requirement

The boilers at each CPP are predefined and cannot be changed in the short term. Each boiler have some requirements for the coal quality, especially for volatile matter content, heat rate, ash content, moisture content, and sulfur content [Dai et al., 2014, Samimi and Zarinabadi, 2012]. If the coal to be burned in the boilers does not meet requirements, there may be serious consequences. Therefore, CPPs must ensure that the blended coal quality meets the boiler requirements. In this chapter, five main characteristics were selected to describe the blended coal quality; $w = 1$ represents volatile matter content, $w = 2$ represents heat rate, $w = 3$ represents ash content, $w = 4$ represents moisture content, and $w = 5$ represents sulfur content. Any deviations on these specific requirements are not permitted, so this constraint can be written as:

$$E[\widetilde{LR_{iw}}] \leq \frac{\sum_{j=1}^{m} E[\widetilde{R_{jw}}]x_{ij}}{\sum_{j=1}^{m} x_{ij}} \leq E[\widetilde{UR_{iw}}] \tag{8.12}$$

where $E[\widetilde{R_{jw}}]$ is the expected value of $\widetilde{R_{jw}}$, which represents the wth quality of the component coal j and LR_{iw} and UR_{iw} represents the upper and the lower requirements for the wth quality at coal in the CPP i. The $E[\widetilde{LR_{iw}}]$ and $E[\widetilde{UR_{iw}}]$ represent the expected value of the lower and upper bounds for coal quality w to meet operations at power plant i.

8.1.3.6 Global Optimization Model

By integrating Eqs. (8.1)–(8.12), the global optimization model is formulated as shown in Eq. (8.13). The optimization process of the model can be summarized: first, based on the historical data and its own objectives, the authority decides on the initial carbon emissions allowances allocation scheme and then informs each CPP, after which each CPP develops their own coal blending plan under the limitation of the carbon emissions allowance, social responsibility, market conditions, and its own technical constraints. The CPPs' plans are submitted to the authority, which then adjusts its initial decisions in consideration of each CPP's emissions performance, after which an improved carbon emissions allowances allocation scheme is sent to the CPPs. The CPPs again consider their previous coal blending plan based on the new allowance and submit an improved coal blending plan to the authority. This process is repeated until an equilibrium solution is arrived at when all stakeholders reach the consensus.

$$\max F_1 = \varphi \sum_{i=1}^{n} y_i + \lambda \sum_{i=1}^{n} \sum_{j=1}^{m} E[\widetilde{T_{ij}}]x_{ij}$$

$$\min F_2 = \pi \sum_{i=1}^{n} E[\widetilde{P_i}]y_i + (1 - \pi) \sum_{i=1}^{n} \sum_{j=1}^{m} E[\widetilde{S_{j3}}]x_{ij}$$

$$\min F_3 = \sum_{i=1}^{n} y_i$$

$$\text{s.t.}\begin{cases} M_i^L \leq y_i \leq M_i^U, \quad \forall i \in \Psi \\ (1-\gamma)\sum_{i=1}^{n} r_i^1 Y_i + \gamma \sum_{i=1}^{n} r_i^2 Y_i - Q \geq 0 \\ \text{where } x_{ij} \text{ solves :} \\ \max G_i = (\mu - \lambda)\sum_{j=1}^{m} E[\widetilde{T_{ij}}]x_{ij} - \sum_{j=1}^{m} C_j x_{ij} - \sum_{j=1}^{m}\sum_{k=1}^{t} E[\widetilde{S_{jk}}]N_{ik}x_{ij} - \varphi y_i \\ \text{s.t.}\begin{cases} \sum_{j=1}^{m} E[\widetilde{V_{ij}}]x_{ij} \leq y_i, \quad \forall i \in \Psi \\ \sum_{j=1}^{m} E[\widetilde{T_{ij}}]x_{ij} \geq U_i, \quad \forall i \in \Psi \\ 0 \leq x_{ij} \leq O_{ij}^U, \quad \forall i \in \Psi, \forall j \in \Omega \\ E[\widetilde{LR_{iw}}] \leq \frac{\sum_{j=1}^{m} E[\widetilde{R_{jw}}]x_{ij}}{\sum_{j=1}^{m} x_{ij}} \leq E[\widetilde{UR_{iw}}], \quad \forall i \in \Psi, \forall w \in \Theta \end{cases} \end{cases} \tag{8.13}$$

The above bilevel model was developed to determine a suitable coal blending ratio for each CPP in the region to improve the atmospheric environment. To make a comprehensive consideration of all the stakeholders' interests and allow the authority has the ability to impact the CPPs' decision, carbon emission allowance allocation method was employed. To enhance the applicability of the model for similar situations, the following considerations are presented:

(1) In the proposed model, π and γ are policy control parameters which were determined by the authority. Employing a parameter setting method means that the authority's decisions are more flexible. Under certain situations, the authority is able to adjust the policy control parameters to identify the most satisfactory decision.
(2) This model can be used across a wider range by changing, adding, or removing the authority's objective functions. For example, in some developed regions, the economy is already well developed. In this case, the authority could focus greater attention on the atmospheric environment, so can remove relative objective function to better control the pollutant emissions.
(3) Changing, adding, or removing some constraints in the proposed model can make it more pervasive. For instance, for CPPs that have relatively wider coal quality requirements, the CPP can add specific constraints to match this model to the actual situation.

8.2 Case Study

To demonstrate the efficiency of the proposed method, a real-world case study is given in this section.

8.2.1 Presentation of the Case Region

Sichuan is one of the most important provinces in southwest China. Like most other province, CPPs are the major electricity power suppliers in Sichuan, providing over 60% of the power demand. However, the coal combustion causes air pollution. Sichuan Provincial Department of ecological environment reported that the number of days of air pollution in 2018 accounted for 15.2%. Atmospheric environmental assessment results also suggested that Sichuan Province has emitted over 13 million tonnes of carbon emissions, ranking eighth in China [He et al., 2002, Wang et al., 2010]. Under such situation, the authority has determined that carbon emissions decreased by 19.5% in five years. Among many carbon emission industries, more than 60% carbon emissions in Sichuan Province come from CPPs. Therefore, reducing the carbon emissions of CPPs is vital to achieve the carbon reduction target.

Not only carbon emission, the CPPs also are the main source of PM_{10} emission from the perspective of industrial structure of Sichuan Province. The annual average concentration of PM_{10} in the Sichuan Province is 62.6 μg per stere in 2018, which is about 50% higher than the current level I standard annual average concentration limit of PM_{10} in China. Therefore, it is imperative to attach great importance to CPPs in Sichuan Province, so as to reduce PM_{10} and carbon emissions.

Grounded on the above situation, Sichuan Province was chosen as the case region to test the efficiency of the proposed combined carbon and PM_{10} emissions reduction optimization method. Besides, considering the feasibility of computer calculation and data acquisition, three main power plants in Sichuan Province were chosen, which are the Jintang power plant, the Neijiang power plant, and the Nanchong power plant.

8.2.2 Model Transformation and Solution Approach

To ensure the sustainable development of Sichuan Province, the authority has to give priority to economic objectives; however, the environmental protecting pressure cannot be ignored. Therefore, to limit pollution into an acceptable range is necessary. Encouraged by Zeng et al.'s Zeng et al. [2014] research, we introduce two parameters α and β to represent the authority's attitudes toward CO_2 emissions and PM_{10} emissions, then the multiobjective model can be transformed into Eq. (8.14).

The proposed bilevel decision model is complex and is an non-deterministic polynomial (NP)-hard problem [Colson et al., 2007], so we designed an interactive algorithm. Details are as follows:

Step 1: Set all constraints of the upper level model in Matlab 7.0 and build a feasible region.
Step 2: Generate a vector of y_i^1 as an initial solution for the upper level.
Step 3: Input y_i^1 into the lower level model, which was transformed into a single-level linear programming.

Step 4: Calculate the lower level model using mathematical toolbox inserted in Matlab 7.0. and represent it as x_{ij}^1, then feedback the x_{ij}^1 to the upper level model.

Step 5: Calculate the optimal solution of the upper level model and send the improved y_i^2 to the lower level model.

Step 6: Repeat the Steps 4 and 5 until the termination condition is reached.

Here the termination condition was set as $\left|y_i^q - y_i^{q-1}\right| / y_i^{q-1} \leq 1\%$, where q represented the number of interactions. Although this interactive algorithm may not be the optimal solution to the bilevel model; using this algorithm can save time as the results are very close to the optimal solution as the allowable error is set at only 1%.

$$
\begin{aligned}
&\max F_1 = \varphi \sum_{i=1}^{n} y_i + \lambda \sum_{i=1}^{n} \sum_{j=1}^{m} E[\widetilde{T_{ij}}]x_{ij} \\
&\text{s.t.} \begin{cases}
\pi \sum_{i=1}^{n} E[\widetilde{P}_i]y_i + (1-\pi) \sum_{i=1}^{n} \sum_{j=1}^{m} E[\widetilde{S_{j3}}]x_{ij} \leq \alpha PE \\
\sum_{i=1}^{n} y_i \leq \beta CE \\
M_i^L \leq y_i \leq M_i^U, \quad \forall i \in \Psi \\
(1-\gamma) \sum_{i=1}^{n} r_i^1 Y_i + \gamma \sum_{i=1}^{n} r_i^2 Y_i - Q \geq 0 \\
\text{where } x_{ij} \text{ solves :} \\
\max G_i = (\mu - \lambda) \sum_{j=1}^{m} E[\widetilde{T_{ij}}]x_{ij} - \sum_{j=1}^{m} C_j x_{ij} - \sum_{j=1}^{m} \sum_{k=1}^{t} E[\widetilde{S_{jk}}]N_{ik}x_{ij} - \varphi y_i \\
\text{s.t.} \begin{cases}
\sum_{j=1}^{m} E[\widetilde{V_{ij}}]x_{ij} \leq y_i, \quad \forall i \in \Psi \\
\sum_{j=1}^{m} E[\widetilde{T_{ij}}]x_{ij} \geq U_i, \quad \forall i \in \Psi \\
0 \leq x_{ij} \leq O_{ij}^U, \quad \forall i \in \Psi, \forall j \in \Omega \\
E[\widetilde{LR_{iw}}] \leq \frac{\sum_{j=1}^{m} E[\widetilde{R_{jw}}]x_{ij}}{\sum_{j=1}^{m} x_{ij}} \leq E[\widetilde{UR_{iw}}], \quad \forall i \in \Psi, \forall w \in \Theta
\end{cases}
\end{cases}
\end{aligned}
\tag{8.14}
$$

8.2.3 Data Collection

Data collected for calculation are divided into uncertain category and crisp category. Crisp data shown in shown in Tables 8.1 and 8.2 were obtained from the Chinese Southwest Electric Power Design Institute and field research. And micro-crisp data such as the actual carbon emission amount in the last production cycle is collected Statistical Yearbook of Sichuan Province, China.

The uncertain parameter (as shown in Tables 8.3 and 8.4) were shown in fuzzy form, the detailed collecting process can be summarized as follows:

Table 8.1 Certain parameters of each CPP.

	Jintang CPP	Neijiang CPP	Langzhong CPP
Emission reduction measure			
For SO_2	LDS	LDS	LDS
For NO_x	SCR	SCR	SCR
For PM_{10}	EP and FF	EP and CED	EP and FF
Emission reduction cost: N_{ik}			
For SO_2 (RMB/kg)	2	1.8	2.3
For NO_x (RMB/kg)	14.5	14.7	14.5
For PM_{10} (RMB/kg)	3.5	2.5	3.3
Emission reduction efficiency			
For SO_2 (%)	95.5	95.7	96.0
For NO_x (%)	84.5	84.9	85.1
For PM_{10} (%)	95.8	94.9	95.0
Others			
Basic electric supply: U_i (kWh)	6×10^9	3.6×10^9	4.3×10^9
Utmost production capacity (kWh)	9.6×10^9	5.7×10^9	6.8×10^9
Basic demand on carbon emission quota: M_i^L (tonne)	6.07×10^6	3.6×10^6	4.1×10^6
Utmost carbon emission amount: M_i^U (tonne)	9.7×10^6	5.7×10^6	6.45×10^6

LDS, lime spray dryer system; SCR, selective catalytic reduction; EP, electrostatic precipitator; CDE, cyclone dust extractor; and FF, fabric filter.

Table 8.2 Other parameters used in the proposed model.

Carbon emission quota price: φ (RMB/tonne)	30
Added-value tax: λ (RMB/kWh)	0.01
Total basic electric supply: Q (kWh)	1.39×10^{10}
Price of unit electric: μ (RMB/kWh)	0.45
Actual carbon emission amount in the last production cycle: CE (tonne)	2.3×10^7
Actual PM_{10} emission amount in the last production cycle: PE (kg)	7.2626×10^6

(1) Ask experts and engineers from each supplier coal mine to give a range for each uncertain parameter;
(2) Analyze the initial collected data and eliminate the extreme values;
(3) Set lower and upper bound of each uncertain parameter using the minimum and maximum value of the remaining data;
(4) Using the values with the highest frequency to represent the most possible range of each uncertain parameter (i.e. form r_{ij}^2 to r_{ij}^3 and form r_i^2 to r_i^3).

Due to the fact that Sichuan Province is developing rapidly, we set the attitude of the authority toward the risk of electric power supply as 1 ($\gamma = 1$), indicating that the authority had a complete strict attitude toward the meeting of the power demands.

8.3 Results and Discussion

By inputting the collected data into the proposed optimization model (i.e. Eq. (8.14)) and run the solution approach on Matlab 7.0, the results of the proposed model can be calculated.

8.3.1 Results Under Different Scenarios

In this section, results under different attitude of the decision-maker (i.e. different policy control parameter: α, β, π) will be shown.

In scenarios 1 and 2, the authority is neutral about historical data and forecast data ($\pi = 0.5$), and the carbon emissions and PM_{10} emissions are set to equal to the actual emissions amount of the last production period ($\beta = 1$, $\alpha = 1$), respectively. By changing authority's attitude toward PM_{10} (α) reduction or carbon emissions (β) with a step length of 0.05, it can be found that total PM_{10} emissions can be controlled at 65% of the actual emissions amount, and total carbon emissions can be controlled at 60% from the actual emissions amount in the last production period.

In scenarios 3 and 4, the authority has a preference for historical records ($\pi = 0.75$), and the carbon emissions and PM_{10} emissions are set to equal to the actual emissions amount of the last production period ($\beta = 1$, $\alpha = 1$), respectively. By changing authority's attitude toward PM_{10} (α) reduction or carbon emissions (β) with a step length of 0.05, it can be found that total PM_{10} emissions can be controlled at 80% of the actual emissions amount, and total carbon emissions can be controlled at 60% from the actual emissions amount in the last production period.

In scenarios 5 and 6, the authority has a preference for the forecasting data ($\pi = 0.35$), and the carbon emissions and PM_{10} emissions are set to equal to the actual emissions amount of the last production period ($\beta = 1$, $\alpha = 1$), respectively. By changing authority's attitude toward PM_{10} (α) reduction or carbon emissions (β) with a step length of 0.05, it can be found that total PM_{10} emissions can be controlled at 60% of the actual emissions amount, and total carbon emissions can be controlled at 60% from the actual emissions amount in the last production period.

Table 8.3 Parameters of component coals in fuzzy form.

	Jincheng	Shenhua	Qujing	Panzhihua
Coal properties: $\widetilde{R_{jw}}$				
Volatile matter (% weight)	(6.08, 6.11, 6.18, 6.22)	(27.5, 28.0, 28.6,p 29.2)	(27.6, 28.2, 28.6, 29.0)	(13.5, 14.0, 14.6, 15.1)
Heat rate (Gj/tonne)	(26.4, 26.8, 29.0, 29.2)	(22.9, 23.1, 23.5, 23.9)	(21.5, 21.8, 22.2, 22.9)	(20.5, 21.0, 21.2, 21.5)
Ash content (% weight)	(19.1, 19.5, 19.8, 20.1)	(8.7, 9.1, 9.3, 9.5)	(9.0, 9.3, 9.5, 9.9)	(13.5, 13.9, 14.2, 14.5)
Moisture content (% weight)	(4.2, 4.4, 4.7, 4.9)	(5.3, 5.5, 5.9, 6.2)	(7.1, 7.3, 7.6, 8.1)	(4.7, 4.9, 5.3, 5.5)
Sulfur content (% weight)	(0.57, 0.59, 0.61, 0.64)	(0.22, 0.25, 0.28, 0.31)	(0.34, 0.37, 0.41, 0.43)	(0.44, 0.47, 0.51, 0.56)
Pollutants emission factors: $\widetilde{S_{jk}}$				
For SO_2 (kg/tonne)	(8.1, 8.3, 8.6, 8.8)	(13.5, 13.8, 14.2, 14.7)	(9.6, 9.8, 10.2, 10.5)	(7.6, 7.8, 8.1, 8.5)
For NO_x (kg/tonne)	(6.3, 6.5, 6.8, 7.1)	(8.9, 9.1, 9.3, 9.6)	(7.6, 7.9, 8.2, 8.6)	(2.2, 2.6, 2.9, 3.2)
For PM_{10} (kg/tonne)	(4.7, 4.9, 5.1, 5.5)	(4.9, 5.3, 5.5, 5.8)	(5.2, 5.5, 5.8, 6.1)	(4.1, 4.5, 5.2, 5.6)
Others				
Procurement price: $\widetilde{C_j}$ (RMB/tonne)	(680, 695, 705, 715)	(650, 665, 680, 700)	(620, 635, 650, 670)	(605, 620, 630, 645)

Table 8.4 Uncertain parameters of each CPP.

	Jintang CPP		Neijiang CPP		Langzhong CPP	
	Lower bound	Upper bound	Lower bound	Upper bound	Lower bound	Upper bound
$(\widetilde{LR_{iw}}, \widetilde{UR_{iw}})$						
Volatile matter (%)	(5.9, 6.1, 6.3, 6.8)	(26.1, 26.5, 26.9, 27.2)	(6.9, 7.3, 7.7, 8.2)	(26.3, 26.6, 27.0, 27.2)	(9.1, 9.3, 9.7, 10.2)	(36.4, 36.8, 37.2, 37.5)
Heat rate (GJ/tonne)	(25.8, 26.0, 26.3, 26.6)	—	(24.5, 24.9, 25.2, 25.6)	—	(22.7, 22.9, 23.3, 23.7)	—
Ash content (%)	—	(19.8, 20.2, 20.5, 21.0)	—	(20.8, 21.3, 21.7, 22.1)	—	(21.8, 22.2, 22.6, 23.0)
Moisture content (%)	—	(4.7, 4.9, 5.1, 5.4)	—	(5.5, 5.7, 6.0, 6.3)	—	(5.0, 5.2, 5.6, 5.9)
Sulfur content (%)	—	(0.6, 0.8, 1.1, 1.2)	—	(0.8, 0.9, 1.1, 1.3)	—	(1.0, 1.2, 1.4, 1.5)
$\widetilde{T}_{ij}$						
Jincheng (kWh/tonne)	(2325, 2350, 2410, 2450)		(2550, 2580, 2600, 2630)		(2570, 2595, 2620, 2650)	
Shenhua (kWh/tonne)	(2205, 2210, 2230, 2250)		(2200, 2220, 2250, 2280)		(2190, 2210, 2260, 2290)	
Qujing (kWh/tonne)	(2420, 2450, 2470, 2510)		(2095, 2130, 2160, 2200)		(2050, 2080, 2110, 2150)	
Panzhihua (kWh/tonne)	(1980, 2000, 2030, 2050)		(2020, 2050, 2070, 2095)		(2000, 2030, 2050, 2080)	
$\widetilde{V}_{ij}$						
Jincheng (kg/tonne)	(2188, 2205, 2228, 2235)		(2205, 2215, 2226, 2239)		(2165, 2186, 2195, 2205)	
Shenhua (kg/tonne)	(1952, 1965, 1978, 1986)		(1965, 1976, 1985, 1998)		(1940, 1955, 1967, 1980)	
Qujing (kg/tonne)	(1940, 1953, 1967, 1980)		(1965, 1980, 1993, 2005)		(1920, 1931, 1943, 1965)	
Panzhihua (kg/tonne)	(1915, 1925, 1936, 1948)		(1935, 1951, 1968, 1981)		(1898, 1909, 1925, 1941)	
Others						
$\widetilde{P}_i$ (kg/tonne)	(0.42, 0.45, 0.48, 0.51)		(0.41, 0.43, 0.46, 0.49)		(0.40, 0.42, 0.46, 0.49)	
$\widetilde{E}_i$ (kWh/tonne)	(950, 970, 1000, 1020)		(960, 990, 1010, 1035)		(975, 1010, 1040, 1065)	

8.3.2 Propositions and Analysis

A discussion on the results under the different scenarios shown above is given in this section.

First of all, it can be concluded that equilibrium strategy-based coal blending method has potential to improve the environment. From an effectiveness test on the policy control α and β, which represented the authority's attitude toward the carbon emissions reduction target and the PM_{10} emissions reduction target, respectively, the lowest possible emissions level for the two kinds of pollutant were determined, as shown in Tables 8.5–8.10. It can be seen that regardless of the scenario, the proposed coal blending method has the potential to improve the environment. Take Tables 8.5 and 8.6 as examples; under this scenario, the lowest possible PM_{10} emission level is 0.65, which means that by using the equilibrium strategy-based coal blending method, the total PM_{10} emissions in this region can be reduced by about 35% compared with the actual PM_{10} emissions amount from the last production cycle and on the other hand, the lowest possible carbon emissions level was 0.6, which indicates that the proposed coal blending method could result in a 40% reduction in the actual carbon emissions from the last production cycle. Similar situation can also be found in other scenarios. This phenomenon is not occasional. Actually, when the equilibrium strategy-based coal blending method is employed, the carbon emissions allowance is the key factor that used by the authority to control and impact the CPPs' decisions, so due to the modeling intention of the authority that the CPPs which have relative better emissions performance will be allocated more allowance, to pursue as a great allowance as possible, each CPP will try their best to adjust the coal blending ratio to improve their own emissions performance even though such behavior may lead to an increase in the unit production costs. This results in the total emission amount reduction in the region.

Furthermore, from the results shown from Tables 8.5–8.10, it can be seen that the marginal emissions growth rate is larger than the financial revenue when the environmental protection constraints are relaxed. With the relaxing of the environmental protection constraints (i.e. changing the policy control parameter α or β from 1 to their possible lowest levels), both the emissions amount and the financial revenue increase; however, the increasing ratio of the former is larger than the latter. Take scenario 3 as an example, the results for which are shown in Table 8.7; the lowest possible PM_{10} emissions control level is 0.8 (i.e. $\alpha = 0.8$) and at this time, the total carbon emission amount is 14.65×10^6 tonnes, the total financial revenue is 6.05×10^8 RMB and the total PM_{10} emission amount is 5.81×10^6 kg. When the authority relaxes its attitude toward environmental protection constraints (i.e. sets $\alpha = 0.85$), the total carbon emissions increase to 15.58×10^6 tonnes, the total financial revenue is 6.42×10^8 RMB, and the total PM_{10} emission amount is 6.17×10^6 kg. The ratio has increased by 6.4%, 6.1%, and 6.2% for these three factors, respectively, with the emissions speed of increase being greater than that of financial revenue. When the authority relaxes its environmental protection attitude again (i.e. sets $\alpha = 0.9$), such situation takes place again, the increase ratio of the three factors being 5.4%, 5.1%, and 5.9%, respectively, with the financial revenue

Table 8.5 Sensitivity analysis on the local authority's attitude toward PM_{10} emission reduction targets.

Policy candidate (π, α, β)	Total benefit: $F(x_{ij}, y_i)$ (10^8 RMB)	CPP	Carbon emission allowance: y_i (10^6 tonnes)	Jincheng: x_{i1} (10^6 tonnes)	Shenhua: x_{i2} (10^6 tonnes)	Qujing: x_{i3} (10^6 tonnes)	Panzhihua: x_{i4} (10^6 tonnes)	Benefit: $G_i(x_{ij}, y_i)$ (10^8 RMB)
$(\pi = 0.5,$	8.86	Jintang CPP	9.7	3.5	0.13	0.51	0	6.89
$\alpha = 1,$		Neijiang CPP	5.7	0.65	0.88	0	1.22	3.39
$\beta = 1)$		Langzhong CPP	6.45	0.66	2.35	0	0	3.59
$(\pi = 0.5,$	8.43	Jintang CPP	8.6261	3.2	0.12	0.46	0	6.12
$\alpha = 0.95,$		Neijiang CPP	5.7	0.65	0.88	0	1.22	3.39
$\beta = 1)$		Langzhong CPP	6.45	0.66	2.35	0	0	3.59
$(\pi = 0.5,$	7.99	Jintang CPP	7.5336	2.72	0.11	0.39	0	5.35
$\alpha = 0.9,$		Neijiang CPP	5.7	0.65	0.88	0	1.22	3.39
$\beta = 1)$		Langzhong CPP	6.45	0.66	2.35	0	0	3.59
$(\pi = 0.5,$	7.55	Jintang CPP	6.4411	2.32	0.09	0.34	0	4.57
$\alpha = 0.85,$		Neijiang CPP	5.7	0.65	0.88	0	1.22	3.39
$\beta = 1)$		Langzhong CPP	6.45	0.66	2.35	0	0	3.59

(Continued)

Table 8.5 (Continued)

Policy candidate (π, α, β)	Total benefit: $F(x_{ij}, y_i)$ (10^8 RMB)	CPP	Carbon emission allowance: y_i (10^6 tonnes)	Jincheng: x_{i1} (10^6 tonnes)	Shenhua: x_{i2} (10^6 tonnes)	Qujing: x_{i3} (10^6 tonnes)	Panzhihua: x_{i4} (10^6 tonnes)	Benefit: $G_i(x_{ij}, y_i)$ (10^8 RMB)
$(\pi = 0.5,$	7.11	Jintang CPP	6.07	2.19	0.08	0.32	0	4.31
$\alpha = 0.8,$		Neijiang CPP	4.982	0.56	0.77	0	1.08	2.96
$\beta = 1)$		Langzhong CPP	6.45	0.66	2.35	0	0	3.59
$(\pi = 0.5,$	6.66	Jintang CPP	6.07	2.19	0.08	0.32	0	4.31
$\alpha = 0.75,$		Neijiang CPP	3.8948	0.44	0.61	0	0.84	2.32
$\beta = 1)$		Langzhong CPP	6.45	0.66	2.35	0	0	3.59
$(\pi = 0.5,$	6.21	Jintang CPP	6.07	2.19	0.01	0.32	0	4.31
$\alpha = 0.7,$		Neijiang CPP	3.6	0.41	0.57	0	0.78	2.14
$\beta = 1)$		Langzhong CPP	5.648	0.58	2.1	0	0	3.14
$(\pi = 0.5,$	5.77	Jintang CPP	6.07	2.19	0.01	0.32	0	4.31
$\alpha = 0.65,$		Neijiang CPP	3.6	0.41	0.57	0	0.78	2.14
$\beta = 1)$		Langzhong CPP	4.5476	0.46	1.66	0	0	2.53

Table 8.6 Sensitivity analysis on the local authority's attitude toward carbon emission reduction targets.

Policy candidate (π, α, β)	Total benefit: $F(x_{ij}, y_i)$ (10^8 RMB)	CPP	Carbon emission allowance: y_i (10^6 tonnes)	Jincheng: x_{i1} (10^6 tonnes)	Shenhua: x_{i2} (10^6 tonnes)	Qujing: x_{i3} (10^6 tonnes)	Panzhihua: x_{i4} (10^6 tonnes)	Benefit: $G_i(x_{ij}, y_i)$ (10^8 RMB)
$(\pi = 0.5,$	8.86	Jintang CPP	9.7	3.5	0.13	0.51	0	6.89
$\alpha = 1,$		Neijiang CPP	5.7	0.65	0.88	0	1.22	3.39
$\beta = 0.95)$		Langzhong CPP	6.45	0.66	2.35	0	0	3.59
$(\pi = 0.5,$	8.39	Jintang CPP	9.6962	3.50	0.13	0.51	0	6.88
$\alpha = 1,$		Neijiang CPP	4.6885	0.53	0.72	0	1.01	2.79
$\beta = 0.9)$		Langzhong CPP	6.3153	0.65	2.31	0	0	3.51
$(\pi = 0.5,$	7.92	Jintang CPP	9.5992	3.45	0.13	0.450	0	6.82
$\alpha = 1,$		Neijiang CPP	5.694	0.65	0.88	0	1.22	3.39
$\beta = 0.85)$		Langzhong CPP	4.2569	0.44	1.55	0	0	2.37
$(\pi = 0.5,$	7.45	Jintang CPP	9.5404	3.44	0.13	0.50	0	6.77
$\alpha = 1,$		Neijiang CPP	4.6907	0.53	0.76	0	1.01	2.79
$\beta = 0.8)$		Langzhong CPP	4.1689	0.43	1.51	0	0	2.32

Table 8.6 (Continued)

Policy candidate (π, α, β)	Total benefit: $F(x_{ij}, y_i)$ (10^8 RMB)	CPP	Carbon emission allowance: y_i (10^6 tonnes)	Jincheng: x_{i1} (10^6 tonnes)	Shenhua: x_{i2} (10^6 tonnes)	Qujing: x_{i3} (10^6 tonnes)	Panzhihua: x_{i4} (10^6 tonnes)	Benefit: $G_i(x_{ij}, y_i)$ (10^8 RMB)
$(\pi = 0.5,$	6.98	Jintang CPP	9.4686	3.42	0.12	0.49	0	6.72
$\alpha = 1,$		Neijiang CPP	3.666	0.42	0.57	0	0.79	2.18
$\beta = 0.75)$		Langzhong CPP	4.1153	0.42	1.49	0	0	2.29
$(\pi = 0.5,$	6.52	Jintang CPP	8.2747	2.98	0.11	0.47	0	5.88
$\alpha = 1,$		Neijiang CPP	3.703	0.42	0.57	0	0.81	2.20
$\beta = 0.7)$		Langzhong CPP	4.1223	0.42	1.50	0	0	2.29
$(\pi = 0.5,$	6.06	Jintang CPP	7.0462	2.55	0.09	0.37	0	5.01
$\alpha = 1,$		Neijiang CPP	3.7717	0.43	0.58	0	0.82	2.23
$\beta = 0.65)$		Langzhong CPP	4.132	0.43	1.50	0	0	2.30
$(\pi = 0.5,$	5.61	Jintang CPP	6.0908	2.20	0.08	0.32	0	4.32
$\alpha = 1,$		Neijiang CPP	3.6013	0.41	0.57	0	0.78	2.14
$\beta = 0.6)$		Langzhong CPP	4.1079	0.42	1.49	0	0	2.28

Table 8.7 Sensitivity analysis on the local authority's attitude toward carbon emission reduction targets.

Policy candidate (π, α, β)	Total benefit: $F(x_{ij}, y_i)$ (10^8 RMB)	CPP	Carbon emission allowance: y_i (10^6 tonnes)	Jincheng: x_{i1} (10^6 tonnes)	Shenhua: x_{i2} (10^6 tonnes)	Qujing: x_{i3} (10^6 tonnes)	Panzhihua: x_{i4} (10^6 tonnes)	Benefit: $G_i(x_{ij}, y_i)$ (10^8 RMB)
$(\pi = 0.75,$	7.45	Jintang CPP	6.192	2.2	0.09	0.33	0	4.4
$\alpha = 1,$		Neijiang CPP	5.7	0.65	0.88	0	1.22	3.39
$\beta = 1)$		Langzhong CPP	6.45	0.66	2.35	0	0	3.59
$(\pi = 0.75,$	7.09	Jintang CPP	6.07	2.1	0.08	0.32	0	4.31
$\alpha = 0.95,$		Neijiang CPP	4.9086	0.56	0.76	0	1.1	2.92
$\beta = 1)$		Langzhong CPP	6.45	0.66	2.35	0	0	3.59
$(\pi = 0.75,$	6.75	Jintang CPP	6.07	2.1	0.08	0.32	0	4.31
$\alpha = 0.9,$		Neijiang CPP	3.9939	0.55	0.62	0	0.86	2.38
$\beta = 1)$		Langzhong CPP	6.45	0.66	2.35	0	0	3.59
$(\pi = 0.75,$	6.42	Jintang CPP	6.07	2.1	0.08	0.32	0	4.31
$\alpha = 0.85,$		Neijiang CPP	3.6	0.41	0.56	0	0.78	2.14
$\beta = 1)$		Langzhong CPP	5.9199	0.61	2.2	0	0	3.29
$(\pi = 0.75,$	6.05	Jintang CPP	6.07	2.1	0.08	0.32	0	4.31
$\alpha = 0.8,$		Neijiang CPP	3.6	0.41	0.56	0	0.78	2.14
$\beta = 1)$		Langzhong CPP	4.9888	0.51	1.9	0	0	2.78

Table 8.8 Sensitivity analysis on the local authority's attitude toward carbon emission reduction targets.

Policy candidate (π, α, β)	Total benefit: $F(x_{ij}, y_i)$ (10^8 RMB)	CPP	Carbon emission allowance: y_i (10^6 tonnes)	Jincheng: x_{i1} (10^6 tonnes)	Shenhua: x_{i2} (10^6 tonnes)	Qujing: x_{i3} (10^6 tonnes)	Panzhihua: x_{i4} (10^6 tonnes)	Benefit: $G_i(x_{ij}, y_i)$ (10^8 RMB)
$(\pi = 0.75,$	7.45	Jintang CPP	6.192	2.2	0.08	0.33	0	4.40
$\alpha = 1,$		Neijiang CPP	5.7	0.65	0.88	0	1.22	3.39
$\beta = 0.95)$		Langzhong CPP	6.45	0.66	2.35	0	0	3.59
$(\pi = 0.75,$	7.45	Jintang CPP	6.192	2.2	0.08	0.33	0	4.40
$\alpha = 1,$		Neijiang CPP	5.7	0.65	0.88	0	1.22	3.39
$\beta = 0.9)$		Langzhong CPP	6.45	0.66	2.35	0	0	3.59
$(\pi = 0.75,$	7.45	Jintang CPP	6.192	2.2	0.08	0.33	0	4.40
$\alpha = 1,$		Neijiang CPP	5.7	0.65	0.88	0	1.22	3.39
$\beta = 0.85)$		Langzhong CPP	6.45	0.66	2.35	0	0	3.59
$(\pi = 0.75,$	7.45	Jintang CPP	6.192	2.2	0.08	0.33	0	4.40
$\alpha = 1,$		Neijiang CPP	5.7	0.65	0.88	0	1.22	3.39
$\beta = 0.8)$		Langzhong CPP	6.45	0.66	2.35	0	0	3.59
$(\pi = 0.75,$	6.99	Jintang CPP	8.5967	3.2	0.12	0.46	0	6.29
$\alpha = 1,$		Neijiang CPP	3.6183	0.41	0.56	0	0.78	2.16
$\beta = 0.75)$		Langzhong CPP	5.035	0.50	1.8	0	0	2.63

($\pi = 0.75$,	6.53	Jintang CPP	7.5304	2.72	0.11	0.40	0	5.35
$\alpha = 1$,		Neijiang CPP	4.381	0.50	0.68	0	0.94	2.61
$\beta = 0.7$)		Langzhong CPP	4.1886	0.43	1.52	0	0	2.33
($\pi = 0.75$,	6.06	Jintang CPP	7.0926	2.56	0.09	0.37	0	5.04
$\alpha = 1$,		Neijiang CPP	3.6023	0.41	0.57	0	0.78	2.14
$\beta = 0.65$)		Langzhong CPP	4.2552	0.43	1.55	0	0	2.37
($\pi = 0.75$,	5.60	Jintang CPP	6.0928	2.20	0.08	0.32	0	4.33
$\alpha = 1$,		Neijiang CPP	3.6056	0.41	0.56	0	0.77	2.13
$\beta = 0.6$)		Langzhong CPP	4.1016	0.42	1.49	0	0	2.28

Table 8.9 Sensitivity analysis on the local authority's attitude toward carbon emission reduction targets.

Policy candidate (π, α, β)	Total benefit: $F(x_{ij}, y_i)$ (10^8 RMB)	CPP	Carbon emission allowance: y_i (10^6 tonnes)	Jincheng: x_{i1} (10^6 tonnes)	Shenhua: x_{i2} (10^6 tonnes)	Qujing: x_{i3} (10^6 tonnes)	Panzhihua: x_{i4} (10^6 tonnes)	Benefit: $G_i(x_{ij}, y_i)$ (10^8 RMB)
$(\pi = 0.35,$	8.86	Jintang CPP	9.7	3.5	0.13	0.51	0	6.89
$\alpha = 1,$		Neijiang CPP	5.7	0.65	0.88	0	1.22	3.39
$\beta = 1)$		Langzhong CPP	6.45	0.66	2.35	0	0	3.59
$(\pi = 0.35,$	8.86	Jintang CPP	9.7	3.5	0.13	0.51	0	6.89
$\alpha = 0.95,$		Neijiang CPP	5.7	0.65	0.88	0	1.22	3.39
$\beta = 1)$		Langzhong CPP	6.45	0.66	2.35	0	0	3.59
$(\pi = 0.35,$	8.86	Jintang CPP	9.7	3.5	0.13	0.51	0	6.89
$\alpha = 0.9,$		Neijiang CPP	5.7	0.65	0.88	0	1.22	3.39
$\beta = 1)$		Langzhong CPP	6.45	0.66	2.35	0	0	3.59
$(\pi = 0.35,$	8.53	Jintang CPP	8.8923	3.2	0.12	0.47	0	6.31
$\alpha = 0.85,$		Neijiang CPP	5.7	0.65	0.88	0	1.22	3.39
$\beta = 1)$		Langzhong CPP	6.45	0.66	2.35	0	0	3.59
$(\pi = 0.35,$	8.03	Jintang CPP	7.645	2.7	0.10	0.40	0	5.43
$\alpha = 0.8,$		Neijiang CPP	5.7	0.65	0.88	0	1.22	3.39
$\beta = 1)$		Langzhong CPP	6.45	0.66	2.35	0	0	3.59

($\pi = 0.35$,	7.53	Jintang CPP	6.3974	2.3	0.09	0.34	0	4.54
$\alpha = 0.75$,		Neijiang CPP	5.7	0.65	0.88	0	1.22	3.39
$\beta = 1$)		Langzhong CPP	6.45	0.66	2.35	0	0	3.59
($\pi = 0.35$,	7.03	Jintang CPP	6.07	2.2	0.08	0.32	0	4.31
$\alpha = 0.7$,		Neijiang CPP	4.7957	0.55	0.74	0	1.03	2.85
$\beta = 1$)		Langzhong CPP	6.45	0.66	2.35	0	0	3.59
($\pi = 0.35$,	6.53	Jintang CPP	6.07	2.2	0.08	0.32	0	4.31
$\alpha = 0.65$,		Neijiang CPP	3.6	0.41	0.56	0	0.78	2.14
$\beta = 1$)		Langzhong CPP	6.4195	0.65	2.33	0	0	3.57
($\pi = 0.35$,	6.02	Jintang CPP	6.07	2.2	0.08	0.32	0	4.31
$\alpha = 0.6$,		Neijiang CPP	3.6	0.41	0.56	0	0.78	2.14
$\beta = 1$)		Langzhong CPP	5.1844	0.53	1.89	0	0	2.88

Table 8.10 Sensitivity analysis on the local authority's attitude toward carbon emission reduction targets.

Policy candidate (π, α, β)	Total benefit: $F(x_{ij}, y_i)$ (10^8 RMB)	CPP	Carbon emission allowance: y_i (10^6 tonnes)	Jincheng: x_{i1} (10^6 tonnes)	Shenhua: x_{i2} (10^6 tonnes)	Qujing: x_{i3} (10^6 tonnes)	Panzhihua: x_{i4} (10^6 tonnes)	Benefit: $G_i(x_{ij}, y_i)$ (10^8 RMB)
$(\pi = 0.35,$	8.86	Jintang CPP	9.7	3.5	0.13	0.51	0	6.89
$\alpha = 1,$		Neijiang CPP	5.7	0.65	0.88	0	1.22	3.39
$\beta = 0.95)$		Langzhong CPP	6.45	0.66	2.35	0	0	3.59
$(\pi = 0.35,$	8.39	Jintang CPP	9.4858	3.4	0.13	0.50	0	6.74
$\alpha = 1,$		Neijiang CPP	4.8846	0.56	0.76	0	1.05	2.91
$\beta = 0.9)$		Langzhong CPP	6.3297	0.65	2.30	0	0	3.52
$(\pi = 0.35,$	7.92	Jintang CPP	9.6239	3.5	0.14	0.52	0	6.83
$\alpha = 1,$		Neijiang CPP	5.2418	0.60	0.81	0	1.13	3.12
$\beta = 0.85)$		Langzhong CPP	4.6843	0.48	1.71	0	0	2.60
$(\pi = 0.35,$	7.45	Jintang CPP	9.6857	3.5	0.13	0.51	0	6.88
$\alpha = 1,$		Neijiang CPP	4.5747	0.52	0.71	0	1.00	2.72
$\beta = 0.8)$		Langzhong CPP	4.1397	0.43	1.60	0	0	2.30
$(\pi = 0.35,$	6.98	Jintang CPP	9.5039	3.4	0.12	0.48	0	6.75
$\alpha = 1,$		Neijiang CPP	3.6308	0.41	0.56	0	0.78	2.16
$\beta = 0.75)$		Langzhong CPP	4.1153	0.42	1.49	0	0	2.29

($\pi = 0.35$,	6.52	Jintang CPP	8.3556	3.1	0.11	0.44	0	5.93
$\alpha = 1$,		Neijiang CPP	3.6308	0.41	0.56	0	0.78	2.16
$\beta = 0.7$)		Langzhong CPP	4.1153	0.42	1.49	0	0	2.29
($\pi = 0.35$,	6.06	Jintang CPP	7.2063	2.6	0.09	0.38	0	5.12
$\alpha = 1$,		Neijiang CPP	3.6308	0.41	0.56	0	0.78	2.16
$\beta = 0.65$)		Langzhong CPP	4.1153	0.42	1.49	0	0	2.29
($\pi = 0.35$,	5.60	Jintang CPP	6.0986	2.2	0.08	0.32	0	4.32
$\alpha = 1$,		Neijiang CPP	3.6308	0.41	0.56	0	0.78	2.16
$\beta = 0.6$)		Langzhong CPP	4.1153	0.42	1.49	0	0	2.29

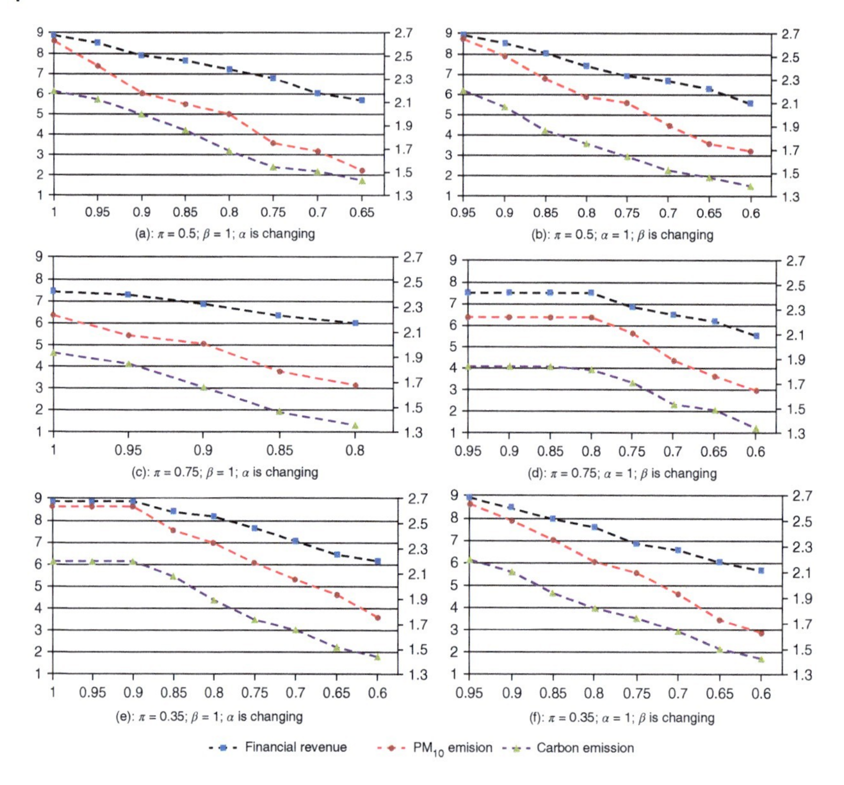

Figure 8.2 Changing trends of sensitive analysis under different dimensions. The financial revenue belongs to the left vertical axis and its unit is 10^8 RMB; the PM_{10} emission and the carbon emission are belong to the right vertical axis and the unit of the PM_{10} emission is 10^6 kg, the unit of the carbon emission is 10^7 tonnes.

increase ratio also being lower than that of the emissions. Similar situations can also be seen other scenarios, which are shown graphically in Figure 8.2. In Figure 8.2, it can be seen that the slope rate for the changing total financial revenue trend is smaller than that of the emissions, indicating that the financial revenue marginal growth rate is lower than the emissions marginal growth rate. From this analysis, it can be concluded that while relatively relaxed environmental protection constraints increase revenue, more damage is caused to the environment.

In addition, we further found that CPPs show different sensitivities toward the changing environmental protection constraints. When the authority changes its attitude toward the environmental protection constraints, the CPPs have different feedback. Take Table 8.6 as an example; in this scenario, if the authority sets the strictest environmental protection constraints (i.e. $\beta = 0.6$), the carbon emissions allowance that each CPP receives is 6.0908×10^6 tonnes, 3.6013×10^6 tonnes, and 4.1079×10^6 tonnes, respectively. When the authority relaxes its attitude toward environmental protection (i.e. sets $\beta = 0.65$), all the three CPPs' carbon emissions allowances will be increased by 15.5%, 5.6%, and 0.7%, respectively. Therefore, the individual CPP sensitivities toward these changes are quite different. When the value of β changes from 0.65 to 0.7, a similar situation can be seen, with the carbon emissions allowances

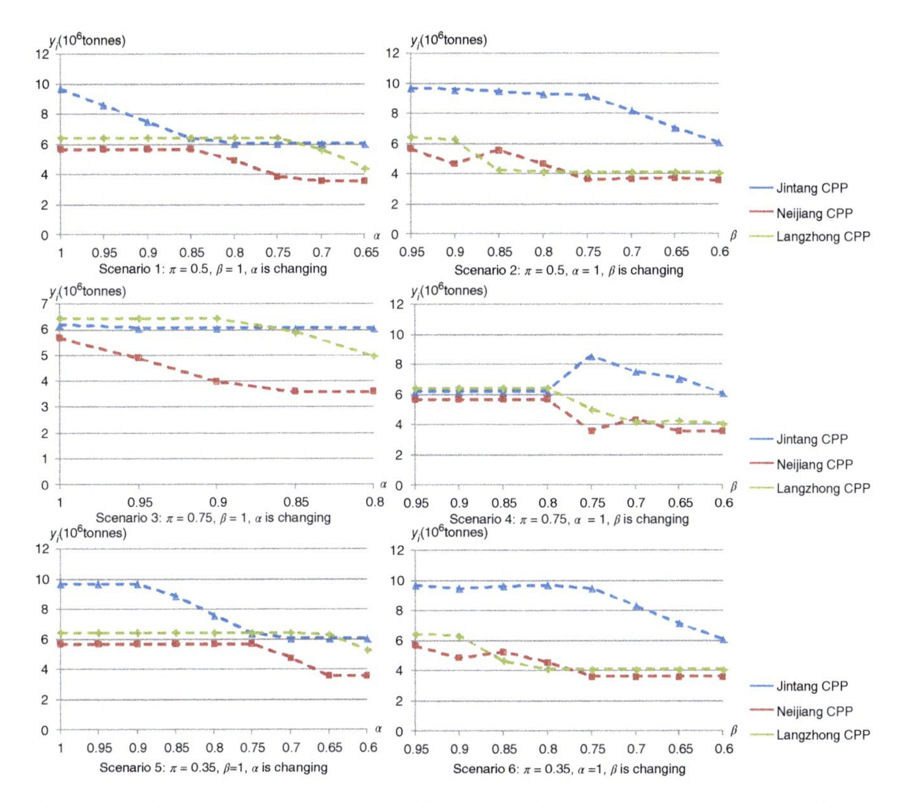

Figure 8.3 CPPs' sensitive toward the changing policy control parameter under different scenarios.

ratio increase at the three CPPs being 17.4%, −1.8%, and −0.2%, respectively. When facing such changes, therefore, the carbon emissions allowances at Neijiang CPP and Langzhong CPP decrease, which demonstrates that when an authority has such an attitude (i.e. $\pi = 0.5$, $\alpha = 1$, and $\beta = 0.7$), the Jintang CPP has an advantage when competing for carbon emissions allowances. This situation can also be found in the other scenarios, as shown graphically in Figure 8.3.

The results of different scenarios demonstrated that the robustness of the proposed method is satisfied and the coal blending ratio of each CPP fluctuates within a tiny range. When the authority changes its attitudes toward the environmental protection constraints under a certain scenario, the coal blending ratio at each CPP fluctuates; however, the fluctuation range is very small, as shown in Tables 8.5–8.10. Take scenario 6 as an example (also shown in Table 8.10); it can be seen that under this scenario, no matter what environmental protection preferences are the set by the authority, the Jintang CPP always mainly depends on coal from the Jincheng colliery which blends with coal from the Shenhua colliery and the Qujing colliery, but not coal from the Panzhihua colliery. The blending ratios for coal from Jincheng, Shenhua, and Qujing collieries fluctuate between 84.5% and 85% and 2.9% to 3.1% and 12.1% to 12.3%, respectively. The Neijiang CPP purchases coal from the Jincheng, Shenhua, and Panzhihua collieries but not from the Qujing colliery. The blending ratios of the coal from these three collieries at Neijiang CPP fluctuate from 23.3%

Figure 8.4 Blending ratio of each CPP under different scenarios.

to 23.6%, 31.8% to 32.1%, and 44.3% to 44.8%, respectively. he Langzhong CPP only depends on coal from the Jincheng and Shenhua collieries with the blending ratio fluctuating from 21.2% to 22.1% and 77.9% to 78.1%, respectively. Similar situation can also be found in other scenarios (as shown in Figure 8.4). These results demonstrate that the proposed method in this chapter has relatively satisfactory robustness.

8.3.3 Management Recommendations

Based on what has been discussed and analyzed above, in this section, some management recommendations are proposed to work as the foundation of the actions to control the CPPs emissions.

First of all, for regions that largely depend on the electricity generated by CPPs, carbon emissions allocation competition mechanisms should be established. Without such a scheme, the CPPs would pursue the highest possible benefit and only ensure that their emissions performance fulfilled the minimum environmental protection regulations, as they would lack the motivation to continuously improve their emissions performance. Using the proposed methodology in this chapter, a carbon emissions allowance quota competition mechanism could be built. As discussed above, such a mechanism could guide CPPs to conduct environmental-friendly production. In addition, carbon emissions should not be the only criterion that is highlighted when the authority buildingthe carbon emission allowance competition mechanisms, other air pollutants should be paid enough attention. In this chapter, PM_{10} emissions reduction was set as the objective of the authority and the NO_x and SO_2 emissions were set as constraints, which made it is possible for the proposed methodology to combine a reduction in both carbon and PM_{10} emissions as well as keeping the other air pollutant emissions within environmental regulations.

Second, the authority can select their own carbon emissions quota allocation mechanism using the proposed model; that is to say, the authority can fully consider their actual situation and choose a desired air pollutant reduction scheme. However, a relaxed environmental protection strategy could result in higher marginal environmental costs and more serious environmental damage. Therefore, we recommend that for developed regions, the authority should have the strictest attitude when using the proposed model (i.e. set the environmental protection policy control parameter α, β, and γ at the strictest level). For developing regions, the authority can gradually tighten up the environmental protection attitude. In other words, the authority in developing regions should have a dynamic attitude when using the model to solve the CPPs emissions problems. For example, to encourage economic development, the authority can set the relative loose emissions reduction targets at the beginning of using the proposed methodology; however, after a period of time, they should change their attitude toward the environmental protection and upgrade the requirements by tightening up the environmental protection policy control parameter. In the next period, the requirements should be upgraded again, finally reaching the most strict environmental protection requirements.

References

Abou-Kandil, H. and Bertrand, P. (1987). Government-private sector relations as a Stackelberg game: a degenerate case. *Journal of Economic Dynamics and Control* 11 (4): 513–517.

Angulo, E., Castillo, E., García-Ródenas, R., and Sánchez-Vizcaíno, J. (2014). A continuous bi-level model for the expansion of highway networks. *Computers & Operations Research* 41: 262–276.

Baek, S.H., Park, H.Y., and Ko, S.H. (2014). The effect of the coal blending method in a coal fired boiler on carbon in ash and NO_x emission. *Fuel* 128: 62–70.

BP Global (2015). BP Statistical Review of World Energy.

Colson, B., Marcotte, P., and Savard, G. (2007). An overview of bilevel optimization. *Annals of Operations Research* 153 (1): 235–256.

Dai, C., Cai, X.H., Cai, Y.P., and Huang, G.H. (2014). A simulation-based fuzzy possibilistic programming model for coal blending management with consideration of human health risk under uncertainty. *Applied Energy* 133: 1–13.

Dubois, D. and Prade, H. (1983). Ranking fuzzy numbers in the setting of possibility theory. *Information Sciences* 30 (3): 183–224.

European Environment Agency (2010). European Union Emission Inventory Report 1990 to 2008 Under the Unece Convention on Long-Range Transboundary Air Pollution.

Franco, A. and Diaz, A.R. (2009). The future challenges for "clean coal technologies": joining efficiency increase and pollutant emission control. *Energy* 34 (3): 348–354.

Goto, K., Yogo, K., and Higashii, T. (2013). A review of efficiency penalty in a coal-fired power plant with post-combustion CO_2 capture. *Applied Energy* 111: 710–720.

He, K., Huo, H., and Zhang, Q. (2002). Urban air pollution in China: current status, characteristics, and progress. *Annual Review of Energy and the Environment* 27 (1): 397–431.

Heinberg, R. and Fridley, D. (2010). The end of cheap coal. *Nature* 468 (7322): 367.

Howell, R.A. (2012). Living with a carbon allowance: the experiences of carbon rationing action groups and implications for policy. *Energy Policy* 41: 250–258.

Hu, Y., Li, H., and Yan, J. (2014). Numerical investigation of heat transfer characteristics in utility boilers of oxy-coal combustion. *Applied Energy* 130: 543–551.

International Energy Agency (2015). Key trends in CO_2 emissions from fuel combustion.

Kauppinen, E.I. and Pakkanen, T.A. (1990). Coal combustion aerosols: a field study. *Environmental Science and Technology* 24 (12): 1811–1818.

Li, W., Huang, G.H., Dong, C., and Liu, Y. (2013). An inexact fuzzy programming approach for power coal blending. *Journal of Environmental Informatics* 21 (2): 112–118.

Li, S., Xu, T., Hui, S., and Wei, X. (2009). NO_x emission and thermal efficiency of a 300 MWe utility boiler retrofitted by air staging. *Applied Energy* 86 (9): 1797–1803.

Liu, Y., Zhang, J., Sheng, C. et al. (2010). Simultaneous removal of NO and SO_2 from coal-fired flue gas by UV/H_2O_2 advanced oxidation process. *Chemical Engineering Journal* 162 (3): 1006–1011.

Lonsdale, C.R., Stevens, R.G., Brock, C.A. et al. (2012). The effect of coal-fired power-plant SO_2 and NO_x control technologies on aerosol nucleation in the source plumes. *Atmospheric Chemistry and Physics* 12 (23): 11519–11531.

Mao, X.Q., Zeng, A., Hu, T. et al. (2014). Co-control of local air pollutants and CO_2 from the Chinese coal-fired power industry. *Journal of Cleaner Production* 67: 220–227.

Meij, R. and Te Winkel, B. (2004). The emissions and environmental impact of PM_{10} and trace elements from a modern coal-fired power plant equipped with ESP and wet FGD. *Fuel Processing Technology* 85 (6–7): 641–656.

Mukherjee, S. and Borthakur, P.C. (2001). Chemical demineralization/desulphurization of high sulphur coal using sodium hydroxide and acid solutions. *Fuel* 80 (14): 2037–2040.

Nowak, W. (2003). Clean coal fluidized-bed technology in Poland. *Applied Energy* 74 (3–4): 405–413.

Pavlish, J.H., Sondreal, E.A., Mann, M.D. et al. (2003). Status review of mercury control options for coal-fired power plants. *Fuel Processing Technology* 82 (2–3): 89–165.

Samimi, A. and Zarinabadi, S. (2012). Reduction of greenhouse gases emission and effect on environment. *Journal of American Science* 8 (8): 1011–1015.

Sherali, H.D., Soyster, A.L., and Murphy, F.H. (1983). Stackelberg-Nash-Cournot equilibria: characterizations and computations. *Operations Research* 31 (2): 253–276.

Shih, J.-S. and Frey, H.C. (1995). Coal blending optimization under uncertainty. *European Journal of Operational Research* 83 (3): 452–465.

Sinha, A., Malo, P., Frantsev, A., and Deb, K. (2014). Finding optimal strategies in a multi-period multi-leader–follower Stackelberg game using an evolutionary algorithm. *Computers & Operations Research* 41: 374–385.

Smoot, L.D. and Smith, P.J. (2013). *Coal Combustion and Gasification*. Springer Science & Business Media.

Sun, R., Liu, G., Zheng, L., and Chou, C.-L. (2010). Characteristics of coal quality and their relationship with coal-forming environment: a case study from the Zhuji exploration area, Huainan coalfield, Anhui, China. *Energy* 35 (1): 423–435.

Tang, L., Wu, J., Yu, L., and Bao, Q. (2015). Carbon emissions trading scheme exploration in China: a multi-agent-based model. *Energy Policy* 81: 152–169.

Tian, H., Wang, Y., Cheng, K. et al. (2012). Control strategies of atmospheric mercury emissions from coal-fired power plants in China. *Journal of the Air and Waste Management Association* 62 (5): 576–586.

Tola, V. and Pettinau, A. (2014). Power generation plants with carbon capture and storage: a techno-economic comparison between coal combustion and gasification technologies. *Applied Energy* 113: 1461–1474.

Tu, Y., Zhou, X., Gang, J. et al. (2015). Administrative and market-based allocation mechanism for regional water resources planning. *Resources, Conservation and Recycling* 95: 156–173.

U.S. Environmental Protection Agency (2014). Profile of the 2011 National Air Emissions Inventory.

Vicente, L.N. and Calamai, P.H. (1994). Bilevel and multilevel programming: a bibliography review. *Journal of Global Optimization* 5 (3): 291–306.

Wang, L., Jang, C., Zhang, Y. et al. (2010). Assessment of air quality benefits from national air pollution control policies in China. Part II: Evaluation of air quality predictions and air quality benefits assessment. *Atmospheric Environment* 44 (28): 3449–3457.

Wang, Q., Zhang, L., Sato, A. et al. (2008). Effects of coal blending on the reduction of PM_{10} during high-temperature combustion 1. Mineral transformations. *Fuel* 87 (13–14): 2997–3005.

Wolde-Rufael, Y. (2010). Coal consumption and economic growth revisited. *Applied Energy* 87 (1): 160–167.

Wu, M., Nakano, M., and She, J.-H. (1999). A model-based expert control strategy using neural networks for the coal blending process in an iron and steel plant. *Expert Systems with Applications* 16 (3): 271–281.

Xia, J., Chen, G., Tan, P., and Zhang, C. (2014). An online case-based reasoning system for coal blends combustion optimization of thermal power plant. *International Journal of Electrical Power & Energy Systems* 62: 299–311.

Xu, J. and Zhou, X. (2011). *Fuzzy-Like Multiple Objective Decision Making*, vol. 263. Springer.

Xu, X., Chen, C., Qi, H. et al. (2000). Development of coal combustion pollution control for SO_2 and NO_x in China. *Fuel Processing Technology* 62 (2–3): 153–160.

Xu, J., Song, X., Wu, Y., and Zeng, Z. (2015). GIS-modelling based coal-fired power plant site identification and selection. *Applied Energy* 159: 520–539.

Yang, C.-J., Xuan, X., and Jackson, R.B. (2012). China's coal price disturbances: observations, explanations, and implications for global energy economies. *Energy Policy* 51: 720–727.

Yue, L. (2012a). Dynamics of clean coal-fired power generation development in China. *Energy Policy* 51: 138–142.

Yue, L. (2012b). Dynamics of clean coal-fired power generation development in China. *Energy Policy* 51: 138–142.

Zeng, Z., Xu, J., Wu, S., and Shen, M. (2014). Antithetic method-based particle swarm optimization for a queuing network problem with fuzzy data in concrete transportation systems. *Computer-Aided Civil and Infrastructure Engineering* 29 (10): 771–800.

9

Equilibrium Biomass–Coal Blending Method Toward Carbon Emissions Reduction

The 2017 BP statistical review of world energy stated that coal still occupied the proportion of 28.1%, situated in the second place after oil in global primary energy consumption. However, heavy reliance on coal results in a great amount of carbon emissions and gives rise to global climate change [Pain, 2017, Basu, 1999, Li et al., 2009]. 2017 U.N. environment meeting report claimed that atmospheric environment existed the largest global environmental risk, as the global carbon emissions have risen from 16 gigatonnes in 1970 to 36.25 gigatonnes in 2015, almost increased by 126.6 % [UNE Programme, 2017, Olivier et al., 2012]. Although fossil fuel power accounts for two-thirds of the global power, they are the major sources of carbon emissions with coal-fired power plant (CPP) being the biggest sources [BP Global, 2015, Service, 2017]. Therefore, the carbon emissions from CPPs are in urgent need of reduction.

9.1 Background Review

Relevant carbon emissions reduction studies are mainly from hard-technology concentrated on technologies, and from soft-technology focused on optimization methods or policy controls to reduce carbon emission [Lv et al., 2016]. As for hard-technology, Mao claimed that front-end control measures such as coal washing, in-the-process control measures such as retrofitting the stream turbine flow passage, and end-of-pipe control measures such as carbon capture and storage/sequestration were all technical reduction measures after investgations in different Chinese CPPs [Mao et al., 2014]. Amitava reported on an effective chemical solvent scrubbing method to remove carbon dioxide from the flue gases emitted from the power plants, which can minimize energy requirements, equipment size as well as corrosion [Bandyopadhyay, 2011]. Low suggested using TiO_2 to achieve carbon reduction due to its ability to increase the carbon dioxide adsorption and activation ability of TiO_2 for carbon emissions reduction [Low et al., 2017]. Although the carbon emissions reduction effect of hard technologies is remarkable, it is generally expensive, especially for the developing countries who heavily depend on coal for power generation [Xu et al., 2017]. Therefore, soft-technology could be a more favored option [Goto et al., 2013, Xu et al., 2015b, Li, 2012].

Innovative Approaches towards Ecological Coal Mining and Utilization, First Edition.
Jiuping Xu, Heping Xie, and Chengwei Lv.

Soft-technology proposed optimization method or policy-making such as blending, carbon tax levies, and carbon emissions trading schemes [Chen et al., 2015, Dai et al., 2014, Shih and Frey, 2007, Yearwood-Lee, 2015, Van Den Broek et al., 1996]. Lv et al. proposed a bilevel programming method-based coal blending method to reduce combined carbon and PM_{10} emissions in large-scale CPPs which can also instruct CPPs to conduct environmental-friendly production [Lv et al., 2016]. Liu explored the development state of the Chinese carbon-trading market and examined its driving and then suggested specific measures which include building a reverse verification mechanism and improving the trading mechanism for the promotion of a Chinese carbon-trading market [Liu et al., 2015a]. As the world is reducing the share of coal in the energy mix to mitigate the carbon emissions, biomass–coal blending method which cannot only achieve economic rationality but also reduce carbon emissions provides the simplest, most practicable, and most cost-effective method of all soft-technology methods [Qi et al., 2016, Baxter, 2005, Li et al., 2012, Sondreal et al., 2001, Tilman et al., 2009]. Eksioglu and and Karimi developed a nonlinear optimization model which captures the additional costs and savings, loss of process efficiencies due to co-firing in support of biomass co-firing decisions for CPPs, and then from the cases study of Mississippi and South Carolina they found that the tax rate should be customized by region to optimize renewable energy production [Eksioglu and Karimi, 2014]. Yilmaz developed a fuzzy multiobjective strategy-based decision-making (DM) model which was already proved the viability to design the most profitable supply chain of biomass by considering inherent uncertainties and can help the authority designate the tax and subsidy strategies toward bioenergy [Yilmaz and Selim, 2015]. However, currently, the greatest challenge associated with biomass–coal blending is technical and logistical issues which involves fuel types, fuel properties, fuel cost, and boiler types [Agbor et al., 2014]. Tillman and Dai investigated the influences of biomass blending during the combustion process and found that any biomass mixed with coal needed to have acceptable physical properties to furnaces without occurrence of fouling and corrosion and impact on performance for the furnaces [Tillman, 2000, Dai et al., 2008]. Sahu and Moon found that a 10% biomass blend ratio was the peak mixing ratio of direct co-firing which could improve combustion efficiency and be acceptable without no irresolvable issues and no effect to combustion reactivities of the individual fuels [Sahu et al., 2014, Moon et al., 2013]. Although these studies have made some progress in reducing carbon emissions, the realistic situation is more complicated and further improvements are still necessary.

Previous biomass–coal blending methodological studies have tended to involve only a single decision-maker, CPP; however in reality, there are several decision entities involved in carbon emissions reduction; the local authority, who has a higher decision-making (DM) position, and the power plants, which occupy the lower DM position, which often have conflicts due to their differing views on financial revenue and enterprise profits [Xu et al., 2016a,2015a]. When making decisions, the local authority is most concerned about environmental protection and total economic development, while the CPPs are seeking their own maximum profits. In these types of situation, a Stackelberg–Nash equilibrium strategy is employed to this intractable problem between several decision-makers. In addition, secure and long-term supplies of biomass feedstock are critical to biomass–coal blending in CPPs [Gielen, 2012]. Previous biomass–coal blending method research has paid greater attention to biomass feedstock supply from a short-term perspective; however, in practice, solving biomass feedstock seasonal variations has proven difficult; that is, in the harvest months, a greater amount of biomass can be blended with coal due to the low price while at other times of the year, less, more expensive biomass is available, which infers that there are inventory management timing conflicts. Under these conditions, a dynamic strategy can be applied, in which optimal decisions are taken in each stage to ensure that enterprises are able to maximize profits from a long-term perspective [Yang and Zhang, 2014]. Therefore, in this chapter, we apply dynamic equilibrium strategy to seek the equilibrium between the authority and the CPPs as well as solving time conflicts in biomass feedstock availability. Dynamic equilibrium strategy has been widely used to solve many troublesome problems. Xu et al. used dynamic equilibrium strategy to solve the conflicts between drought emergency temporary water transfer and allocation management in order to mitigate severe drought and is appropriate for the situation where there are water transfer projects with multiple donor reservoirs and one recipient reservoir [Xu et al., 2016b]. Yang et al. employed dynamic equilibrium strategy to capture the welfare effects of interregional subsidy game for renewable energy investment between two neighboring regions and then found that the optimal subsidy strategies for both regions and the corresponding scale of renewable energy investment relied on plenty of factors after comparative analysis [Yang et al., 2017]. What's more, biomass–coal blending is quite difficult to solve because of the many uncertain, imprecise, and vague factors that influence decision-makers; for example, emissions factors have considerable uncertainties because of the lack of actual measurements [Liu et al., 2015b]. Therefore, uncertainty theory is included to deal with the parameter uncertainty problems [Xu and Zhou, 2011]. Based on the above discussion, this chapter proposes a comprehensive methodology that integrates bilevel programming method, dynamic programming method, and multiobjective programming method to solve conflicts between the authority and the CPPs, settle time conflicts in biomass feedstock availability, and balance the trade-off between environmental protection and economic development under a fuzzy environment.

9.2 Key Problem Statement

Based on a biomass–coal blending method, this chapter uses dynamic equilibrium strategy to solve multiple conflicts and achieve economic and environmental coordination, the specific details, basic background, and descriptions for which are as follows.

In practice, many decision entities are involved in regional CPPs carbon emissions reduction. As different decision-makers make different decisions and have different objectives, conflicts such as the financial revenue and enterprise profits are inevitable [Xu et al., 2015b]. The authority, which has the higher decision-making priority and has the ability to move first, attaches greater importance to the environment to protect the general public, but also needs to assure a certain financial revenue. As there is a trade-off between economic development and environmental protection, the authority develops carbon emissions allowance quota schemes to distribute the allocations to the CPPs. The CPPs, which are independent of each other, focus more on profit maximization. Under the carbon emissions allowance allocated by the authority, each CPP reacts in a rational way to develop their biomass–coal blending schemes. For each possible decision of the authority, each CPP has its own reactions [Zhang et al., 2015]. However, the authority is also affected by the coal-fired power plants (CCPs)' decisions through feedback, at which time it reconsiders its initial decisions and makes some further adjustments [Lv et al., 2016]. This process is repeated several times until a final scheme that is acceptable to the authority and all the CPPs. Therefore, the relationship between the upper-level authority and the lower-level CPPs are similar to a Stackelberg game. Bilevel programming is needed to find the solution to the Stackelberg–Nash equilibrium [Colson et al., 2007]. The key difference between bilevel programming problems and other optimization problems is in their nested structure. A bilevel programming encompasses another optimization task within the constraints denominated as the lower-level problem of the outer problem termed as the upper level [Sinha et al., 2014]. Therefore, a multiobjective bilevel programming method can be used to address the conflicts between the superior authority and the inferior CPPs and seek the trade-off between economic development and environmental protection.

Even though the cumulative carbon emissions from bioenergy are smaller than from traditional fossil fuels, the heavy use of biomass is still difficult because biomass feedstock vary with the months, meaning that there is a fluctuation in the quantities available; therefore, a dynamic strategy can be employed to resolve such timing conflicts [Pehl et al., 2017, Tilman et al., 2009]. In the real world, the CPPs usually decide on fuel resource quantities depending on price and availability and for safety, they usually guarantee a certain amount to be stored in the warehouses; therefore, when the price is lower, the CPPs buy additional fuel resources to store; however, as there is a limit to the amount that can be stored they can only stockpile to the limit of their storage. When the price is higher, the CPPs buy less or none and use the stockpiled fuel resource, as shown in Figure 9.1. Therefore, to achieve lower costs, a dynamic strategy is appropriate for handling the biomass feedstock fluctuations.

From the discussion above, a dynamic equilibrium strategy method is needed in biomass–coal blending method and a bilevel dynamic multiobjective optimization model is developed in Section 9.3 to resolve the conflicts between the authority and the CPPs, address the biomass time conflicts, and search for the economic development and environmental protection trade-off.

9.3 Modeling

The detailed modeling process will be introduced in this section.

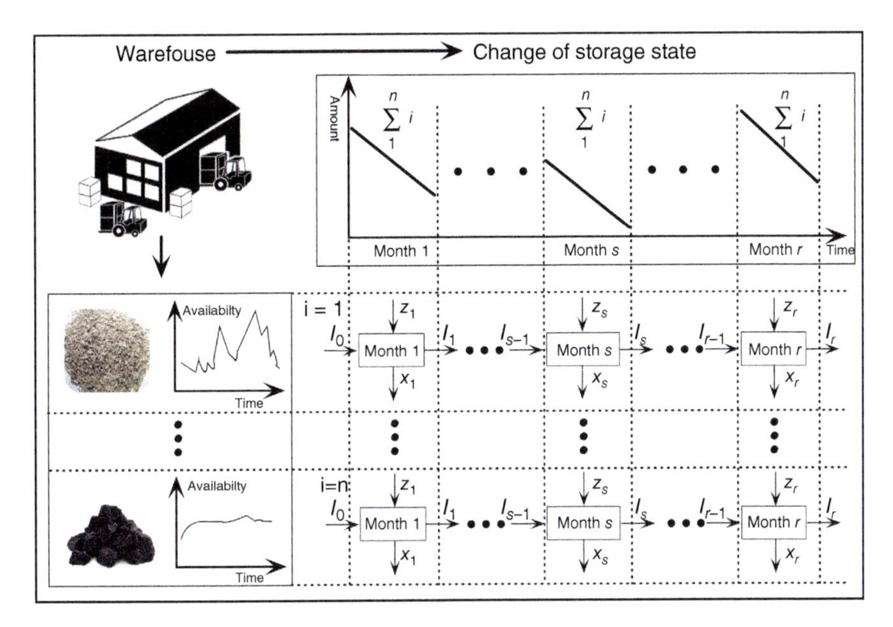

Figure 9.1 Dynamic inventory management.

9.3.1 Assumption

(1) It is assumed that blending at low ratios does not pose any threat or major problems to the boiler operation [Basu et al., 2011].
(2) It is assumed that volatile matter of fuels in the burners can be completely burnt [Sami et al., 2001].
(3) It is assumed that the added-value tax rate that CPPs pay is fixed.

9.3.2 Notations

The following notations will be employed in this chapter.

Indices

j = Fuel index, $j = 1, 2, \dots, l, \dots, J$, where $1 - l$ represent biomass, $l - J$ represent coal.
i = CPP index, $i = 1, 2, \dots, I$.
p = Pollution index, $p = 1, 2, \dots, P$.
q = Fuel quality index, $q = 1, 2, \dots, Q$.
m = Month index, $m = 1, 2, \dots, M$.

Certain parameters

ε = Price of a unit of carbon emissions quota.

(Continued)

(Continued)

Q = Price of a unit of power.
A_{ijs} = Procurement price of a unit of fuel j in s month at CPP i.
ϖ = Added-value tax.
U_{ijs} = Maximum available quantity of fuel j in s month at CPP i.
D_{is} = Amount of power needed to be generated in s month by CPP i.
θ = Upper percent of biomass in blended fuel.
O_i^L, O_i^U = Lower and upper bounds for the annual carbon emissions that CPP i can carry.
V_{ip} = Unit operating costs to reduce pollutant p at CPP i.
K = Storage price monthly per unit.
L_A = Actual carbon emissions in the last production period.
I_i^U = Upper storage ability of CPP i.
I_{ijs} = Inventories of fuel j at the end of s month for CPP i.

Uncertain parameters

$\widetilde{P_{ij}}$ = Conversion parameters from a unit of fuel j to power at CPP i.
$\widetilde{M_{jp}}$ = Pollutant P from a unit of fuel i.
$\widetilde{H_{ij}}$ = Carbon emissions from a unit of fuel i at CPP i.
$\widetilde{Q_{jq}}$ = Quality t of fuel j.
$\widetilde{\eta_j}$ = Burn-out fraction of fuel j char.
$\widetilde{T_{iq}^L}, \widetilde{T_{iq}^U}$ = Lower and upper bounds for fuel blend quality q at CPP i.

Control parameters

β = Attitude of the authority toward carbon emissions reduction.
ψ = Attitude of the authority toward the gap between the quota and actual value.
∂ = Combustion efficiency.

Decision variables

x_{ijs} = Fuel i used in s month for CPP i.
z_{ijs} = Fuel i purchased in s month for CPP i.
y_i = Carbon emissions quota that the authority allocates to CPP i.

9.3.3 Model for the Local Authority

9.3.3.1 Objective 1: Maximizing Financial Revenue

Because it is hard to tackle uncertainties in realistic decision-making problems, uncertainty theory is applied to measure the fuzzy variables. $\widetilde{P_{ij}}$ varies from a minimum value r_{11} to a maximum value r_{14}, with the more likely values being from r_{12} and r_{13}, where $r_{11} \leq r_{12} \leq r_{13} \leq r_{14}$. As these fuzzy variables cannot be calculated directly, the expected value operation presented by Xu and Zhou as Figure 9.2 is used [Xu and Zhou, 2011]. In this chapter, uncertain parameters can be described as:

$$\widetilde{P_{ij}} \to E[\widetilde{P_{ij}}] = \frac{1-\theta}{2}(r_{11} + r_{12}) + \frac{\theta}{2}(r_{13} + r_{14}) \tag{9.1}$$

The authority imposes financial revenue on the CPPs to exercise its responsibility. In this chapter, financial revenue consists of added-value tax income and fees on carbon emissions quotas. Let ϖ be added-value tax rate and $\sum_{j=1}^{J}\sum_{s=1}^{S} QE[\widetilde{P_{ij}}]x_{ijs}$ be the profits of CPP i, so the total added-value tax that the authority receives is $\sum_{i=1}^{I}\sum_{j=1}^{J}\sum_{s=1}^{S} \varpi QE[\widetilde{P_{ij}}]x_{ijs}$. Then let ε be price of a unit of carbon emissions quotas and $\sum_{i=1}^{I} y_i$ be the total carbon emissions allowances, so fees on carbon emissions quotas are $\varepsilon \sum_{i=1}^{I} y_i$. Combining these, this objective can be described as:

$$\max F_1 = \sum_{i=1}^{I}\sum_{j=1}^{J}\sum_{s=1}^{S} \varpi QE[\widetilde{P_{ij}}]x_{ijs} + \varepsilon \sum_{i=1}^{I} y_i \tag{9.2}$$

9.3.3.2 Objective 2: Minimizing Carbon Emissions

The authority has the responsibility to protect the environment as it stands for the public. Let y_i be carbon emissions quota that the authority allocates to CPP i. Thus, the total carbon emission quotas are $\sum_{i=1}^{I} y_i$.

$$\min F_2 = \sum_{i=1}^{I} y_i \tag{9.3}$$

9.3.3.3 Limitation on the CPPs' Operations

Each CPP's operation rights should be guaranteed. Let O_i^L be the minimal carbon emissions when CPP i achieves balance of payment, and y_i should exceed this minimum value. Let O_i^L be the maximal carbon emissions when CPP i is under full-load production, and y_i should be below this maximum value. Therefore, the constraints can be formulated as:

$$O_i^L \leq y_i \leq O_i^U \tag{9.4}$$

9.3.3.4 Power Supply Demand Restriction

The power supply is supposed to meet the needs in the region. Let C_i be conversion parameter for a unit of carbon emissions to power at CPP j, and $C_i y_i$ be the power provided by CPP i. The total power supply is $\sum_{i=1}^{I} C_i y_i$ and that must be more than the power needs in the region D. This constraint is formed as:

$$\sum_{i=1}^{I} C_i y_i \geq D \tag{9.5}$$

9.3.3.5 Limitation on the Different Between the Quota and the Actual Emission

To ensure rational and equal allocation, the authority pays attention to the difference between the quota and the actual emission. Let $y_i - \sum_{j=1}^{J}\sum_{s=1}^{S} E[\widetilde{H_{ij}}]x_{ijs}$ be the

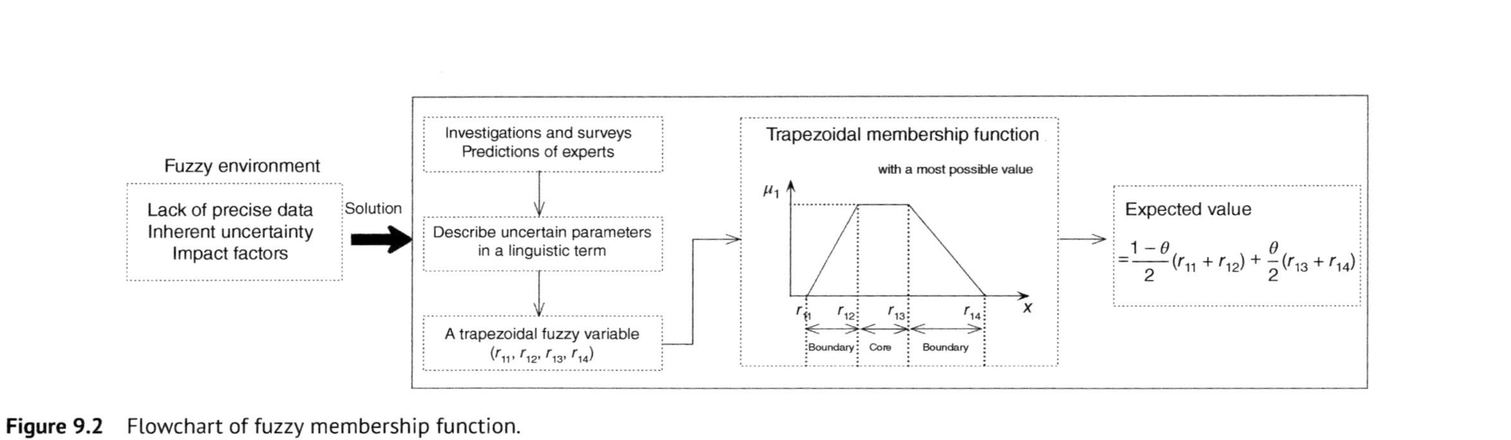

Figure 9.2 Flowchart of fuzzy membership function.

difference between the quota and the actual emission; therefore, this constraints can be written as follows:

$$\frac{y_i - \sum_{j=1}^{J} \sum_{s=1}^{S} E[\widetilde{H_{ij}}]x_{ijs}}{y_i} \leq \psi \tag{9.6}$$

9.3.4 Model for CPPs

9.3.4.1 Objective: Maximizing Economic Benefits

The major objective of CPPs is to obtain maximum economic benefits; that is, they prefer to generate more power using the cheapest fuels under the limited carbon emissions quota. Let Q be price of a unit of power and $\sum_{j=1}^{J} \sum_{s=1}^{S} E[\widetilde{P_{ij}}]x_{ijs}$ be the power generated by CPP i, so the income from power generation is $\sum_{j=1}^{J} \sum_{s=1}^{S} QE[\widetilde{P_{ij}}]x_{ijs}$. Then after-tax income is $\sum_{j=1}^{J} \sum_{s=1}^{S}(1 - \varpi)QE[\widetilde{P_{ij}}]x_{ijs}$. As for cost, it can be classified into fuel procurement costs, fuel storage costs, pollutants treatment costs, and fees on carbon emissions quotas. Let z_{ijs} be purchase quantity and A_{ijs} be procurement price of fuel j in s month, so the fuel procurement costs are $\sum_{j=1}^{J} \sum_{s=1}^{S} A_{ijs}z_{ijs}$. Moreover, let I_{ijs} be inventory in s month and K be the monthly price of a unit of fuel, thus the fuel storage costs are $K \sum_{j=1}^{J} \sum_{s=1}^{S} I_{jis}$. Additionally, let $\widetilde{M_{jp}}$ be pollutant p from fuel j and V_{ip} be operating costs for treating pollutant p at CPP i, so the total pollutants treatment costs are $\sum_{j=1}^{J} \sum_{s=1}^{S} \sum_{p=1}^{P} E[\widetilde{M_{jp}}]V_{ip}x_{ijs}$. The fees on carbon emissions quota are εy_i. Therefore, the profit of CPP i can be formulated as:

$$\begin{aligned} \max f = &\sum_{j=1}^{J} \sum_{s=1}^{S} (1 - \varpi)QE[\widetilde{P_{ij}}]x_{ijs} - \sum_{j=1}^{J} \sum_{s=1}^{S} A_{ijs}z_{ijs} - K\sum_{j=1}^{J} \sum_{s=1}^{S} I_{jis} \\ &- \sum_{j=1}^{J} \sum_{s=1}^{S} \sum_{p=1}^{P} E[\widetilde{M_{jp}}]V_{ip}x_{ijs} - \varepsilon y_i \end{aligned} \tag{9.7}$$

9.3.4.2 Combustion Efficiency Constraint

The best combustion efficiently retrieves energy from the fuels with the least amount of loss. Due to relatively lower heat value and higher moisture content of biomass, possible loss in overall efficiency may occur to biomass–coal blend. Sami et al. investigated that combustion efficiency is typically limited by the extent of char combustion as biomass have higher volatile matter that is completely burnt [Sami et al., 2001]. Let $\widetilde{Q_{jq}}(q = 1)$ be the volatile matter content and $\widetilde{\eta_j}$ be the burn-out fraction of fuel j char, so $\sum_{j=1}^{J} E[\widetilde{Q_{jq}}]x_{ijs} + \sum_{j=1}^{J} E[\widetilde{\eta_j}](1 - E[\widetilde{Q_{jq}}])x_{ijs}$ is the burnt content. ∂ is combustion efficiency, the co-combustion efficiency of the blended fuel should exceed ∂. Therefore, this restriction is described as:

$$\frac{\sum_{j=1}^{J} E[\widetilde{Q_{jq}}]\, x_{ijs} + \sum_{j=1}^{J} E[\widetilde{\eta_j}](1 - E[\widetilde{Q_{jq}}])\, x_{ijs}}{\sum_{j=1}^{J} x_{jis}} \geq \partial \tag{9.8}$$

9.3.4.3 Limitations on Fuel Quantities and Qualities

Let U_{ijs} be maximum supply quantity of fuel i in s month at CPP i, so CPP i can buy U_{ijs} at most. In every month, the usage amount of fuel i cannot exceed the purchase amount. The usage amount and the purchase amount cannot be negative. Therefore, this constraint is formulated as:

$$0 \leq x_{ijs} \leq z_{ijs} \leq U_{ijs} \tag{9.9}$$

The qualities of biomass dramatically differ from coal in the aspect of physical and chemical qualities, composition, and energy content. Biomass contain more volatile materials, more ash, less sulfur and nitrogen, less heat value, and more moisture content and have larger and a wider range of particle sizes. When biomass is co-fired with coal, the qualities of blended fuel should be controlled to be accepted by the boilers. Let q be the fuel quality, $q = 1$ be the volatile matter content, $q = 2$ be the heat rate, $q = 3$ be the ash content, $q = 4$ be the moisture content, and $q = 5$ be the sulfur content [Lv et al., 2016].

$$E\left[\widetilde{T_{iq}^{L}}\right] \leq \frac{\sum_{j=1}^{J} E[\widetilde{Q_{jq}}]\, x_{ijs}}{\sum_{j=1}^{J} x_{ijs}} \leq E\left[\widetilde{T_{iq}^{U}}\right] \tag{9.10}$$

9.3.4.4 Technical Constraint

Biomass contains less carbon, more hydrogen, and more oxygen than coal, which indicates biomass contains less energy density. The fuel density of a typical biomass contains only around one-tenth of that of coal. It has been found that blending coal with a certain ratio percentage of biomass can improve combustion efficiency [Agbor et al., 2014]. Let $1 - l$ be the biomass and $l - n$ be the coal. Therefore, this constraint can be written as:

$$\frac{\sum_{j=1}^{l} x_{ijs}}{\sum_{j=1}^{J} x_{ijs}} \leq \theta \tag{9.11}$$

9.3.4.5 Social Responsibility Limitation

Each CPP undertakes the responsibility to meet the electricity demand in the region in terms of its power generation capabilities. Let D_{is} be power demand in s month at CPP i, and the actual power generation $\sum_{j=1}^{J} E[\widetilde{P_{ij}}]\, x_{ijs}$ must be more than D_{is}.

$$\sum_{j=1}^{J} E[\widetilde{P_{ij}}]\, x_{ijs} \geq D_{is} \tag{9.12}$$

9.3.4.6 Carbon Emissions Quota Constraint

Decisions are made sequentially from the upper level to the lower level. As a lower-level decision-maker, CPP i must obey the decisions from the authority; that is, the tactual carbon emissions at CPP i cannot exceed the quota allocated by the authority. Let $\widetilde{H_{ij}}$ be the carbon emissions from a unit of fuel j. Therefore, this mathematical formulation is described as:

$$\sum_{j=1}^{J} \sum_{s=1}^{S} E[\widetilde{H_{ij}}]\, x_{ijs} \leq y_i \tag{9.13}$$

9.3.4.7 Fuel Resources Storage Limitation

As biomass is a renewable carbon dioxide neutral energy source, it is considered as an attractive renewable fuel to supplement coal combustion in boilers. However, the available quantity of biomass heavily relies upon the months. Therefore, dynamic programming is employed to characterize the fuel resources storage. The model is a dynamic multi-stage decision-making model, with the process-based control being the essential difference from previous work. The CPPs monitor the fuel resource storage and adjust fuel purchase strategies each month. In addition, the storage quantity at the terminal cannot be negative.

$$I_{ijs} = I_{ij(s-1)} + z_{ijs} - x_{ijs} \tag{9.14}$$

$$0 \leq x_{ijs} \leq z_{ijs} + I_{ij(s-1)} \tag{9.15}$$

Let I_j^U be the upper boundary for the warehousing capability. CPPs cannot allow fuel inventory to exceed this upper boundary. Therefore, storage quantities at the terminal in the previous month and the purchase quantities in the present month are restricted by the upper boundary.

$$\sum_{j=1}^{J} I_{ij(s-1)} + \sum_{j=1}^{J} z_{ijs} \leq I_i^U \tag{9.16}$$

9.3.5 Global Model

To reduce carbon emissions, there are multiple conflicts to be addressed; decision-makers' conflicts, objectives' conflicts, and time conflicts. To address these complicated conflicts, an equilibrium strategy-based bilevel multiobjective dynamic model is developed. In the proposed model, the authority acts as the leader in the upper level while the CPPs are the followers in the lower level. The authority aims to obtain maximum financial revenue and achieve minimum carbon emissions, then develops the initial strategy on carbon emissions quota allocation. Following that, under the limited carbon emissions quota, CPPs aim to maximize their economic benefits and develop the biomass–coal blending ratio schemes considering the fuel resource quantity and quality limitations, the technical constraint, carbon emissions allowance, and social responsibility constraints. The CPPs feedbacks to the authority and urges it to adjust the initial strategy, after which the authority updates the carbon emissions quota allocation and the CPPs modify the biomass–coal blending ratios. After several interactive processes, an optimal solution that both the authority and the CPPs satisfy is reached. From an integration of the objectiveS and constraints as Eqs. (9.2)– (9.16), the full equilibrium strategy-based multiobjective bilevel dynamic model Eq. (9.17) can be formulated.

Compared with previous biomass–coal blending models, this model considers the interaction between the authority and the CPPs, the trade-off between the economy and the environment, and time conflict in biomass availability; therefore, it is suitable for actual production and can assist CPPs make reasonable production schemes under the authority's policy. In this model, as β presents attitude of the authority toward carbon emissions reduction and λ presents the authority's attitude toward the difference between the quota and actual emissions, this model is flexible and robustness. Under various situations, the authority is able to adjust the policy control parameters to achieve the most satisfactory results.

9.4 Case Study

In this section, a real-world application of the proposed model will be given to show its effectiveness.

9.4.1 Case Description

China is entering a post-coal growth stage. In this stage, coal consumption accompanied by carbon emission is projected to reach the peak, with a proportion of nearly 50% of global coal consumption in 2035 Qi et al. [2016]. Besides, China's market of coal is regarded as the largest all over the world. Figure 9.4 is not only an economically prosperous province ranking second of all Chinese provinces, but also a populous province with an estimated 2016 population at more than 78.66 million. Jiangsu Province, locating in southeast China, is a major carbon emitter in China at 20.3 million tonnes per annum due to the large population and its economic superiority [Zhao et al., 2008]. It is known that CPP is the most important end user of coal, and correspondingly, coal consumption is a vital source for carbon emissions. Above all, Jiangsu's authority has been making efforts to reduce carbon emissions while maintaining economic growth for a long time. However, through the lens of economic development and environmental sustainability, there is a need for developing cleaner and lower-carbon fuel sources. The Chinese government's Energy Development Strategy Action Plan states to reduce coal's share to less than 62% of the energy mix by 2020 BP Global [2017]. Fortunately, Jiangsu Province is known as being rich in agricultural bioenergy. Since 2009, Jiangsu Province has produced an annual bioenergy output of about 40 million tonnes and had already considered biomass–coal co-firing to deal with a rising trend in energy demand and carbon emission [Zhang et al., 2013]. Since 2003, several biomass–coal co-firing CPPs have been built due to the abundant biomass resources in Jiangsu Province. In this research, three main biomass–coal co-firing CPPs are examined in this case: Fengxian CPP, Ganyu CPP, and Baoying CPP, all of which use straw (S), wood waste (W), and coal as fuel. As coal from different places has different properties, Yaoqiao coal (Y) and Zhangshuanghe coal (Z) are used in these CPPs.

9.4.2 Model Transformation and Solution Approach

Aiming to develop provincial economy, Jiangsu's authority has to balance financial revenue and environmental protection. Let β be the authority's attitude toward carbon emissions reduction, according to Lu et al., the objective of minimizing carbon emissions could be transformed into a constraint controlled in the accepted range [Lv et al., 2016]. Therefore, Eq. (9.17) can be transformed as shown in Eq. 9.18.

Within the hierarchical framework, there is a trade-off between the upper-level decision-maker and the lower-level decision-makers because of the conflicting objectives. Further, the following algorithm in Matlab is employed to solve the linear bilevel model. The specific process is that: first, an initial solution to financial revenue maximization at the upper level is determined. Then, the initial optimal solution is put into the lower-level model and an optimal solution is calculated based on the lower-level objectives. Next, the solutions at the lower level are regarded as feedback and put into the upper-level model, and then an updated solution is obtained. This process is repeated until an optimal solution to the bilevel programming is found [Lv et al., 2016]. The procedure for this solution approach is shown in Figure 9.3

$$
\begin{array}{l}
\max F_1 = \sum\limits_{i=1}^{I}\sum\limits_{j=1}^{J}\sum\limits_{s=1}^{S} \varpi Q E[\widetilde{P_{ij}}]\, x_{ijs} + \varepsilon \sum\limits_{i=1}^{I} y_i \\
\min F_2 = \sum\limits_{i=1}^{I} y_i \\
\text{s.t.} \left\{
\begin{array}{l}
O_i^L \le y_i \le O_i^U, \quad \forall i \\
\sum\limits_{i=1}^{I} C_i y_i \ge D \\
\frac{y_i - \sum_{j=1}^{J}\sum_{s=1}^{S} E[\widetilde{H_{ij}}] x_{ijs}}{y_i} \le \psi, \quad \forall i \\
\max f = \sum\limits_{j=1}^{J}\sum\limits_{s=1}^{S} (1-\varpi) Q E[\widetilde{P_{ij}}]\, x_{ijs} - \sum\limits_{j=1}^{J}\sum\limits_{s=1}^{S} A_{ijs} z_{ijs} - K \sum\limits_{j=1}^{J}\sum\limits_{s=1}^{S} I_{jis} \\
\qquad - \sum\limits_{j=1}^{J}\sum\limits_{s=1}^{S}\sum\limits_{p=1}^{P} E[\widetilde{M_{jp}}] V_{ip} x_{ijs} - \varepsilon y_i \\
\text{s.t.} \left\{
\begin{array}{l}
\frac{\sum_{j=1}^{J} E[\widetilde{Q_{jq}}] x_{ijs} + \sum_{j=1}^{J} E[\widetilde{\eta_j}](1 - E[\widetilde{Q_{jq}}]) x_{ijs}}{\sum_{j=1}^{J} x_{jis}} \ge \partial, \quad \forall i, s \\
0 \le x_{ijs} \le z_{ijs} \le U_{ijs}, \quad \forall i, s \\
E\left[\widetilde{T_{iq}^{L}}\right] \le \frac{\sum_{j=1}^{J} E[\widetilde{Q_{jq}}] x_{ijs}}{\sum_{j=1}^{J} x_{ijs}} \le E\left[\widetilde{T_{iq}^{U}}\right], \quad \forall i, s, q \\
\frac{\sum_{j=1}^{l} x_{ijs}}{\sum_{j=1}^{J} x_{ijs}} \le \theta, \quad \forall i, s \\
\sum\limits_{j=1}^{J} E[\widetilde{P_{ij}}]\, x_{ijs} \ge D_{is}, \quad \forall i \\
\sum\limits_{j=1}^{J}\sum\limits_{s=1}^{S} E[\widetilde{H_{ij}}]\, x_{ijs} \le y_i, \quad \forall i \\
I_{ijs} = I_{ij(s-1)} + z_{ijs} - x_{ijs}, \quad \forall i \\
0 \le x_{ijs} \le z_{ijs} + I_{ij(s-1)} \\
\sum\limits_{j=1}^{J} I_{ij(s-1)} + \sum\limits_{j=1}^{J} z_{ijs} \le I_i^U, \quad \forall i, s
\end{array}
\right.
\end{array}
\right.
\end{array}
\tag{9.17}
$$

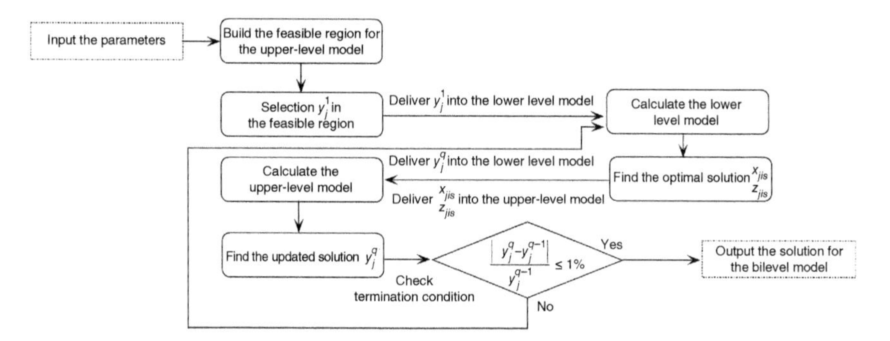

Figure 9.3 Flowcharts of the algorithm.

$$
\max F = \sum_{i=1}^{I}\sum_{j=1}^{J}\sum_{s=1}^{S} \varpi Q E[\widetilde{P_{ij}}]\, x_{ijs} + \varepsilon \sum_{i=1}^{I} y_i
$$
$$
\text{s.t.}\begin{cases}
\sum_{i=1}^{I} y_i \leq \beta L_A \\
O_i^L \leq y_i \leq O_i^U, \quad \forall i \\
\sum_{i=1}^{I} C_i y_i \geq D \\
\frac{y_i - \sum_{j=1}^{J}\sum_{s=1}^{S} E[\widetilde{H_{ij}}] x_{ijs}}{y_i} \leq \psi, \quad \forall i \\
\max f = \sum_{j=1}^{J}\sum_{s=1}^{S} (1-\varpi) Q E[\widetilde{P_{ij}}]\, x_{ijs} - \sum_{j=1}^{J}\sum_{s=1}^{S} A_{ijs}\, z_{ijs} - K \sum_{j=1}^{J}\sum_{s=1}^{S} I_{jis} \\
\qquad - \sum_{j=1}^{J}\sum_{s=1}^{S}\sum_{p=1}^{P} E[\widetilde{M_{jp}}] V_{ip}\, x_{ijs} - \varepsilon y_i \\
\text{s.t.}\begin{cases}
\frac{\sum_{j=1}^{J} E[\widetilde{Q_{jq}}] x_{ijs} + \sum_{j=1}^{J} E[\widetilde{\eta_j}](1 - E[\widetilde{Q_{jq}}]) x_{ijs}}{\sum_{j=1}^{J} x_{jis}} \geq \partial, \quad \forall i, s \\
0 \leq x_{ijs} \leq z_{ijs} \leq U_{ijs}, \quad \forall i, s \\
E\left[\widetilde{T_{iq}^L}\right] \leq \frac{\sum_{j=1}^{J} E[\widetilde{Q_{jq}}] x_{ijs}}{\sum_{j=1}^{J} x_{ijs}} \leq E\left[\widetilde{T_{iq}^U}\right], \quad \forall i, s, q \\
\frac{\sum_{j=1}^{l} x_{ijs}}{\sum_{j=1}^{J} x_{ijs}} \leq \theta, \quad \forall i, s \\
\sum_{j=1}^{J} E[\widetilde{P_{ij}}]\, x_{ijs} \geq D_{is}, \quad \forall i \\
\sum_{j=1}^{J}\sum_{s=1}^{S} E[\widetilde{H_{ij}}]\, x_{ijs} \leq y_i, \quad \forall i \\
I_{ijs} = I_{ij(s-1)} + z_{ijs} - x_{ijs}, \quad \forall i \\
0 \leq x_{ijs} \leq z_{ijs} + I_{ij(s-1)} \\
\sum_{j=1}^{J} I_{ij(s-1)} + \sum_{j=1}^{J} z_{ijs} \leq I_i^U, \quad \forall i, s
\end{cases}
\end{cases}
\tag{9.18}
$$

9.4.3 Data Collection

The basic necessary data for Fengxian CPP, Ganyu CPP, and Baoying CPP are shown in Tables 9.1–9.6 and are divided into the fixed data and the uncertain data. The

Table 9.1 Straw resource quantities, prices, and power demand.

	Fengxian CPP		Ganyu CPP		Baoying CPP	
Month	Quantity: O^U_{jis} (10^4 tonnes)	Power: U_{js} (10^7 kWh)	Quantity: O^U_{jis} (10^4 tonnes))	Power: U_{js} (10^7 kWh)	Quantity: O^U_{jis} (10^4 tonnes)	Power: U_{js} (10^7 kWh)
1	4.2	22.4	0.6	10.07	0.6	14.47
2	3.7	15.24	1.1	5.8	0.9	8.46
3	11.7	21.56	1.4	8.61	1.2	13.1
4	18.5	20.26	2	7.97	2	12.52
5	29.6	21.97	81.9	9.07	92.9	13.3
6	116.5	22.3	91.3	9.38	95.9	13.44
7	38.2	25.23	19.9	11.06	65.7	16.78
8	36.8	23.26	28.7	9.93	34.3	15.18
9	177.5	16.61	142.3	7.36	132.6	9.67
10	77.2	17.34	58.2	7.98	60.2	11.28
11	22.1	21.5	9.5	10.82	11.3	13.64
12	10.3	22.57	1.3	11.95	1.7	14.78

Table 9.2 Fuzzy straw prices.

	Fengxian CPP	Ganyu CPP	Baoying CPP
Month	Price: $\widetilde{P_{jis}}$ (CNY/tonne)	Price: $\widetilde{P_{jis}}$ (CNY/tonne)	Price: $\widetilde{P_{jis}}$ (CNY/tonne)
1	(715,725,735,740)	(725,730,740,750)	(727,732,748,755)
2	(710,720,730,740)	(690,695,705,710)	(720,725,735,740)
3	(705,715,725,735)	(670,675,680,685)	(710,715,720,725)
4	(690,695,700,705)	(670,680,690,705)	(685,695,705,710)
5	(663,670,676,685)	(625,630,640,645)	(565,570,575,580)
6	(630,635,640,650)	(590,598,606,615)	(528,535,543,650)
7	(640,645,656,660)	(633,638,645,650)	(630,635,640,645)
8	(630,635,645,650)	(620,626,632,638)	(655,660,665,670)
9	(665,675,685,790)	(577,585,594,600)	(580,590,600,610)
10	(615,620,630,635)	(615,620,625,631)	(620,635,645,660)
11	(645,655,665,670)	(652,657,665,670)	(675,685,695,700)
12	(705,710,720,730)	(690,695,705,710)	(710,715,720,725)

Table 9.3 Crisp parameters.

	Fengxian CPP	Ganyu CPP	Baoying CPP
Least carbon emissions: Q_j^L (tonnes)	5.76×10^6	4.67×10^6	6.47×10^6
Utmost carbon emissions: Q_j^U (tonnes)	1.05×10^7	1.034×10^7	1.13×10^7
Max storage ability: I_j^U (tonnes)	7.85×10^5	5.91×10^5	6.15×10^5
Procurement price: V_k (CNY/kg)			
For SO_2	1.8	2	2.2
For NO_x	14.4	14.7	14.5

Table 9.4 Other parameters.

Carbon emissions quota price: u (CNY/tonne)	30
Value-added tax: S (%)	1
Price of unit electric: Q (CNY/kWh)	0.45
Amount of power needed in a region: U(kWh)	5.1686×10^9
Upper percentage of biomass in fuel blend: UR	0.1
Storage price monthly per unit: K (CNY/10^4 tonnes)	0.125
Carbon emissions last year: LCE (tonnes)	2.37×10^7

fixed data shown in Tables 9.1, 9.3, and 9.4 were taken from the White Paper on straw power generation project construction management in Jiangsu Province, the Jiangsu Province Statistical Yearbook, and the official CPP websites (Jiangsu Statistical Bureau). Tables 9.1 and 9.3 show the requirements and responsibilities at the three main CPPs, and the other needed parameters for the proposed model are shown in Table 9.4. The uncertain data collected from the Jiangsu Province Statistical Yearbook and the official websites of each CPP are shown in Tables 9.5, and 9.6. Table 9.2 shows straw price over 12 months, Table 9.5 shows the fuel characteristics of the various bunkers, and Table 9.6 gives the boiler requirements at the three main CPPs.

As Jiangsu Province is a relatively developed region, carbon emissions quota allocation satisfaction and regional fairness are paramount. Therefore, the control parameter λ is set at 0, indicating that the authority has a strict attitude toward the gap between the quota and the actual value.

9.5 Results and Discussion

Detailed results and related discussion will be given in this section.

Table 9.5 Fuzzy uncertain parameters for each fuel.

	Straw	**Wood waste**	**Yaoqiao coal**	**Zhangshuanghe coal**
Fuel properties: $\widetilde{B_{it}}$				
Volatile matter (wt%)	(64.2,64.7,65.4,65.9)	(44.7,45.03,45.32,45.6)	(25.5,25.8,26.2,26.5)	(31.6,31.9,32.2,32.5)
Heat rate (G_j/tonne)	(19.3,19.6,20.2,20.9)	(18.35,18.7,19.1,19.81)	(27.8,28.2,28.55,28.8)	(21.8,22.2,22.75,23.1)
Ash content (wt%)	(27.28,28.07,28.4,28.88)	(19.76,20,20.4,20.56)	(8.86,9.1,9.5,9.8)	(19,19.3,19.7,20)
Moisture content (wt%)	(12.5,12.8,13.2,13.5)	(8.6,8.93,9.2,9.55)	(6.75,6.9,7.2,7.4)	(4.6,4.8,5.22,5.45)
Sulfur content (wt%)	(0.17,0.18,0.21,0.24)	(0.09,0.1,0.12,0.14)	(0.33,0.4,0.47,0.52)	(0.59,0.62,0.66,0.69)
Pollutants emission				
Factor: $\widetilde{E_{ik}}$ (kg/tonne)				
For SO_2	(1.5,1.8,2.13,2.4)	(3.95,4.1,4.4,4.72)	(2.7,3,3.35,3.7)	(8.25,8.6,8.9,9.2)
For NO_x	(4.6,4.8,5.1,5.34)	(0.92,1.1,1.3,1.45)	(9.34,9.6,9.86,10.2)	(9.9,10.2,10.52,10.7)
Char burn-out fraction: $\widetilde{\eta_i}$(%)	(44,48,52,56)	(36,38,40,43)	(17,19,21,23)	(17,20,22,24)

Table 9.6 Fuzzy CPP requirements and factors.

	Fengxian CPP		Ganyu CPP		Baoying CPP	
	Lower bound	Upper bound	Lower bound	Upper bound	Lower bound	Upper bound
Volatile matter (wt%)	(5.5,5.9,6.2,6.6)	(35.9,36.2,36.5,36.9)	(7.1,7.3,7.6,7.8)	(37.6,37.9,38.1,38.4)	(8.7,8.9,9.2,9.4)	(38.6,38.8,39,39.3)
Heat rate (G_j/tonne)	(20.9,21.3,21.6,21.9)	—	(21.6,21.9,22.1,22.4)	—	(21.4,21.7,21.9,22.1)	—
Ash content (wt%)	—	(25.6,26,26.3,26.5)	—	(27.7,27.9,28.2,28.4)	—	(28.6,29,29.2,29.5)
Moisture content (wt%)	—	(8.6,8.9,9.1,9.4)	—	(9.6,9.9,10.1,10.4)	—	(12.6,12.9,13.2,13.7)
Sulfur content (wt%)	—	(0.91,0.96,0.99,1.15)	—	(1.1,1.3,1.8,2.2)	—	(1.7,1.9,2.2,2.5)
Carbon emissions						
$\widetilde{CE}_{ji}$ (kg/tonne)						
Straw	(1375,1500,1600,1750)		(1200,1350,1500,1650)		(1350,1400,1500,1600)	
Wood waste	(1700,1850,2070,2290)		(1750,1900,2000,2150)		(1350,1500,1700,1850)	
Yaoqiao coal	(2325,2450,2500,2600)		(2125,2350,2500,2630)		(2250,2300,2450,2600)	
Zhangshuanghe coal	(1970,2030,2070,2100)		(2150,2200,2375,2450)		(2150,2250,2300,2400)	
Fuel to power						
$\widetilde{C}_{ji}$ (kWh/tonne)						
Straw	(1250,1400,1550,1700)		(1500,1650,1850,1950)		(1300,1500,1700,1900)	
Wood waste	(1550,1700,1900,2050)		(1450,1650,1870,2090)		(1200,1400,1550,1750)	
Yaoqiao coal	(2050,2200,2350,2500)		(2100,2200,2350,3450)		(2200,2300,2450,2550)	
Zhangshuanghe coal	(1950,2100,2250,2350)		(1900,2050,2130,2250)		(2010,2150,2300,2450)	
Carbon to power						
$\widetilde{H}_j$ (kWh/tonne)	(935,945,955,965)		(1003,1007,1013,1016)		(960,966,978,993)	

9.5.1 Results Under Different Scenarios

The results under different scenarios are shown in Tables 9.7–9.9. α and β, respectively, represent combustion efficiency and attitude of the authority toward carbon emissions reduction. The results under scenario 1 are shown in Table 9.7 demonstrates that when $\alpha = 0.44$ and β reduces from 1 to 0.75. Scenario 2 in Table 9.8 shows the results when $\alpha = 0.435$ and β reduces from 1 to 0.75. Scenario 3 in Table 9.9 shows the results when $\alpha = 0.43$ and β reduces from 1 to 0.75. After analyzing these scenarios, some findings are drawn in the following.

With β reducing, revenue, the total carbon emissions and CPPs profits reduce. For example, as shown in Table 9.7, when β reduces from 1 to 0.75, the revenue reduces from 824 124 154 to 619 531 368 CNY, and the total carbon emissions reduce from 23 700 000 to 17 775 000 tonnes. The profits of Fengxian CPP, Ganyu CPP, and Baoying CPP, respectively, reduce from 4 451 759 826 to 3 906 769 677 tonnes, from 4 168 156 720 to 2 552 713 223 tonnes, and from 4 436 308 120 to 3 566 293 572 tonnes. Similar situation can be seen in scenarios 2 and 3.

When there is no environmental protection, the CPPs make larger profits. For example, as what was listed in Table 9.8, when the carbon emission constraint is considered and set as 1, the profits of Fengxian CPP, Ganyu CPP, and Baoying CPP, respectively, are 4 454 356 818, 4 183 966 990, and 4 452 589 908 CNY. When carbon emission constraint is ignored, the profits of Fengxian CPP, Ganyu CPP, and Baoying CPP are 4 604 992 957, 4 162 462 747, and 4 488 139 326 CNY, respectively. Similar situations can also be found in scenarios 1 and 3.

Even though the CPP profits vary with combustion efficiency, the carbon emissions allowances do not. For example, as shown in Tables 9.7 and 9.8, when β is set as 1 and α reduces from 0.44 to 0.435, the profits of Fengxian CPP, Ganyu CPP, and Baoying CPP all slightly increase, but the carbon emission allowances that the three CPPs receive are not changed. Similar situations can be seen in comparison between scenarios 1 and 3 as well as between scenarios 2 and 3.

9.5.2 Propositions and Analyses

In this section, the results under diverse scenarios (Tables 9.7–9.9) are discussed.

First, dynamic equilibrium strategy-based coal and biomass blending method can be important significance for carbon emission reduction. β represents the authority's attitude toward carbon emission reduction, and the lowest potential carbon emission level is 0.9. This means using the dynamic equilibrium strategy-based coal and biomass blending method has potential to reduce carbon emission by about 10% compared with carbon emission of last production period. In actual fact, when the dynamic equilibrium strategy-based coal and biomass blending method is applied, the local authority allocates carbon emission allowance quota to different CPPs and it prefers to allocate more carbon emission which shows better carbon emission performance. As for CPPs, when allocated limitative carbon emission, they will adjust coal and biomass blend ratio and make fuel purchase schemes to improve their carbon emission performance. Therefore, dynamic equilibrium strategy-based coal and biomass blending method can reduce carbon emission as much as possible.

Table 9.7 Results when $\alpha = 0.44$ and β is changing.

	The authority	Fengxian CPP		Ganyu CPP		Baoying CPP	
	Revenue (CNY)	Profits (CNY)	Carbon emissions allowances (tonnes)	Profits (CNY)	Carbon emissions allowances (tonnes)	Profits (CNY)	Allowances (tonnes)
$\beta = 1$	824 124 154	4 451 759 826	7 560 926	4 168 156 720	7 746 840	4 436 308 120	8 392 234
$\beta = 0.95$	783 256 321	4 306 500 002	7 261 291	3 726 845 015	6 899 931	4 419 829 892	8 353 778
$\beta = 0.9$	742 526 138	4 148 066 003	6 935 241	3 421 066 080	6 318 592	4 298 714 817	8 076 167
$\beta = 0.85$	701 780 815	3 939 222 482	6 509 867	3 332 479 147	6 150 287	3 993 437 743	7 484 847
$\beta = 0.8$	660 556 261	3 914 166 707	6 459 159	2 554 275 247	4 674 309	4 170 160 156	7 826 532
$\beta = 0.75$	619 531 368	3 906 769 677	6 444 189	2 552 713 223	4 671 356	3 566 293 572	6 659 454
No carbon emission constraint	837 332 385	4 603 086 052	7 875 000	4 133 640 676	7 755 000	4 471 773 541	8 475 000

Table 9.8 Results when $\alpha = 0.435$ and β is changing.

	The authority	Fengxian CPP		Ganyu CPP		Baoying CPP	
	Revenue (CNY)	Profits (CNY)	Carbon emissions allowances (tonnes)	Profits (CNY)	Carbon emissions allowances (tonnes)	Profits (CNY)	Allowances (tonnes)
$\beta = 1$	824 017 103	4 454 356 818	7 560 926	4 183 966 990	7 746 840	4 452 589 908	8 392 234
$\beta = 0.95$	783 123 956	4 309 906 907	7 261 291	3 739 566 760	6 899 931	4 436 072 652	8 353 778
$\beta = 0.9$	742 550 083	4 152 222 662	6 935 241	3 432 827 618	6 318 592	4 316 834 319	8 076 167
$\beta = 0.85$	701 798 004	3 944 814 098	6 509 867	3 343 889 217	6 150 287	4 017 184 370	7 484 847
$\beta = 0.8$	660 588 047	3 919 885 299	6 459 159	2 562 826 418	4 674 309	4 194 894 359	7 826 532
$\beta = 0.75$	619 548 106	3 912 525 755	6 444 189	2 561 258 417	4 671 356	3 587 622 374	6 659 454
No carbon emission constraint	837 640 608	4 604 992 957	7 875 000	4 162 462 747	7 755 000	4 488 139 326	8 475 000

Table 9.9 Results when $\alpha = 0.43$ and β is changing.

	The authority	Fengxian CPP		Ganyu CPP		Baoying CPP	
	Revenue (CNY)	Profits (CNY)	Carbon emissions allowances (tonnes)	Profits (CNY)	Carbon emissions allowances (tonnes)	Profits (CNY)	Allowances (tonnes)
$\beta = 1$	823 832 277	4 454 356 817	7 560 926	4 196 291 295	7 746 840	4 466 603 158	8 392 234
$\beta = 0.95$	783 005 652	4 309 906 906	7 261 291	3 751 866 731	6 899 931	4 450 064 080	8 353 778
$\beta = 0.9$	742 436 605	4 152 222 660	6 935 241	3 444 181 222	6 318 592	4 330 668 228	8 076 167
$\beta = 0.85$	701 851 371	3 944 814 099	6 509 867	3 354 888 571	6 150 287	4 040 413 096	7 484 847
$\beta = 0.8$	660 654 846	3 919 885 301	6 459 159	2 570 950 912	4 674 309	4 219 174 655	7 826 532
$\beta = 0.75$	619 629 466	3 912 525 758	6 444 189	2 557 310 004	4 671 356	3 608 335 619	6 659 454
No carbon emission constraint	837 639 036	4 604 992 957	7 875 000	4 181 655 784	7 755 000	4 502 188 550	8 475 000

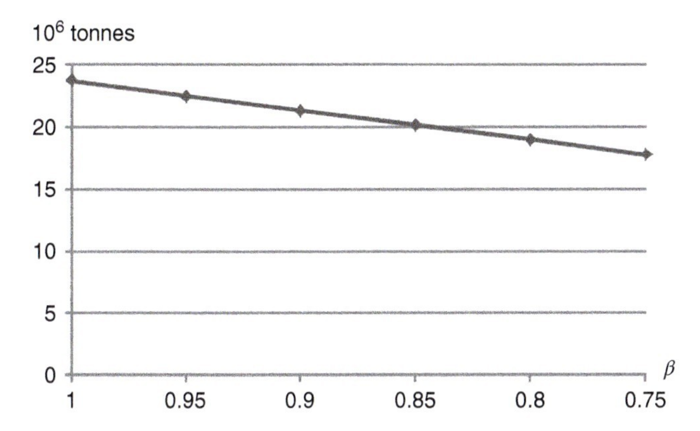

Figure 9.4 Total carbon emissions.

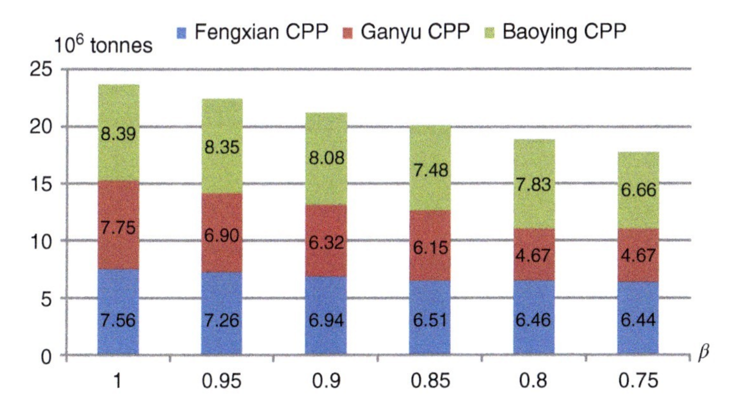

Figure 9.5 CPP carbon emissions.

Second, combustion efficiency improvement can achieve economy–environment coordinated sustainable development. In this chapter, combustion efficiency is mainly restricted by volatile matter content and char combustion. From the results, it can be seen that with combustion efficiency increasing CPPs gain more economical benefits and discharge less carbon emission. When α_1, α_2, and α_3 change from 0.459, 0.469, and 0.449 to 0.468, 0.478, and 0.458, respectively, the carbon emission allowances of Fengxian CPP, Ganyu CPP, and Baoying CPP are reduced while their economic benefits increase. In terms of CPPs, their carbon emission allowance quotas decrease but their economical profits all increase. Actually, combustion efficiency improvement is necessary for green development. Therefore, combustion efficiency improvement can help to build resource-saving, environment-friendly society.

Further, the reduction ratio of carbon emission amount is larger than that of financial revenue when carbon emission and combustion efficiency constraints are stricter. The financial revenue and the total carbon emission amount both decrease when carbon emission and combustion efficiency constraints are stricter. However,

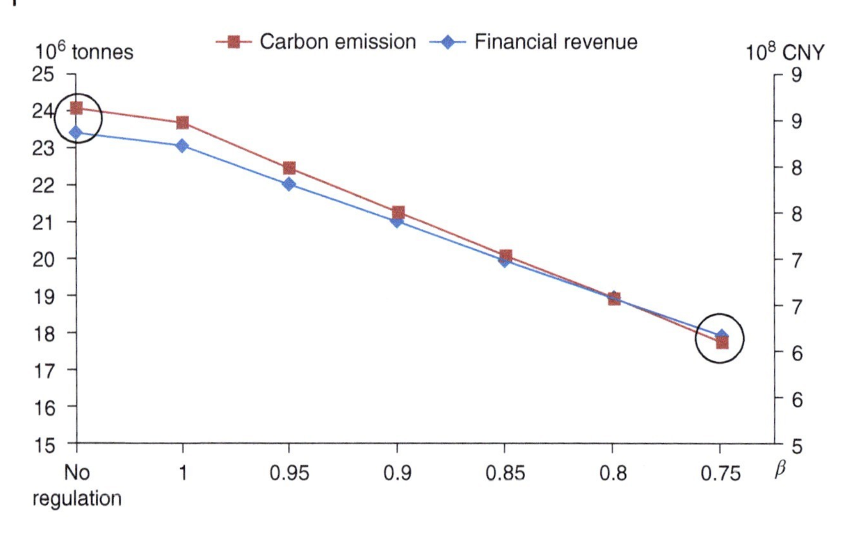

Figure 9.6 Revenue and carbon emissions under different β values when $\alpha = 0.44$.

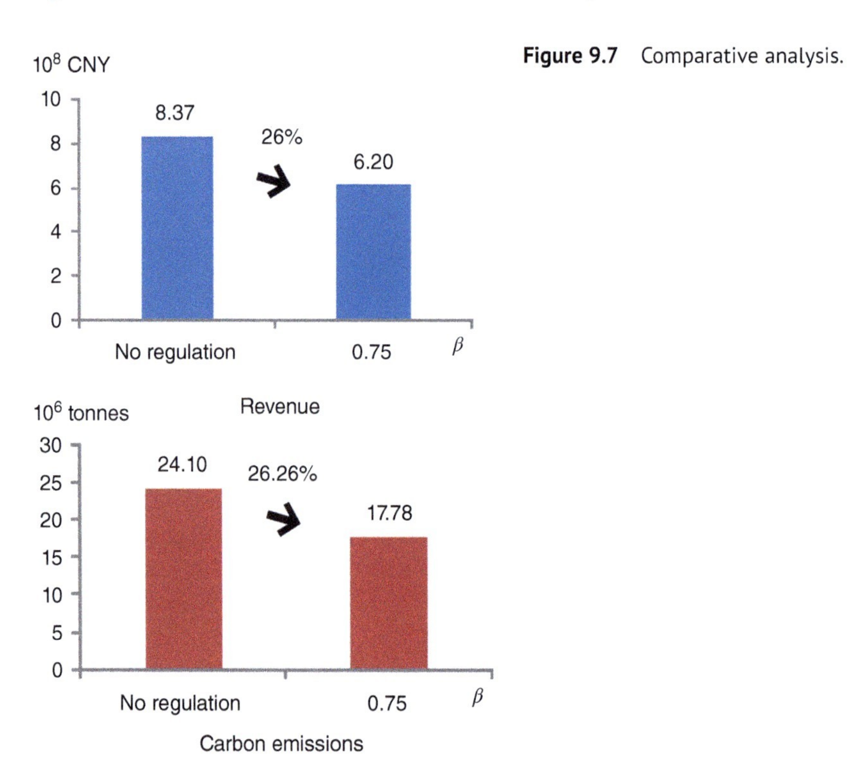

Figure 9.7 Comparative analysis.

the marginal carbon emission reduction rate is larger than financial revenue. It can be inferred that although relatively stricter environmental protection and combustion efficiency decreases financial revenue, it would cause much less carbon emission and bring much more benefits to the environment.

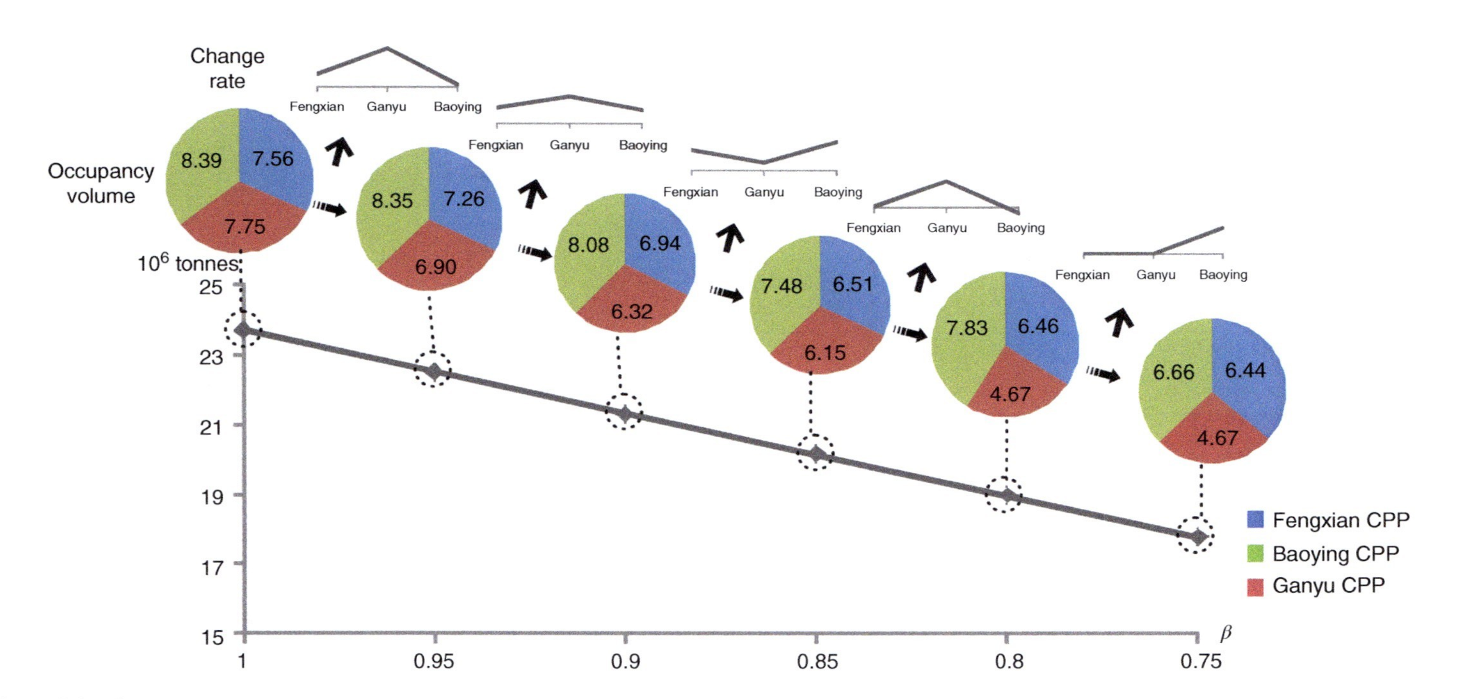

Figure 9.8 Carbon emissions under different β values.

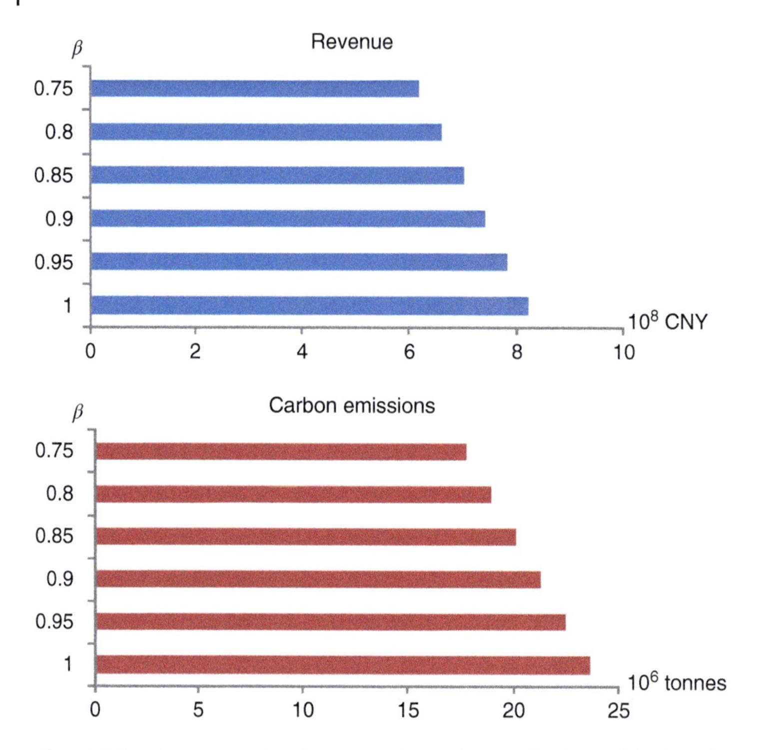

Figure 9.9 Revenue and carbon emissions when α=0.44 and β is changing.

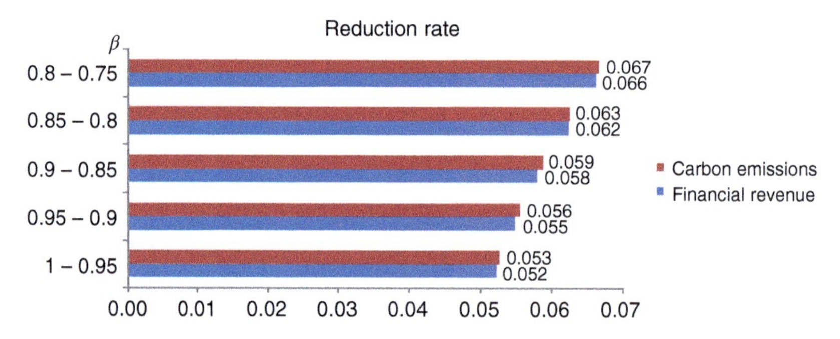

Figure 9.10 Reduction rate.

Finally, combustion efficiency does not significantly affect CPP profits.From the results, it can be seen that when the combustion efficiency increases, there is little effect on CPP profits. When β remains unchanged and α increases, profits reduce. As α increases again to 0.44, profits still decrease. The reason is that combustion efficiency does not change the carbon emissions allowances which are the key factors influencing profits. It just impacts biomass–coal co-firing ratio and slightly effects profits. Therefore, CPPs would be willing to sacrifice a small fall in profits to improve combustion efficiency and reduce emissions.

9.5.3 Policy Implications

Based on the above discussions and analyses, some management recommendations are proposed. First, a carbon emission allocation competition mechanism which involves the authority and CPPs is recommended. In such a mechanism, the authority allocates carbon emission allowances to multiple CPPs and its decision is conversely affected by the decision of multiple CPPs. Each CPP makes the biomass and coal blending ratio schemes for each decision from the authority and is impelled to conduct environmental-friendly power generation. The authority would allocate more carbon emission allowances to the CPP which has better carbon emission performance. CPPs compete against each other to acquire more carbon emission allowances. Second, combustion efficiency gradual growth should be emphasized. Considering energy conservation and emission reduction, combustion efficiency is of important significance. In this chapter, combustion efficiency is limited by the extent of char combustion in biomass and coal co-firing method. In the pursuit of high combustion efficiency, CPPs cautiously select biomass feedbacks and coal, and properly burn them, making their combustion process cheaper and discharging less carbon emission. Third, a changing attitude toward carbon emission reduction should be taken. It is recommended that the authority should choose a changing carbon emission reduction scheme according to the level of economic development. When the authority makes its decisions, a changing attitude should be taken. The authority should take the actual situation into consideration and choose an optimal carbon emission reduction scheme.

References

Agbor, E., Zhang, X., and Kumar, A. (2014). A review of biomass co-firing in north America. *Renewable and Sustainable Energy Reviews* 40: 930–943.

Bandyopadhyay, A. (2011). Amine versus ammonia absorption of CO as a measure of reducing GHG emission: a critical analysis. *Clean Technologies and Environmental Policy* 13 (2): 269–294.

Basu, P. (1999). Combustion of coal in circulating fluidized-bed boilers: a review. *Chemical Engineering Science* 54 (22): 5547–5557.

Basu, P., Butler, J., and Leon, M.A. (2011). Biomass co-firing options on the emission reduction and electricity generation costs in coal-fired power plants. *Renewable Energy* 36 (1): 282–288.

Baxter, L. (2005). Biomass–coal co-combustion: opportunity for affordable renewable energy. *Fuel* 84 (10): 1295–1302.

BP Global (2015). BP Energy Outlook.

BP Global (2017). BP Energy Outlook.

Chen, Z.M., Liu, Y., Qin, P. et al. (2015). Environmental externality of coal use in China: welfare effect and tax regulation. *Applied Energy* 156: 16–31.

Colson, B., Marcotte, P., and Savard, G. (2007). An overview of bilevel optimization. *Annals of Operations Research* 153 (1): 235–256.

Dai, C., Cai, X.H., Cai, Y.P., and Huang, G.H. (2014). A simulation-based fuzzy possibilistic programming model for coal blending management with consideration of human health risk under uncertainty. *Applied Energy* 133 (10): 1–13.

Dai, J., Sokhansanj, S., Grace, J.R. et al. (2008). Overview and some issues related to co-firing biomass and coal. *Canadian Journal of Chemical Engineering* 86 (3): 367–386.

Eksioglu, S.D. and Karimi, H. (2014). An optimization model in support of biomass co-firing decisions in coal fired power plants. *Ecoproduction* 111–123.

Gielen, D. (2012). Renewable energy technologies: cost analysis series biomass for power generation.

Goto, K., Yogo, K., and Higashii, T. (2013). A review of efficiency penalty in a coal-fired power plant with post-combustion CO_2 capture. *Applied Energy* 111 (11): 710–720.

Jiangsu Statistical Bureau (2018). http://tj.jiangsu.gov.cn/col/col4009/index.html (accessed 28 May 2021).

Li, Y. (2012). Dynamics of clean coal-fired power generation development in China. *Energy Policy* 51 (6): 138–142.

Li, J., Brzdekiewicz, A., Yang, W., and Blasiak, W. (2012). Co-firing based on biomass torrefaction in a pulverized coal boiler with aim of 100% fuel switching. *Applied Energy* 99 (6): 344–354.

Li, S., Xu, T., Hui, S., and Wei, X. (2009). NO_x emission and thermal efficiency of a 300 MWe utility boiler retrofitted by air staging. *Applied Energy* 86 (9): 1797–1803.

Liu, L., Chen, C., Zhao, Y., and Zhao, E. (2015a). China's carbon-emissions trading: overview, challenges and future. *Renewable and Sustainable Energy Reviews* 49: 254–266.

Liu, Z., Guan, D., Wei, W. et al. (2015b). Reduced carbon emission estimates from fossil fuel combustion and cement production in China. *Nature* 524 (7565): 335.

Low, J., Cheng, B., and Yu, J. (2017). Surface modification and enhanced photocatalytic CO_2 reduction performance of TIO_2: a review. *Applied Surface Science* 392: 658–686.

Lv, C., Xu, J., Xie, H. et al. (2016). Equilibrium strategy based coal blending method for combined carbon and PM_{10} emissions reductions. *Applied Energy* 183: 1035–1052.

Mao, X.Q., Zeng, A., Hu, T. et al. (2014). Co-control of local air pollutants and CO_2 from the Chinese coal-fired power industry. *Journal of Cleaner Production* 67 (67): 220–227.

Moon, C., Sung, Y., Ahn, S. et al. (2013). Effect of blending ratio on combustion performance in blends of biomass and coals of different ranks. *Experimental Thermal & Fluid Science* 47 (5): 232–240.

Olivier, J., Janssens-Maenhout, G., Muntean, M., and Peters, J. (2012). Trends in global CO_2 emissions: 2015 report.

Pain, S. (2017). Power through the ages. *Nature* 551: 134–137.

Pehl, M., Arvesen, A., Humpender, F. et al. (2017). Understanding future emissions from low-carbon power systems by integration of life-cycle assessment and integrated energy modelling. *Nature Energy* 2 (12): 939–945.

Qi, Y., Stern, N., Wu, T. et al. (2016). China's post-coal growth. *Nature Geoscience* 9: 564–566.

Sahu, S.G., Chakraborty, N., and Sarkar, P. (2014). Coal-biomass co-combustion: an overview. *Renewable and Sustainable Energy Reviews* 39 (6): 575–586.

Sami, M., Annamalai, K., and Wooldridge, M. (2001). Co-firing of coal and biomass fuel blends. *Progress in Energy and Combustion Science* 27 (2): 171–214.

Service, R.F. (2017). Cleaning up coal–cost-effectively. *Science* 356: 798.

Shih, J.S. and Frey, H.C. (2007). Coal blending optimization under uncertainty. *European Journal of Operational Research* 83 (3): 452–465.

Sinha, A., Malo, P., Frantsev, A., and Deb, K. (2014). Finding optimal strategies in a multi-period multi-leader–follower Stackelberg game using an evolutionary algorithm. *Computers & Operations Research* 41 (1): 374–385.

Sondreal, E.A., Benson, S.A., Hurley, J.P. et al. (2001). Review of advances in combustion technology and biomass cofiring. *Fuel Processing Technology* 71 (1–3): 7–38.

Tillman, D.A. (2000). Biomass cofiring: the technology, the experience, the combustion consequences. *Biomass and Bioenergy* 19 (6): 365–384.

Tilman, D., Socolow, R., Foley, J.A. et al. (2009). Energy beneficial biofuels–the food, energy, and environment trilemma. *Science* 325 (5938): 270.

UNE Programme (2017). Towards a pollution-free planet.

Van Den Broek, R., Faaij, A., and Van Wijk, A. (1996). Biomass combustion for power generation. *Biomass and Bioenergy* 11 (4): 271–281.

Xu, J. and Zhou, X. (2011). Fuzzy-like multiple objective decision making. *Studies in Fuzziness & Soft Computing* 263: 227–294.

Xu, J., Lv, C., Zhang, M., Yao, L., and Zeng, Z. (2015a). Equilibrium strategy-based optimization method for the coal-water conflict: a perspective from China. *Journal of Environmental Management* 160: 312–323.

Xu, J., Song, X., Wu, Y., and Zeng, Z. (2015b). GIS-modelling based coal-fired power plant site identification and selection. *Applied Energy* 159: 520–539.

Xu, J., Lv, C., Zuo, J. et al. (2016a). A seasonal changes-based equilibrium strategy for coal-water conflict: a case study at the Yanzhou coal field. *Environmental Earth Sciences* 75 (8): 1–18.

Xu, J., Ma, N., and Lv, C. (2016b). Dynamic equilibrium strategy for drought emergency temporary water transfer and allocation management. *Journal of Hydrology* 539: 700–722.

Xu, J., Ma, N., and Xie, H. (2017). Ecological coal mining based dynamic equilibrium strategy to reduce pollution emissions and energy consumption. *Journal of Cleaner Production*. 167: 514–529

Yang, N. and Zhang, R. (2014). Dynamic pricing and inventory management under inventory-dependent demand. *Operations Research* 62 (5): 177–194.

Yang, Y.C., Nie, P.Y., Liu, H.T., and Shen, M.H. (2017). On the welfare effects of subsidy game for renewable energy investment: toward a dynamic equilibrium model. *Renewable Energy*. 121: 420–428.

Prepared By Yearwood-Lee, E. (2015). Carbon Emission Trading.

Yilmaz, B. and Selim, H. (2015). A decision model for cost effective design of biomass based green energy supply chains. *Bioresource Technology* 191: 97–109.

Zhang, G., Lu, J., and Gao, Y. (2015). *Multi-Level Decision Making*. Berlin, Heidelberg: Springer-Verlag.

Zhang, Q., Zhou, D., Zhou, P., and Ding, H. (2013). Cost analysis of straw-based power generation in Jiangsu Province, China. *Applied Energy* 102 (2): 785–793.

Zhao, Y., Wang, S., Duan, L. et al. (2008). Primary air pollutant emissions of coal-fired power plants in China: current status and future prediction. *Atmospheric Environment* 42 (36): 8442–8452.

10

Carbon Emission Reduction-Oriented Equilibrium Strategy for Thermal–Hydro–Wind Generation System

As the International Energy Agency statistics revealed, the world's carbon emissions from fuel combustion in 2014 had reached 32 billion tonnes in which coal-combusted electricity generation contributes a quite remarkable ratio [IEA, 2016]. It is also reported that the developing countries which rely on the coal-dominant energy structures are the main carbon emissions sources [IEA, 2016]. Take China as an example, as one of the most popular developing country, coal-combusted electricity generation emission was the largest at 3.6 billion tonnes, followed by manufacturing (31.64%) and transport (8.63%) [IEA, 2016]. Besides, BP revealed in the World Energy Statistics Yearbook 2016 that if China's coal continued to be mined at the 2015 level, the remaining reserves would last only 31 years, compared to the United States' 292 years and Japan's 296 years [BP, 2016]. However, because of China's energy mix and the ever-increasing economic development, the demand for coal-combusted electricity is expected to remain high for the next 20–30 years [Li et al., 2011]. Therefore, the energy savings and emission reductions being achieved by Chinese coal-combusted electricity generation have attracted extensive global attention. This situation should not only be focused in China, but also all the developing countries which rely on the coal-dominant energy structures.

10.1 Background Introduction

Many previous studies have examined emissions mitigation and the environmental impacts of coal-fired power plants (CPPs) [Mao et al., 2014, Lonsdale et al., 2012]. Some tended to focus on clean technology improvements to transform dirty black coal into clean fuels and chemicals, of which an integral part has been the integration of carbon dioxide capture and storage installations which is also known as carbon capture, utilization, and storage [Haszeldine, 2009, Skorek-Osikowska et al., 2017, Xu et al., 2017]. Skorek-Osikowska, for instance, analyzed the environmental effect of systems integrated with carbon capture installations and found that annual emissions in the post-combustion system decreased from 735 kg CO_2/mWh to 100.77 kg/mWh [Skorek-Osikowska et al., 2017]. These solutions have been considered the most useful methods, because it can achieve high efficiencies, but the capital cost and the additional operation cost are so expensive to deploy on large

Innovative Approaches towards Ecological Coal Mining and Utilization, First Edition.
Jiuping Xu, Heping Xie, and Chengwei Lv.

scale [Tapia et al., 2016]. Others focused on management methods, such as optimal economic-environmental dispatch, carbon emissions trading, and unit commitment strategies [Lv et al., 2016, Wang et al., 2012, Chen et al., 2016a]. Of these, proposing different unit commitment strategies for power plant to determine reasonable scheduling has proven to be effective in controlling emissions [Sun et al., 2017, Jiang et al., 2012]. For example, Chandrasekaran et al. Chandrasekaran et al. [2012] proposed a novel methodology that employs fuzzified binary real coded artificial bee colony algorithm for solving multi-objective unit commitment problem, in which three conflicting functions such as fuel cost, emission, and reliability level of the system are simultaneously considered. However, with more and more variable renewable energy integrated to the traditional generation network, significant operating challenges also occur. To propose more appropriate scheduling strategies, world energy system is becoming more and more important.

Actually, renewable energy sources haven been actively studied and investigated by global researchers and operation engineers to search new methods in integrating them into traditional power network [Squalli, 2017, Best, 2017]. Among all these sources, wind power, with nearly zero carbon emissions and the largest market share, has become an competitive integrating source [Wang et al., 2012, Garciagonzalez, 2008, Huang et al., 2015]. Many researches have been conducted to integrate this source. For example, Laia et al. Laia et al. [2016] presented a stochastic mixed-integer linear programming approach for solving the self-scheduling problem of a price-taker thermal and wind power producer taking part in a pool-based electricity market. Sun et al. Sun et al. [2017] proposed an optimal day-ahead wind–thermal unit commitment strategy considering statistical and predicted features of wind speeds to achieve secure power system operations and economic scheduling. These efforts have been significant important in solving the challenges bring about by integrating wind energy into power networks, but some thermal units at low loads have low efficiency, are not environmentally friendly, and are a safety risk [Gu et al., 2016, Davidson et al., 2016]; therefore, new coordination operation sources need to be integrated.

Hydropower, which can be turn on and off rapidly, is regarded as one promising complementary energy source to mitigate wind power fluctuation and reduce dependence on thermal power for reserve. Therefore, combining hydro, wind, and thermal energy sources into a hybrid system and operating the system coordinately is attracting more and more attention [Dubey et al., 2016, Chen et al., 2016b]. To enhance the flexibility and reliability of power system operations, Chen et al. proposed a distributionally robust hydro–thermal–wind economic dispatch method which described the uncertain wind power output and optimized expected operating costs for the worst distribution [Chen et al., 2016b]. For example, Zhou et al. Zhou et al. [2016] constructs a unit commitment model that coordinates hydro and thermal power generation to support secure and economic wind power integration. Chen et al. Chen et al. [2017a] presented a short-term hydro–thermal–wind economic emission dispatching problem, which aimed to distribute the load among hydro, thermal, and wind power units to simultaneously minimize economic cost and pollutant emission. These researchs have made very important contributions in modeling the coordinated operation of hybrid wind–hydro–thermal power generation system, but few have considered the carbon emission reduction and reliable power supplier of this system under the influence of stochastic wind and water flows.

In the optimal hybrid wind–hydro–thermal generation system scheduling problem, methods to handle renewable energy-related uncertainties and methods to balance environmental-reliable generation trade-offs need detail statements.

Due to the differences in weather conditions or geographical locations, wind speed changes across seasons, days, hours, and even minutes [Feng and Shen, 2017]. On the other hand, although hydropower has advantages of good peak shaving ability, flexible start/stop operations, and promising reserves, the water flow of power reservoir is high uncertain because of rainfall inconsistencies [De Queiroz, 2016]. These strong natural resource-related uncertainties of wind and hydropower throw significant scheduling complexity to hybrid power operators, making it unreasonable for them to provide only one determined scheduling strategy for different scenarios. To handle these uncertainties, many efforts have been made, for example, Jiang et al. Jiang et al. [2012] proposed a robust optimization approach to accommodate wind output uncertainty, with the objective of providing a robust unit commitment schedule for the thermal generators in the day-ahead market. Séguin et al. Seguin et al. [2016] presented an optimization approach to solve the short-term hydropower unit commitment and loading problem with uncertain inflows, in which the uncertain water flow is built by scenario trees. However, the seasonal properties of wind speed and hydro water flow have not been concurrently considered. In this chapter, 12 typical scheduling scenarios are established based on the wind and hydropower uncertainties. The scheduling scenarios are generated by interacting wind speed scenario set and water flow level set. The wind speed set includes four elements, representing typical daily wind speed data in the four different seasons. The water flow level set includes three elements; high, normal, and low. Details of these scenarios are shown in Figure 10.1.

The dilemma between reliable power supply and carbon emissions reductions is still difficult problem. The integration of renewable wind power makes hybrid power system cleaner in generation, but it also brings about highly randomness and variability, making it difficult to guarantee stable and reliable electricity supplies [Wang et al., 2012]. On the other hand, although coal-combusted thermal power is highly reliable and stable, the serious pollutants and emissions released during coal burning process have led to significant environmental damages [Xu et al., 2017, Schill et al., 2017]. Therefore, the system operator needs to seek a reasonable scheduling strategy that can achieve equilibrium reliable power supply and emission reduction for the hybrid wind–hydro–thermal generation system. Multi-objective method is very efficient in solving trade-off problems among different goals [Tsai and Chen, 2016]. For example, Wang et al. applied the multi-objective method to the microgrids scheduling problems and successfully achieved the effectiveness of this approach in dealing with economic-environmental conflicts [Wang et al., 2017]. Demissie et al. applied the multi-objective method to the natural gas pipeline networks operation problem and successfully achieved the balance between power consumption minimization and gas delivery flow rate

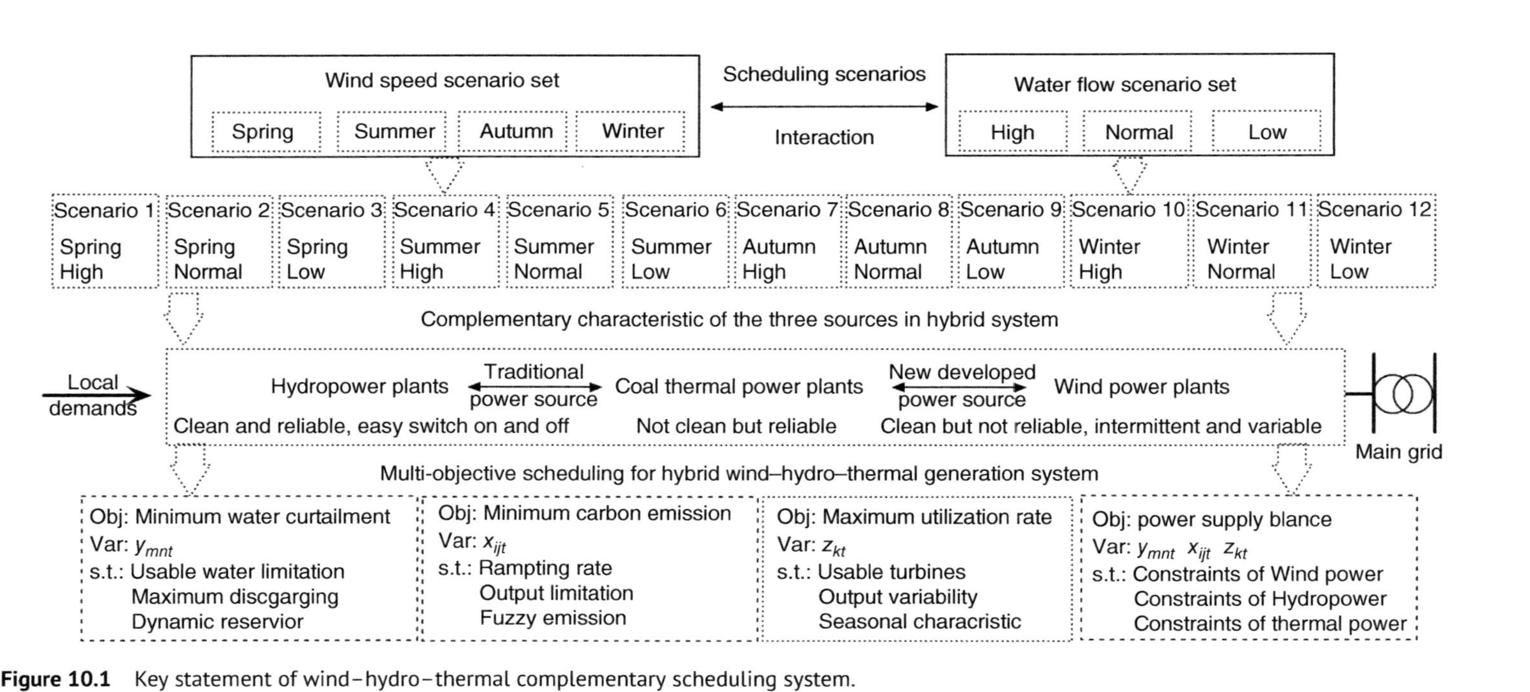

Figure 10.1 Key statement of wind–hydro–thermal complementary scheduling system.

maximization [Demissie et al., 2017]. Encouraged by these successful experiences, this chapter prepares to apply the multi-objective method for system operators to deal with the trade-off between clean and reliable power generation.

10.2 Modeling

This section adopts multi-objective programming to address the trade-offs between the reliable and environmental power generation scheduling in the hydro–thermal–wind complementary system, and the detail scheduling model is built as follows.

10.2.1 Notations

The following notations will be used in the modeling process.

Indices

i = Index for coal fired power plant.
j = index for coal fired units.
k = index for wind power plant.
l = index for conventional wind turbines.
m = index for hydropower station.
n = index for hydro turbines.
t = time period index.

Decision variables

g_{ijt} = the on/off status of thermal unit i in power plant i at time t.
x_{ijt} = the coal to be combusted by the thermal unit i in power plant i at time t.
y_{mnt} = the water flow through turbine n in hydro station m at time t.
z_{kt} = the number of conventional wind turbines to be operated at wind farm k at time t.

Certain parameters

$A_i^{\max}$ = upper bound of coal that can be used by coal fired power plant i.
$A_{ij-\min}$ = minimum output of unit j at power plant i.
$A_{ij-\max}$ = maximum output of unit j at power plant i.
A_{ij}^- = the lower limitation of unit ramping rate.
A_{ij}^+ = the upper limitation of unit ramping rate.
$A_{ct}^{\min}$ = minimum output of thermal power plants.
$A_{ct}^{\max}$ = maximum output of thermal power plants.
B = maximum discharge flowing into the turbine.
B_{mt} = water left in the reservoir m at time t.
$B_{\min,m}$ = the minimum amount of water during the scheduled time period at station m.

(Continued)

(Continued)

$B_{\max,m}$ = the amount of water that can be used to generate electricity during the allocation period.
$B_{ht}^{\max}$ = the maximum output of hydropower plants at time.
$B_{ht}^{\min}$ = the minimum output of hydropower plants at time.
$c_{tur,t}$ = the output of a wind turbine.
c^r = the rated power output of wind turbines.
C = the area of the cylinder swept by the blades of turbine.
C_k^1 = total sample turbines of wind farm k.
C_k^2 = total conventional turbines of wind farm k.
q_{mt} = the water inflow into the reservoir m at time t, $(\mathrm{m}^3/\mathrm{s})$.
v_{kt} = the actual wind speed.
v_{in} = cut-in speed of the wind turbines.
v_{rated} = rated speed of the wind turbines.
v_{out} = cut-out speed of the wind turbines.
D_t = power demand at time t.
$E\left(\tilde{A}_{ij}\right)$ = expected value of uncertain parameter $\tilde{A}_{ij}$.
$E\left(\tilde{T}_{ij}\right)$ = expected value of uncertain parameter $\tilde{T}_{ij}$.
F_t^c = output of the coal-combusted power plants during time t.
F_t^h = output of the hydropower plants during time t.
F_t^w = output of the wind power plants during time t.
K_1 = carbon emissions of previous production period.
K_2 = wind power utilization rate of previous production period.
α = positive and negative attitude parameter of power generator toward parameter $\tilde{A}_{ij}$.
β_m = the comprehensive efficiency of the turbine and generator at plant m.
χ = coefficient of wind turbine.
ρ = represent the air density.
ε = the system load spinning reserve rate.
ϖ_1 = attitude parameter of system operator toward carbon emissions reduction.
ϖ_2 = attitude parameter of system operator toward water resource waste.
ϖ_3 = attitude parameter of system operator toward wind power utilization.

Uncertain parameters
$\tilde{A}_{ij}$ = coal to carbon parameter, which means $\tilde{A}_{ij}$ kilograms of carbon emissions will be released if 1 tonnes of coal is burned.
$\tilde{T}_{ij}$ = coal to electricity parameter, which means $\tilde{A}_{ij}$ megawatt of power will be generated if 1 tonnes of coal is burned.

10.2.2 Objectives

10.2.2.1 Carbon Emissions Reduction

To minimize carbon emissions of electricity generation process is very important for social development, especially for the coal combustion thermal units of hybrid generation system. Here we use x_{ijt} to represent the coal to be combusted by the thermal unit i in power plant i at time t, g_{ijt} represent the on/off status of thermal unit i in power plant i at time t, and $\tilde{A}_{ij}$ to represent the coal to carbon parameter, which means $\tilde{A}_{ij}$ kilograms of carbon emissions will be released if 1 tonnes of coal is burned. However, due to the lack of historical data and immature technology, the exact value of the coal to carbon parameter is very difficult to obtain, but through field research with practical engineers, we found that the value of this parameter always belonging to an interval, which is very similar to the character of fuzzy number. Therefore, to ensure a more accurate and more scientific value, uncertain parameter $\tilde{A}_{ij}$ is represented by a triangular fuzzy number. Then to minimize carbon emissions in the generation process can be written as follows:

$$\min f_1 = \sum_t \sum_i \sum_j E\left(\tilde{A}_{ij}\right) g_{ijt} x_{ijt} \tag{10.1}$$

In which, $E\left(\tilde{A}_{ij}\right) = \frac{1-\alpha}{2}(a_{ij}^1 + a_{ij}^2) + \frac{\alpha}{2}(a_{ij}^3 + a_{ij}^4)$ is the expected value of the uncertain parameter $\widetilde{A_{ij}}$ calculated by the method proposed by Xu and Zhou Xu and Zhou [2011]. In this representation, the parameters a_{ij}^1 and a_{ij}^4 are the minimum and maximum possible value of the uncertain parameter $\tilde{A}_{ij}$, and the most likely value of $\tilde{A}_{ij}$ is between a_{ij}^2 and a_{ij}^3, therefore, $a_{ij}^1 < a_{ij}^2 < a_{ij}^3 < a_{ij}^4$. The parameter α, $(0 \leq \alpha \leq 1)$ is positive and negative attitude parameter of power generator toward parameter $\tilde{A}_{ij}$.

10.2.2.2 Water Resources Wastes

Hydropower is very flexible and clean, if a higher proportion of hydropower integrated in the electricity supply process, environmental pressure and reserve demands can be reduced [Alizadeh et al., 2017]. However, highly flexibility of hydro turbines also bring a certain amount of water waste, because it is usually used as reserve to shave the peak [Singh and Singal, 2017]. Therefore, to maintain sustainable scheduling in the hybrid wind–hydro–thermal generation system, the amount of water that can be used to generate electricity but was wasted should be minimized. Here we use $B_{\max,m}$ to represent the amount of water that can be used to generate electricity during the allocation period, and $y_{mnt}(\mathrm{m}^3/\mathrm{s})$ is the water flow through turbine n in hydro station m at time t to generate electricity, and then the objective of minimized water resources wastes can be represented as follows:

$$\min f_2 = \sum_m (B_{\max,m} - \sum_t \sum_n y_{mnt} \Delta t) \tag{10.2}$$

10.2.2.3 Wind Power Utilization

As Lu et al. [2016] stated, the utilization of the wind power equipment in China is not inefficient enough yet. Many studies have been conducted to reduce wind power curtailment and increase wind power utilization (see [Zhang et al., 2016, Xiong et al., 2016]). Here we use the sample turbines method to calculate the wind energy consumption of the wind farm, and this method was proposed by State Electricity Regulation Commission of China in 2013, see details in BP [2013]. Let C_k^1 be the sample turbines of wind farm k, it accounts 10–20% of the total turbines, z_{kt} be the conventional turbines to be operated at wind farm k in time period t, and $c_{tur,t}$ represent the output of a wind turbine; therefore, the actual output of wind farm k can be written as: $\sum_k c_{tur,t}(z_{kt} + c_k^1)$. Further, let C_k^2 be the total conventional turbines of wind farm k, then the theoretical output at wind farm k is $\sum_k c_{tur,t}(c_k^1 + c_k^2)$; therefore, the wind power utilization rate is equal to the ratio of actual output and theoretical output, which can be written as:

$$\max f_3 = \frac{\sum_t \sum_k c_{tur,t}(z_{kt} + c_k^1)}{\sum_t \sum_k c_{tur,t}(c_k^1 + c_k^2)} \tag{10.3}$$

10.2.2.4 Power Supply Balance

In this chapter, both power supply shortages and excesses are considered when considering power supply balance. Because if there is more power generated than necessary, it is a waste of resources, and if the power supply field to meet demand, it will cause economic losses and social problems [Lv et al., 2016]. Here in the hybrid generation system, hydropower plants, wind power plants, and coal-fired thermal power plants are scheduled simultaneously. Let F_t^c represent the output of the coal-combusted power plants during time t, F_t^h and F_t^w represent the outputs of the hydropower and wind power plants at time t, and D_t is the power demand at time t. Then this objective function can be written as follows:

$$\min f_4 = \sum_t \left(F_t^c + F_t^h + F_t^w - D_t\right)^2 \tag{10.4}$$

For the output of coal-combusted power plants F_t^c:

$$F_t^c = \sum_i \sum_j E\left(\tilde{G}_{ij}\right) x_{ijt} g_{ijt} \tag{10.5}$$

For the output of wind power plants F_t^w:

$$F_t^w = \sum_k c_{tur,t}(z_{kt} + c_k^1) \tag{10.6}$$

For the output of hydropower plants F_t^h:

$$F_t^h = \sum_m \sum_n 9.81 y_{mnt} H_m \beta_m \tag{10.7}$$

In which, x_{ijt} is the coal to be combusted by the thermal unit i in power plant i at time t and is the decision variable for the CPPs. z_{kt} is the number of conventional wind turbines to be operated at wind farm k at time t, the more the turbines operated, the more wind power will be sent to the main grid, and it is the decision variable for the wind farms. The decision variable for hydropower is $y_{mnt}(\mathrm{m}^3/\mathrm{s})$, which is the water flow through turbine n in hydro station m at time t that allows the system operator to provide a reasonable amount of power; H_m is the fall height at hydropower plant m, and β_m is the comprehensive efficiency of the turbine and generator at plant m, and the specific weight of the water is $9.81(\mathrm{kN}/\mathrm{m}^3)$ [Lopes and Borges, 2014].

10.2.3 Constraint

In this section, constraints for the equilibrium strategy-based scheduling of hybrid wind–hydro–thermal generation system are given as follows:

10.2.3.1 Constraints of Wind Power

10.2.3.1.1 Power Output Curve

Wind turbine power output curve is characterized by cut-in, rated, and cut-out speeds [Aliari and Haghani, 2016]. Speeds below the cut-in speed are insufficient to start the turbines, while speeds higher than the cut-out speed may damage turbine components [Shi et al., 2017, Huang et al., 2015]. Here we let $c_{tur,t}$ represent the output of a wind turbine, c^r represent the rated power output of wind turbines, χ represent coefficient of wind turbine, ρ represent the air density, C is the area of the cylinder swept by the blades of turbine, and v_{kt} is the actual wind speed, then the output curve of wind turbines is as follows:

$$c_{tur,t}\left(v_{kt}\right) = \begin{cases} 0, & 0 \leq v_{kt} < v_{in} \\ \frac{1}{2}\chi \rho_k {v_{kt}}^3 C, & v_{in} \leq v_{kt} < v_{rated} \\ c^r, & v_{rated} \leq v_{kt} < v_{out} \\ 0, & v_{out} \leq v_{kt} \end{cases} \tag{10.8}$$

10.2.3.1.2 Useable Wind Turbines

The total installed wind turbines in each wind farms are limited; therefore, the wind turbines that can be operated during the dispatch period have upper limits. Here we use c_k^2 to represent all conventional turbines at wind farm k, which is also the upper level of usable turbines at plant k; therefore, this constraint can be written as:

$$z_{kt} \leq c_k^2 \tag{10.9}$$

10.2.3.2 Constraints of Coal-Combusted Power Plants

10.2.3.2.1 Accessible Coal

The maximum amount of coal available during the scheduling period is limited, that is, the plant cannot purchase unlimited coal. Here we let $A_i^{\max}$ to represent the upper bound of coal that can be used by CPP i for electricity generation, so this constraint can be written as:

$$\sum_t \sum_j x_{ijt} g_{ijt} \leq A_i^{\max} \tag{10.10}$$

10.2.3.2.2 Unit Ramping Rate

To maintain a sustainable and stable utilization at the CPPs, a unit ramp rate must be seriously considered when deciding on the peak shaving unit [Xie et al., 2016]. Here we use A_{ij}^- and A_{ij}^+ to represent the lower and upper limitation of unit ramping rate, so this constraint can be written as:

$$\begin{array}{l} E\left[\tilde{T}_{ij}\right] x_{ijt} - E\left[\tilde{T}_{ij}\right] x_{ij(t-1)} \le g_{ijt} A_{ij}^+ + \left(1 - g_{ij(t-1)}\right) A_{ij-\max} \\ E\left[\tilde{T}_{ij}\right] x_{ij(t-1)} - E\left[\tilde{T}_{ij}\right] x_{ijt} \le g_{ijt} A_{ij}^- + \left(1 - g_{ijt}\right) A_{ij-\max} \end{array} \tag{10.11}$$

10.2.3.2.3 Unit Output Limitation

Thermal unit output limitation ensures output is within the normal range [Laia et al., 2016]. Let $A_{ij-\min}$ be the minimum output of unit j at power plant i, $A_{ij-\max}$ be the maximum output of unit j at power plant i, so this constraint can be written as:

$$A_{ij-\min} g_{ijt} \le E\left[\tilde{T}_{ij}\right] x_{ijt} \le A_{ij-\max} g_{ijt} \tag{10.12}$$

10.2.3.2.4 Minimal On/Off Time

The total number of hours for which plant i unit j has been running (A_{ij}^{on}) must be greater than or equal to the minimum plant up time (A_{up}). Similarly, the total number of hours for which power unit j in power plant i has been turned off (A_{ij}^{off}) must be greater than or equal to the minimum plant down time (A_{down}) [Chen et al., 2017b]. Therefore, when deciding the on/off status of coal-fired unit j at thermal power plant i at time t, this constraint can be written as follows.

$$g_{ijt} = \begin{cases} 1, & A_{ijt,on} \ge A_{ij,up} \\ 0, & A_{ijt,off} \ge A_{ij,up} \end{cases} \tag{10.13}$$

10.2.3.3 Constraint of Hydropower Station

10.2.3.3.1 Accessible Water

These exists lower and upper bound of water available for electricity generation. Here let $B_{\min,m}$ be the minimum amount of water during the scheduled time period at station m, and $B_{\max,m}$ be the maximum amount of water that can be used for electricity generation; therefore, the total amount of water used at station m can be represented as: $\sum_t \sum_n y_{mnt} \Delta t$, this constraint can be written as [Panda et al., 2017]:

$$A_{\min,m} \le \sum_t \sum_n y_{mnt} \Delta t \le A_{\max,m} \tag{10.14}$$

10.2.3.3.2 Maximum Water Discharge

The maximum amount of water flow that can be tolerated by hydro turbine is limited. Let B be the maximum discharge flowing into the turbine, which means that the water used to generate the electricity through the water turbines cannot be larger than this parameter, and this constraint can be written as:

$$y_{mnt} \le B \tag{10.15}$$

10.2.3.3.3 *Dynamic Water Inventory*

In the scheduling period, the total water in the reservoir are changing dynamically [Banerjee et al., 2016]. Let B_{mt} be the water left in the reservoir m at time t, and $q_{mt}(\text{m}^3/\text{s})$ be the water inflow into the reservoir m at time t; therefore, the total water flow into the reservoir can be written as $q_{mt} \times \Delta t$, where Δt is a time conversion parameter. During time period t, the total water out flow quantity by all hydropower plants can be written as $\sum_m \sum_n y_{mnt} \Delta t$; therefore, this constraint can be written as:

$$\begin{aligned} B_{mt} &= B_{m(t-1)} + q_{mt}\Delta t - \sum_m \sum_n y_{mnt}\Delta t \\ B_{m1} &= q_{m1}\Delta t - \sum_m \sum_n y_{mnt}\Delta t \end{aligned} \tag{10.16}$$

10.2.3.4 Constraints of Hybrid Generation System

10.2.3.4.1 *System Spinning Reserve*

The stochastic characteristic of wind and hydropower indicating that thermal power needs to assign spinning reserve for the hybrid generation system. In this chapter, we define the minimum and maximum output of thermal power plants at time t as $A_{ct}^{\min} = \sum_i \sum_j A_{ij-\min}$ and $A_{ct}^{\max} = \sum_i \sum_j A_{ij-\max}$, the maximum and minimum output of hydropower plants at time t as $B_{ht}^{\max} = \sum_m \sum_n B_{mn-\max}\Delta t$ and $B_{ht}^{\min} = \sum_m \sum_n B_{mn-\min}\Delta t$, let ε represent the system load spinning reserve rate, so this constraint can be written as [Wang et al., 2016]:

$$\begin{aligned} &\sum_i \left(A_{ct}^{\max} + B_{ht}^{\max} + C_{wt}\right) \geq D_t(1+\varepsilon) \\ &\sum_i \left(A_{ct}^{\min} + B_{ht}^{\min} + C_{wt}\right) \leq D_t \end{aligned} \tag{10.17}$$

10.2.3.5 Global Model

Integrating the objectives and constraints we stated above, the optimization scheduling model of hybrid hydro–wind–thermal generation system can be obtained, as shown in Eq. (10.18). In this multi-objective model, characteristics of hydro, wind, and thermal power are simultaneously considered to present a reasonable schedule to achieve higher emission reduction and renewable energy utilization. The flexible hydropower can mitigate fluctuations of wind power output, so that dependance on thermal power reserves will be alleviated. The clean wind power integration can reduce a certain amount of thermal power output, so as to achieve the goal of emission reduction. And reliable thermal power will bear the basic load. In the hybrid system generation scheduling problem, objectives of reducing carbon emissions, minimizing water resources waste, maximizing wind power utilization, and minimizing power supply and demand risk are considered at the same time by establishing multi-objective model. Further, in the established model, the wind turbine output piece function is applied to characterize wind turbine power curve; the fuzzy theory is applied to describe the coal to power output parameter and the coal to carbon emission parameter, which are difficult to precisely determine; the dynamic theory is applied in the hydropower plants to describe the water volume in the reservoir.

$$
\min f_1 = \sum_t \sum_i \sum_j E\left(\tilde{A}_{ij}\right) g_{ijt} x_{ijt}
$$
$$
\min f_2 = \sum_m (B_{\max,m} - \sum_t \sum_n y_{mnt} h_{mnt} \Delta t)
$$
$$
\max f_3 = \frac{\sum_t \sum_k c_{tur,t}(z_{kt} + c_k^1)}{\sum_t \sum_k c_{tur,t}(c_k^1 + c_k^2)}
$$
$$
\min f_4 = \sum_t \left(F_t^c + F_t^h + F_t^w - D_t\right)^2
$$
$$
\left\{
\begin{array}{l}
F_t^c = \sum_i \sum_j E\left(\tilde{G}_{ij}\right) x_{ijt} g_{ijt} \\
F_t^w = \sum_k c_{tur,t}(z_{kt} + c_k^1) \\
F_t^h = \sum_m \sum_n 9.81 y_{mnt} H_m \beta_m \\
c_{tur,t}\left(v_{kt}\right) = \left\{
\begin{array}{ll}
0, & 0 \le v_{kt} < v_{in} \\
\frac{1}{2} \chi \rho_k {v_{kt}}^3 C, & v_{in} \le v_{kt} < v_{rated} \\
c^r, & v_{rated} \le v_{kt} < v_{out} \\
0, & v_{out} \le v_{kt}
\end{array}
\right. \\
z_{kt} \le c_k^2 \\
\sum_t \sum_j x_{ijt} g_{ijt} \le A_i^{\max} \\
E\left[\tilde{T}_{ij}\right] x_{ijt} - E\left[\tilde{T}_{ij}\right] x_{ij(t-1)} \le g_{ijt} A_{ij}^{+} + \left(1 - g_{ij(t-1)}\right) A_{ij-\max} \\
E\left[\tilde{T}_{ij}\right] x_{ij(t-1)} - E\left[\tilde{T}_{ij}\right] x_{ijt} \le g_{ijt} A_{ij}^{-} + \left(1 - g_{ijt}\right) A_{ij-\max} \\
A_{ij-\min} g_{ijt} \le E\left[\tilde{T}_{ij}\right] x_{ijt} \le A_{ij-\max} g_{ijt} \\
g_{ijt} = \left\{
\begin{array}{ll}
1, & A_{ijt,on} \ge A_{ij,up} \\
0, & A_{ijt,off} \ge A_{ij,up}
\end{array}
\right. \\
A_{\min,m} \le \sum_t \sum_n y_{mnt} \Delta t \le A_{\max,m} \\
y_{mnt} \le B \\
B_{mt} = B_{m(t-1)} + q_{mt} \Delta t - \sum_m \sum_n y_{mnt} \Delta t \\
B_{m1} = q_{m1} \Delta t - \sum_m \sum_n y_{mnt} \Delta t \\
\sum_i \left(A_{ct}^{\max} + B_{ht}^{\max} + C_{wt}\right) \ge D_t (1 + \varepsilon) \\
\sum_i \left(A_{ct}^{\min} + B_{ht}^{\min} + C_{wt}\right) \le D_{t358}
\end{array}
\right.
\tag{10.18}
$$

Compared with the models in previous complementary generation strategy, this model has a more comprehensive and systematic structure, in which the top three electricity sources in China are coordinated to seek a global optimal solution, and 12 different scheduling scenarios are proposed to ensure better coordination. The model proposed is suitable for scenarios where wind speed and water flow have been determined. For the other proposed scenarios, while the mathematical essence and mathematical forms are the same, the wind speed and water flow data are correspondingly adjusted for calculation. By inserting hydropower plants into traditional wind–thermal complementary system, this model has more extensive peak shaving ability and can therefore assist system operator to have great control over the actual emission reduction and energy-saving effectiveness. In addition, different water flow level and wind speed scenarios are consider to ensure the model is more practicable, thus allowing the system operator to select the most appropriate scheduling strategy to ensure greater precision in the clean and reliable objective.

10.3 Case Study

In this section, a practical daily generation scheduling example from Bijie, Guizhou, China is given to show the practicality and efficiency of the proposed optimization.

10.3.1 Case Description

The Guizhou Provincial Development and Reform Commission reported that standard coal consumption for Guizhou power was 327 g/kWh, which is 12 g/kWh higher than the national average. Also, 174 billion kWh electricity was generated in 2015 by thermal power plants, accompanied by 56 billion tonnes of carbon emissions [Guizhou Province, 2017a]. Therefore, the reduction of carbon emitted from thermal power has become urgent. The International Energy Network reported that installed capacities for hydro, thermal, and wind power in Guizhou province in October 2017 were 18.8, 30.88, and 4.6 thousand megawatts, respectively (in Chinese) [Guizhou Province, 2017b]. It is obvious that Guizhou's installed capacity for thermal, hydro, and wind power is lower than in those in other parts of China, and it is reasonable to establish a complementary power generation system. A successful application of a complementary system in Guizhou's power grid could assist in persuading other provinces to build complementary systems. In this chapter, two CPPs, two wind power plants (WPPs), and one hydro power plant (HPP) in Bijie City were chosen as the studied area, which is helpful to the establishment of complementary system.

10.3.2 Model Transformation

A reliable power supplier is an essential for economic and social development. Meanwhile, with the demand for resource conservation and environmental sustainability, renewable energy needs to be used, and pollution emissions should be minimized. It is known that wind energy utilization enable to control these within an acceptable range. Based on Zeng et al.'s research, three attitude parameters of system operators toward the reduction of CO_2 emissions, the waste water resources, and the wind energy utilization rate are introduced [Zeng et al., 2014]. ϖ_1 is the attitude parameter of system operator toward carbon emissions reduction, to be specific, carbon emissions in this scheduling period is ϖ_1 times less than (or equal) to that of the previous production period. In the same way, ϖ_2 is the attitude parameter toward water resource waste, and ϖ_3 is the attitude parameter toward wind power utilization. K_1 and K_2, respectively, represent the carbon emissions and the waste water resource of previous production period. To solve the proposed model, we transformed it into its equivalent single-objective form as follows.

$$
\min f = \sum_t \left(F_t^c + F_t^h + F_t^w - D_t\right)^2
$$

$$
\left\{
\begin{array}{l}
\sum_t \sum_i \sum_j E\left(\tilde{A}_{ij}\right) g_{ijt} x_{ijt} \leq \varpi_1 K_1 \\
\sum_m (B_{\max,m} - \sum_t \sum_n y_{mnt} h_{mnt} \Delta t) \leq \varpi_2 K_2 \\
\frac{\sum_t \sum_k c_{tur,t}(z_{kt}+c_k^1)}{\sum_t \sum_k c_{tur,t}(c_k^1+c_k^2)} \geq \varpi_3 \\
F_t^c = \sum_i \sum_j E\left(\tilde{G}_{ij}\right) x_{ijt} g_{ijt} \\
F_t^w = \sum_k c_{tur,t}(z_{kt} + c_k^1) \\
F_t^h = \sum_m \sum_n 9.81 y_{mnt} H_m \beta_m \\
c_{tur,t}\left(v_{kt}\right) = \begin{cases} 0, & 0 \leq v_{kt} < v_{in} \\ \frac{1}{2} \chi \rho_k {v_{kt}}^3 C, & v_{in} \leq v_{kt} < v_{rated} \\ c^r, & v_{rated} \leq v_{kt} < v_{out} \\ 0, & v_{out} \leq v_{kt} \end{cases} \\
z_{kt} \leq c_k^2 \\
\sum_t \sum_j x_{ijt} g_{ijt} \leq A_i^{\max} \\
E\left[\tilde{T}_{ij}\right] x_{ijt} - E\left[\tilde{T}_{ij}\right] x_{ij(t-1)} \leq g_{ijt} A_{ij}^+ + \left(1 - g_{ij(t-1)}\right) A_{ij-\max} \\
E\left[\tilde{T}_{ij}\right] x_{ij(t-1)} - E\left[\tilde{T}_{ij}\right] x_{ijt} \leq g_{ijt} A_{ij}^- + \left(1 - g_{ijt}\right) A_{ij-\max} \\
A_{ij-\min} g_{ijt} \leq E\left[\tilde{T}_{ij}\right] x_{ijt} \leq A_{ij-\max} g_{ijt} \\
g_{ijt} = \begin{cases} 1, & A_{ijt,on} \geq A_{ij,up} \\ 0, & A_{ijt,off} \geq A_{ij,up} \end{cases} \\
A_{\min,m} \leq \sum_t \sum_n y_{mnt} \Delta t \leq A_{\max,m} \\
y_{mnt} \leq B \\
B_{mt} = B_{m(t-1)} + q_{mt} \Delta t - \sum_m \sum_n y_{mnt} \Delta t \\
B_{m1} = q_{m1} \Delta t - \sum_m \sum_n y_{mnt} \Delta t \\
\sum_i \left(A_{ct}^{\max} + B_{ht}^{\max} + C_{wt}\right) \geq D_t (1 + \varepsilon) \\
\sum_i \left(A_{ct}^{\min} + B_{ht}^{\min} + C_{wt}\right) \leq D_t
\end{array}
\right.
\tag{10.19}
$$

As the global model has already been transformed into a single-objective model that has mixed-integer nonlinear programming, Lingo was chosen to solve the model directly. Widely used in scientific research and industry, Lingo is characterized by fast execution and convenient inputs to modify, solve, and analyze mathematical optimization problems. In our model, the decision variable z_{kt}, which indicates the number of wind turbines operated at wind farm k during time t, is an integer variable, the wind turbine power output P_{kt} is a piecewise function directly related to wind speed v_{kt}, and the objective function is a nonlinear function. As Lingo can be used to solve linear nonlinear programming with a single objective, and it also allows the decision variables in the model to be integers, Lingo13.0 and a selected global solver was used to solve this transformed model.

10.4 Data Collection

After the presentation of the case region, data and parameters were collected for the calculation. The data in Table 10.1 for the hydropower plants were mainly collected from published documents [Liang, 2006]. The turbine parameters, water head data, and some rated parameters for the hydropower station in Table 10.2 were collected from Qing [2007]. Data for the wind farms in Tables 10.3 and 10.4 were mainly were

Table 10.1 Parameters of hydropower station.

Parameters	Unit	Value	Parameters	Unit	Value
Annual average flow	m^3/s	155	Maximum discharge flowing turbine	m^3/s	163.5
Annual average runoff	Billion m^3	4.89	Comprehensive output coefficient	—	8.4
Normal water level	m	1140	Annual average generating capacity	Billion kWh	1.56
Dead water level	m	1076	Normal reservoir capacity	Billion m^3	4.497
Installed capacity	MW	3×200	Regulating storage capacity	Billion m^3	3.361

Table 10.2 Turbine parameters of hydropower station.

Parameters	Value	Parameters	Value	Parameters	Value
Maximum water head	164 m	Rated water head	135 m	Minimum water head	90 m
Rated power	204.1 MW	Rated flow	164.1 m^3/s		

Table 10.3 Parameters of wind farm 1.

Parameters	Value	Parameters	Unit	Value	Parameters	Unit	Value
Turbines in total	198	Turbine rated capacity	kW	1500	Cut-in wind speed	m/s	3.5
Normal turbines	178	Installed capacity	MW	297	Rated wind speed	m/s	11
Sample turbines	20	Air density	g/cm^3	0.92	Cut-out wind speed	m/s	23

Table 10.4 Parameters of wind farm 2.

Parameters	Value	Parameters	Unit	Value	Parameters	Unit	Value
Turbines in total	122	Turbine rated capacity	kW	1500	Cut-in wind speed	m/s	3.2
Normal turbines	110	Installed capacity	MW	183	Rated wind speed	m/s	11
Sample turbines	12	Air density	g/cm^3	0.95	Cut-out wind speed	m/s	25

Table 10.5 Crisp parameters of coal-combusted power plants.

Parameters	Units of measure	Dafang CPP	Bijie CPP
Parameters	Units	Dafang CPP	Bijie CPP
Installed capacity	MW	4 × 300	2 × 150
Minimum unit output	MW	120	45
Maximum unit output	MW	330	150
Units ramping rate	MW/min	12	5

Table 10.6 Fuzzy parameters of coal-combusted power plants.

Power plants	Units	Coal to power parameter (kW/tonnes)	Carbon emission parameter (kg/tonnes)
Dafang CPP	Dafang No.1	(2205 2215 2235 2250)	(2005 2020 2032 2044)
	Dafang No.2	(2200 2220 2250 2275)	(2013 2025 2033 2045)
	Dafang No.3	(2195 2205 2230 2245)	(1960 1975 1993 2005)
	Dafang No.4	(2175 2200 2210 2245)	(1952 1965 1978 1986)
Bijie CPP	Bijie No.1	(2150 2185 2200 2225)	(1940 1956 1969 1986)
	Bijie No.2	(2165 2180 2200 2230)	(1945 1956 1973 1982)

collected from news reports and government websites. The wind speed data were collected from Guizhou Meteorological Bureau. Data for thermal power plants in Table 10.5 were mainly collected from the Bijie government websites, and data in Table 10.6 were collected from field investigations.

10.5 Result and Discussion

By inputting the collected data into the model, and running the solution approach on Lingo13.0 software, the results were calculated, as follows:

10.5.1 Result Under Different Scenarios

In this chapter, calculation results under our proposed 12 scheduling scenarios are presented, and in each scenario, the min$f$ equals to zero, meaning that all power demand need be satisfied. Figures 10.2–10.4 displayed the hourly wind, hydro and thermal power plant output when the water flow is high, normal, and low. As can be seen in Figure 10.2, the line in orange, gray, and yellow represent the output of hydropower, wind power, and coal-combusted thermal power, and the line in blue is load.

From the four subfigures in Figure 10.2, it is easy to find that hydropower output in summer is obviously higher than that in autumn, while the outputs in spring and winter are relatively lower; however, the output of wind power has a opposite situation, meaning that its output in spring and winter are relatively higher than that of summer and autumn. However, no matter in what season, the base condition of thermal power does not change. The output of hydro, thermal, and wind power of scenario #1 are 1937, 19 302, and 2406 MW, respectively, with the carbon emissions reaching 17 390 kg, water curtailment reaching 322 thousand cubic miters, and wind power online rate reaching 0.73. The output of hydro, thermal, and wind power of scenarios 4, 7, and 10 and corresponding can be seen carbon emissions, water curtailment and wind power online rate can be seen in Table 10.7.

From the four subfigures in Figure 10.3, the same seasonal output situation of hydropower, wind power, and thermal power can be obtained. However, as the water flow decrease, the output of hydropower decreased accordingly. As can be seen in Table 10.8, the hydropower output in scenario #2 is 1616 MW, while that in scenario#1 is 1937 MW. Accordingly, the decreased hydropower outputs are supplied by wind and thermal power, with thermal power increased from 19 302 to 19 599 MW, and wind power output increased from 2406 to 2431 MW. Apart from the changes of water flow, seasonal changes also influenced the output of wind power, as can be seen, the output of wind power in spring, summer, autumn, and winter when the water flow is normal are 2431, 623, 820, and 3505 MW, respectively.

From the four sub-figures in Figure 10.4, the power output of wind, hydro, and thermal in the four seasons when the water flow is low can be obtained. In this four scenario, hydropower output are 1294, 4719, 3356, and 1341 MW, and the output of wind power are 2471, 644, 836, and 3631 MW, while output of thermal power are 19 882, 20 514, 19 022, and 20 458 MW (Table 10.9).

10.5.2 Comprehensive Discussion of Results

By interpreting the objective value for the different scenarios in Tables 10.10 and 10.11, some proportions were obtained as follows.

Based on the calculation results, some core conclusions can be obtained. Firstly, the proposed optimization model can effectively reduce carbon emissions and increase wind power utilization of hydro–thermal–wind generation system, with guarantees of system reliability. Take the results of scenarios 1 in Table 10.10 as an example, in each scenario, power supply reliability is ensured by setting the objective value as zero. Carbon emissions in scenario 1 decreased to 17 390 kg, and wind

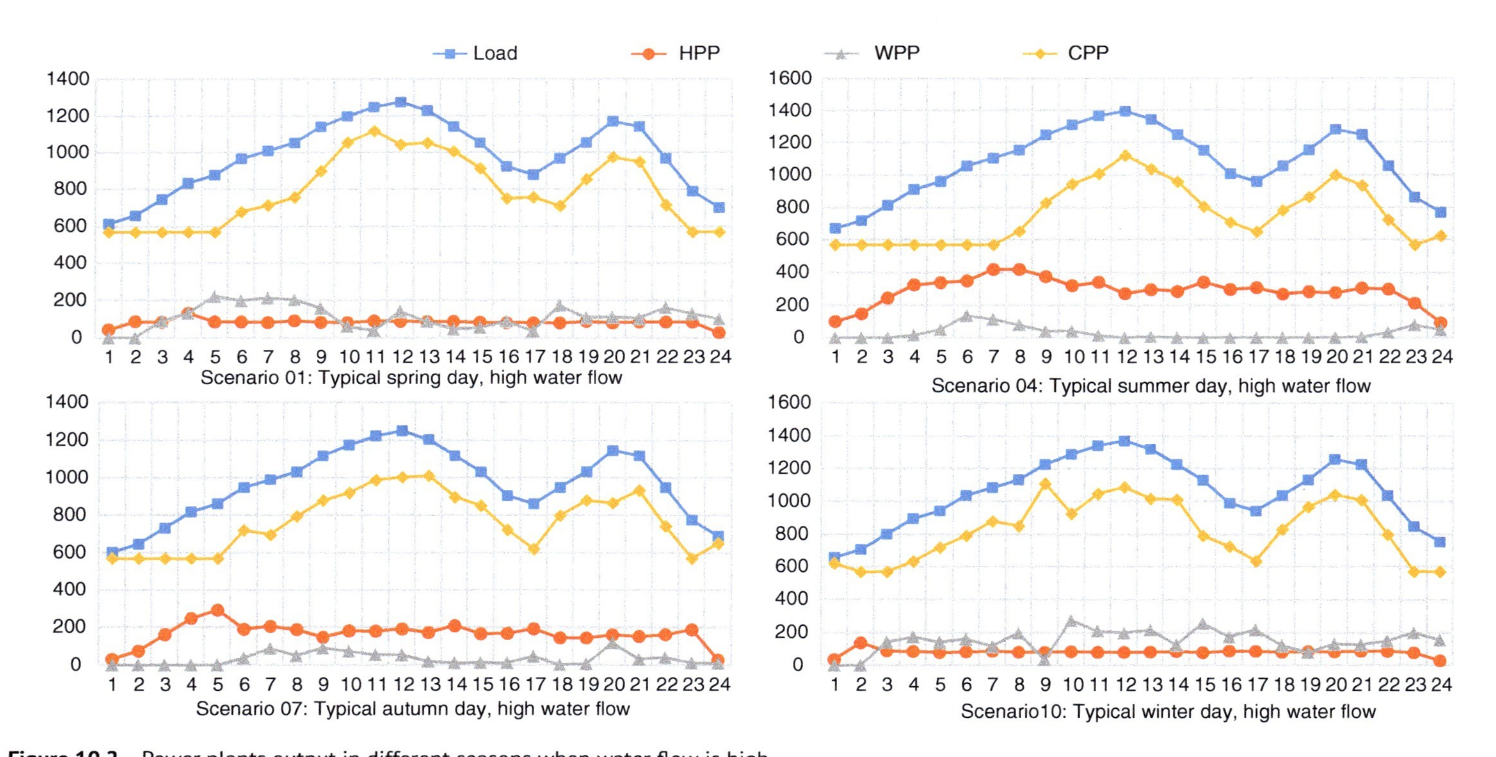

Figure 10.2 Power plants output in different seasons when water flow is high.

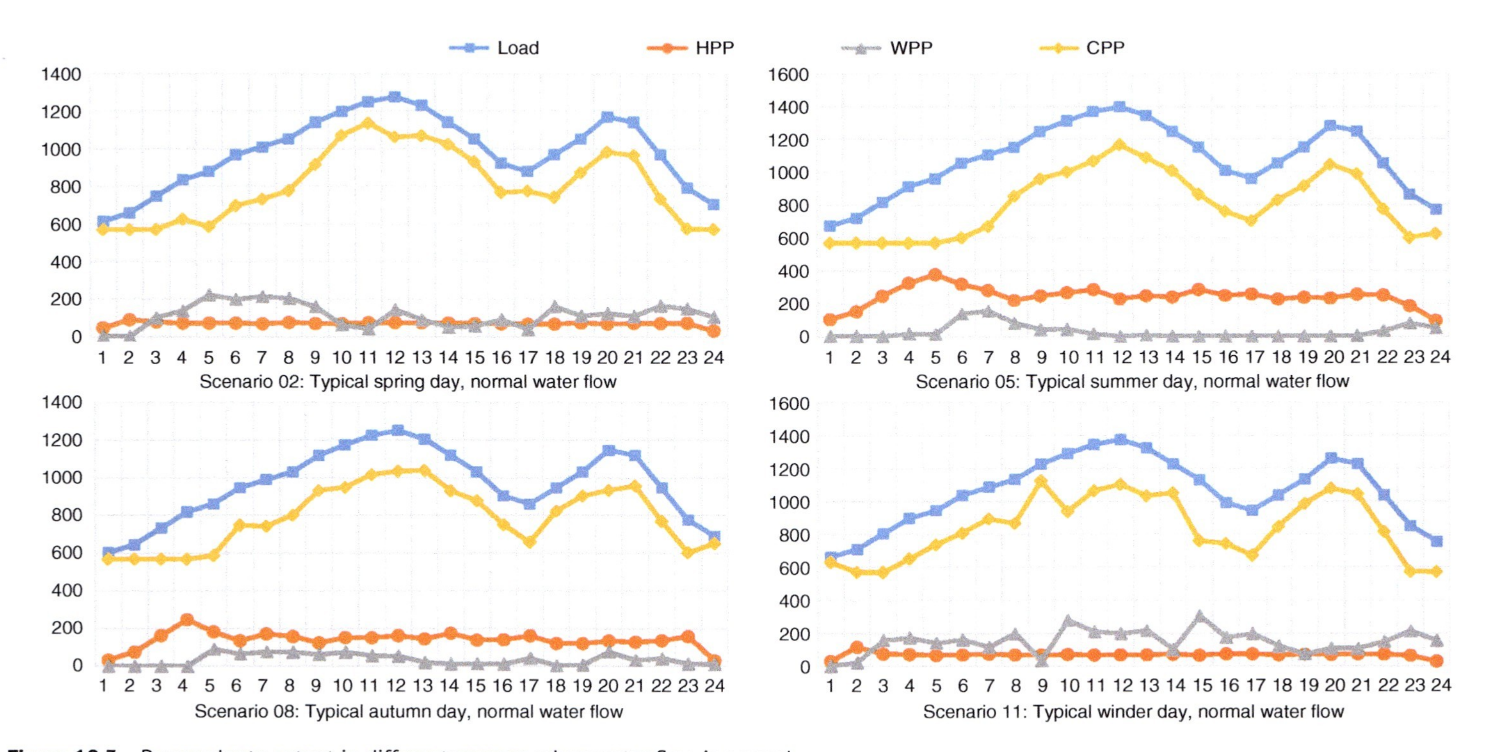

Figure 10.3 Power plants output in different seasons when water flow is normal.

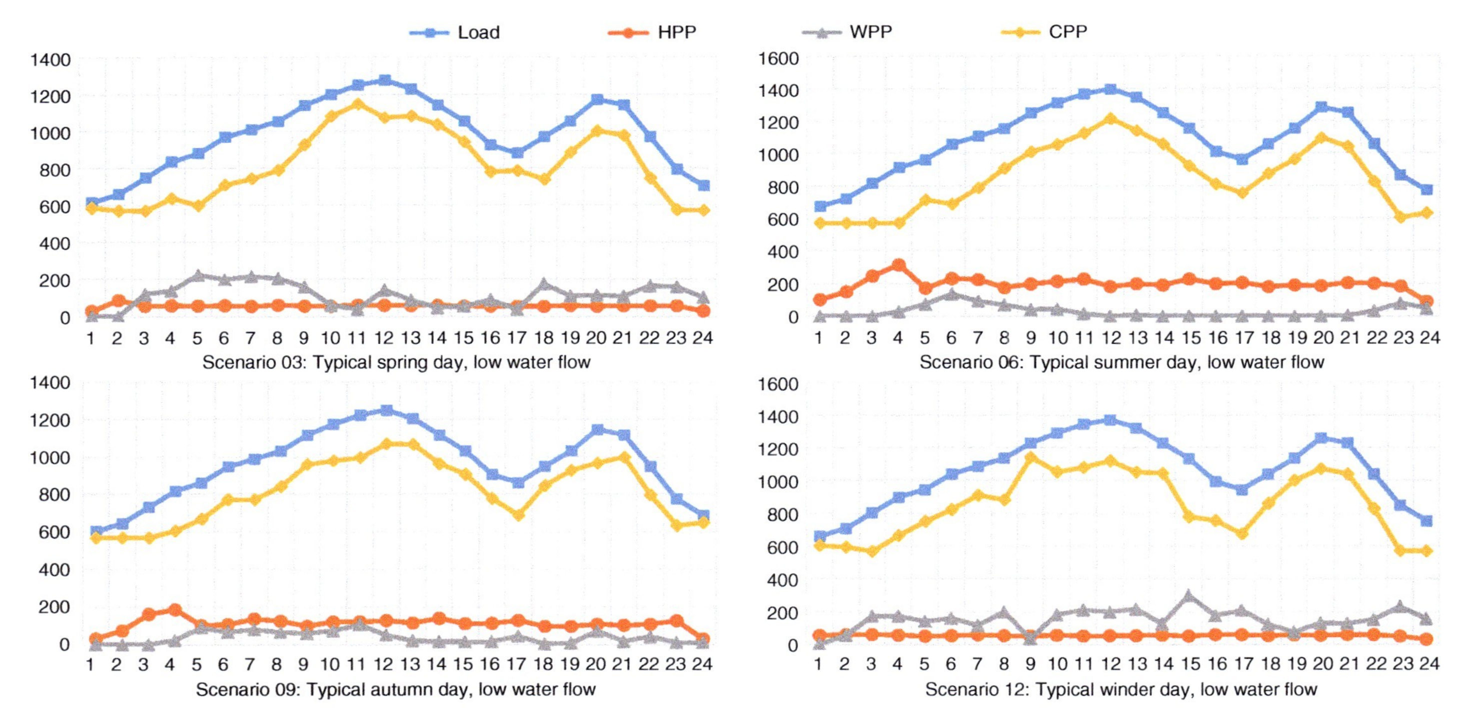

Figure 10.4 Power plants output in different seasons when water flow is low.

Table 10.7 Power plant output in different seasons when water flow is high.

	Plants	Time period																							
		1	2	3	4	5	6	7	8	9	10	11	12	13	14	15	16	17	18	19	20	21	22	23	24
1#	CPP	570	570	570	638	576	664	696	737	887	1084	1142	1058	1134	1047	956	757	757	697	860	990	964	741	570	641
	WPP	21	21	20	107	218	217	233	227	173	36	19	131	11	9	16	85	41	190	108	100	94	143	134	52
	HPP	24	68	156	90	84	85	81	90	82	81	89	87	86	87	81	81	80	79	87	80	84	82	86	10
4#	CPP	570	570	570	570	570	570	570	661	905	943	1008	1122	1042	961	807	709	650	784	868	1001	937	720	588	620
	WPP	8	8	12	11	7	67	115	85	44	48	16	0	5	0	0	0	0	0	0	0	4	34	60	54
	HPP	95	143	235	332	384	420	420	408	299	322	344	275	299	288	345	300	311	273	286	280	309	303	217	95
7#	CPP	570	570	570	570	570	682	651	784	854	934	979	1010	1010	897	863	723	659	799	884	855	952	745	580	657
	WPP	2	4	2	5	9	53	134	60	117	60	65	49	21	12	2	13	9	2	3	130	13	39	4	3
	HPP	31	72	160	243	283	213	207	189	149	183	182	194	175	211	168	170	195	147	147	163	155	164	191	30
10#	CPP	589	570	640	660	755	853	909	833	1132	965	1030	1071	999	1014	777	710	664	820	979	1079	1036	827	570	570
	WPP	16	18	73	149	108	102	88	218	15	238	231	218	239	127	274	192	190	137	68	95	102	123	202	156
	HPP	56	121	90	88	81	84	89	83	81	87	82	83	84	87	81	90	90	82	87	85	90	88	78	30

Table 10.8 Power plant output in different seasons when water flow is normal.

	Plants	Time period																							
		1	2	3	4	5	6	7	8	9	10	11	12	13	14	15	16	17	18	19	20	21	22	23	24
2#	CPP	570	570	570	656	570	684	711	753	902	1097	1158	1072	1147	1062	971	770	771	765	874	1010	977	759	584	596
	WPP	21	21	61	104	238	211	232	227	171	35	18	131	11	8	15	85	41	135	108	93	94	137	135	97
	HPP	24	68	115	74	70	71	67	75	69	68	74	73	72	72	68	67	66	66	72	67	70	69	70	10
5#	PCT	570	570	570	570	570	610	639	886	956	997	1066	1167	1092	1009	865	759	702	830	915	1048	988	787	570	659
	PWT	8	8	12	11	7	109	184	47	44	48	16	0	5	0	0	0	0	0	0	0	4	18	86	15
	PHT	95	143	235	332	384	337	282	221	249	268	286	230	249	240	287	250	259	227	239	233	257	252	209	95
8#	CPP	570	570	570	570	657	765	685	820	879	965	1013	1041	1039	932	891	751	670	824	909	882	977	773	612	657
	WPP	2	4	2	5	9	48	134	56	117	60	62	49	21	12	2	13	30	2	4	130	13	39	3	3
	HPP	31	72	160	243	196	135	172	158	124	153	152	162	146	176	140	141	162	122	122	135	129	136	159	30
11#	CPP	570	570	644	655	769	867	929	851	1142	975	1047	1088	1013	1021	786	724	678	840	993	1088	1025	838	571	570
	WPP	23	59	84	170	108	102	82	214	18	243	228	216	239	134	279	192	191	130	68	100	129	127	214	156
	HPP	68	80	76	73	67	70	75	69	68	72	69	69	70	73	67	75	75	69	73	71	75	74	65	30

Table 10.9 Power plant output in different seasons when water flow is low.

	Plants	Time period																							
		1	2	3	4	5	6	7	8	9	10	11	12	13	14	15	16	17	18	19	20	21	22	23	24
3#	CPP	570	570	649	626	579	692	724	887	914	1140	1187	1087	1162	1076	983	784	783	724	889	992	991	731	570	570
	WPP	21	21	20	151	243	217	233	107	173	6	4	131	11	9	16	85	41	190	108	125	94	179	163	123
	HPP	24	68	77	57	56	57	53	60	55	54	59	58	57	57	54	53	53	52	57	53	57	56	57	10
6#	CPP	570	570	570	570	755	731	696	929	1006	1050	1123	1214	1142	1057	922	809	754	875	963	1094	1040	821	595	660
	WPP	8	8	12	11	23	94	184	49	44	48	16	0	5	0	0	0	0	0	0	0	4	34	86	19
	HPP	95	143	235	332	183	232	225	177	199	215	229	183	200	192	230	200	207	182	191	187	206	201	184	90
9#	CPP	570	570	570	617	698	762	794	822	934	995	1031	1074	1068	967	917	779	682	848	934	909	1003	820	646	658
	WPP	2	4	2	5	61	78	59	86	86	60	74	49	21	12	3	13	50	2	2	130	13	19	1	2
	HPP	31	72	160	195	103	108	138	126	99	123	122	129	117	141	112	113	129	98	98	109	104	109	128	30
12#	CPP	571	570	573	649	783	881	939	861	1158	1021	1058	1101	1028	1031	809	740	693	848	1008	1107	1027	814	618	570
	WPP	33	78	170	190	108	103	88	217	15	211	231	216	239	139	270	192	191	137	68	95	141	166	180	156
	HPP	57	61	60	58	53	56	59	56	55	57	55	56	56	58	53	60	60	55	58	57	60	59	52	30

Table 10.10 Calculation results under different scenarios.

	Units	Scenarios											
		1#	2#	3#	4#	5#	6#	7#	8#	9#	10#	11#	12#
Carbon emission	kg	17 390	17 656	17 911	16 509	17 478	18 485	16 551	17 140	17 729	18 063	18 243	18 426
Water curtailment	10^5 m^3	3.22	2.60	1.99	13.20	9.16	6.05	5.18	4.10	3.06	2.28	1.69	1.08
Hydropower output	MW	1937	1616	1294	6982	5860	4719	4019	3356	2693	1997	1670	1341
Coalpower output	MW	19 302	19 599	19 882	18 317	19 395	20 514	18 367	19 022	19 669	20 055	20 256	20 458
Windpower output	MW	2406	2431	2471	577	623	644	812	820	836	3379	3505	3631
Wind online rate		0.73	0.74	0.75	0.52	0.55	0.57	0.62	0.64	0.65	0.80	0.83	0.86

Table 10.11 Power plants output rate under different scenarios.

Plants / Seasons	Spring			Summer			Autumn			Winter		
Water flow	**High (%)**	**Normal (%)**	**Low (%)**	**High (%)**	**Normal (%)**	**Low (%)**	**High (%)**	**Normal (%)**	**Low (%)**	**High (%)**	**Normal (%)**	**Low (%)**
Hydropower output	8.2	6.8	5.5	27	22.6	18.2	17.3	14.5	11.6	7.9	6.6	5.3
Coalpower output	81.6	82.9	84.1	70.8	74.9	79.3	79.2	82	84.8	78.9	79.7	80.4
Windpower output	10.2	10.3	10.4	2.2	2.4	2.5	3.5	3.5	3.6	13.3	13.8	14.3

power utilization rate increased to 73%. The same increased and decreased situation can also be seen in other scenarios. Secondly, in each season, when the water flow changes from high to normal and from normal to low, hydropower generation will decrease. It is worth noting that when hydropower generation is reduced, the power generation of CPPs and wind power plants increased accordingly. However, the added difference is inconsistent. For example, in Table 10.11, the coal-combusted power plants in summer took up 8.5% and the wind power plants took up only 0.3% of the 8.8% hydropower share decrement. This indicating that wind power has a disadvantage in terms of the connection to the main grid compared to the coal-combusted power plants. Thirdly, wind power and hydropower output have strong seasonal complementary characteristics. As shown in Table 10.11, the output share of wind power plants in spring and winter is slightly higher than that in summer and autumn. The average shares of wind power plant production in spring and winter are 10.3% and 13.8%, while that in summer and autumn are 2.4% and 3.5%, respectively. On the contrary, the average output shares of hydropower plants in summer and autumn are 22.6% and 14.5%, respectively, while that in winter and spring are 6.8% and 6.6%, respectively. It can be clearly seen that cooperation between wind and hydropower plants reduces the peak shaving demand of coal-fired units, thus extending the service life of these units while ensuring cleaner and more economical power generation.

10.5.3 Management Recommendations

In this section, some management suggestions are given to better explore the proposed optimization model. Firstly, for power generation operators, it is worth considering establishing hydro–wind–thermal complementary system to reduce carbon emissions and provide a cleaner energy future. This is because although CPPs can provide the most reliable power for economic development, they are now facing the need to significantly reduce carbon emissions to protect the environment. As a viable energy source, wind energy emits close to zero carbon emissions, but it is random and intermittent. Therefore, under the coordination of hydropower, the three can be complementary. However, because complementary systems are a new generation of cooperation models encouraged by researchers and engineers around the world, they need government policy support to gain wider recognition. For example, targeted policies need to be formulated to encourage the development of complementary power generation systems to achieve a clean power future. Thirdly, the technological improvement of the power generation system also needs to be continued. By integrating hydropower plants into the traditional wind–thermal power generation system, the peak shaving pressure of CPPs can be reduced and the utilization rate of wind energy can be improved; however, the economic performance of hydropower plants and CPPs will be weakened. Moreover, if the wind energy fails to reach its expected output, the hydropower station will immediately adjust its schedule to generate more power to ensure the safety of power supply. Therefore, technical improvements are needed to ensure accurate wind energy prediction and to increase cleaner thermal power technology.

References

Aliari, Y. and Haghani, A. (2016). Planning for integration of wind power capacity in power generation using stochastic optimization. *Renewable and Sustainable Energy Reviews* 59: 907–919.

Alizadeh, M.R., Nikoo, M.R., and Rakhshandehroo, G.R. (2017). Hydro-environmental management of groundwater resources: a fuzzy-based multi-objective compromise approach. *Journal of Hydrology* 551: 540–554.

Banerjee, S., Dasgupta, K., and Chanda, C.K. (2016). Short term hydro–wind–thermal scheduling based on particle swarm optimization technique. *International Journal of Electrical Power & Energy Systems* 81: 275–288.

Best, R. (2017). Switching towards coal or renewable energy? The effects of financial capital on energy transitions. *Energy Economics* 63: 75–83.

State Electricity Regulatory Commission of China (2013). Calculation method for wind energy consumption of wind farms (for trial implementation).

BP Energy (2016). BP World Energy Statistics Yearbook.

Chandrasekaran, K., Hemamalini, S., Simon, S.P., and Padhy, N.P. (2012). Thermal unit commitment using binary/real coded artificial bee colony algorithm. *Electric Power Systems Research* 84 (1): 109–119.

Chen, X., Chan, C.K., and Lee, Y.C.E. (2016a). Responsible production policies with substitution and carbon emissions trading. *Journal of Cleaner Production* 134: 642–651.

Chen, Y., Wei, W., Liu, F., and Mei, S. (2016b). Distributionally robust hydro–thermal–wind economic dispatch. *Applied Energy* 173: 511–519.

Chen, F., Zhou, J., Wang, C. et al. (2017a). A modified gravitational search algorithm based on a non-dominated sorting genetic approach for hydro–thermal–wind economic emission dispatching. *Energy* 121: 276–291.

Chen, J.J., Zhuang, Y.B., Li, Y.Z. et al. (2017b). Risk-aware short term hydro–wind–thermal scheduling using a probability interval optimization model. *Applied Energy* 189: 534–554.

Davidson, M.R., Zhang, D., Xiong, W. et al. (2016). Modelling the potential for wind energy integration on China's coal-heavy electricity grid. *Nature Energy* 1: 16086.

Demissie, A., Zhu, W., and Belachew, C.T. (2017). A multi-objective optimization model for gas pipeline operations. *Computers & Chemical Engineering* 100: 94–103.

De Queiroz, A.R. (2016). Stochastic hydro–thermal scheduling optimization: an overview. *Renewable and Sustainable Energy Reviews* 62: 382–395.

Dubey, H.M., Pandit, M., and Panigrahi, B.K. (2016). Hydro–thermal–wind scheduling employing novel ant lion optimization technique with composite ranking index. *Renewable Energy* 99: 18–34.

Feng, J. and Shen, W.Z. (2017). Wind farm power production in the changing wind: Robustness quantification and layout optimization. *Energy Conversion and Management* 148: 905–914.

Garciagonzalez, J. (2008). Stochastic joint optimization of wind generation and pumped-storage units in an electricity market. *IEEE Transactions on Power Systems* 23 (2): 460–468.

Gu, Y., Xu, J., Chen, D. et al. (2016). Overall review of peak shaving for coal-fired power units in China. *Renewable and Sustainable Energy Reviews* 54: 723–731.

Guizhou Province Development and Reform Commission (2017a). Guizhou Province Energy Development Plan for 13th Five-Year (in Chinese).

(2017b). Rank and Contrast of Thermal Power Installed and Utilization Hours in 1–10 Months of 2017 (2017) (in Chinese).

Haszeldine, R.S. (2009). Carbon capture and storage: how green can black be? *Science* 325 (5948): 1647.

Huang, C., Li, F., and Jin, Z. (2015). Maximum power point tracking strategy for large-scale wind generation systems considering wind turbine dynamics. *IEEE Transactions on Industrial Electronics* 62 (4): 2530–2539.

International Energy Agency (2016). CO_2 *Emissions From Fuel Combustion-2016 Edition*. International Energy Agency.

Jiang, R., Wang, J., and Guan, Y. (2012). Robust unit commitment with wind power and pumped storage hydro. *IEEE Transactions on Power Systems* 27 (2): 800–810.

Laia, R., Pousinho, H.M.I., Melico, R., and Mendes, V.M.F. (2016). Bidding strategy of wind–thermal energy producers. *Renewable Energy* 99: 673–681.

Li, L., Tan, Z., Wang, J. et al. (2011). Energy conservation and emission reduction policies for the electric power industry in China. *Energy Policy* 39 (6): 3669–3679.

Liang, L. (2006). Study on short-term generation optimization dispatching system for reservoirs of cascade hydropower stations in Wujiang. PhD thesis. Xi'an University of Technology (in Chinese).

Lonsdale, C.R., Stevens, R.G., Brock, C.A. et al. (2012). The effect of coal-fired power-plant SO_2 and NO_x control technologies on aerosol nucleation in the source plumes. *Atmospheric Chemistry and Physics* 12 (23): 11519–11531.

Lopes, V.S. and Borges, C.L.T. (2014). Impact of the combined integration of wind generation and small hydropower plants on the system reliability. *IEEE Transactions on Sustainable Energy* 6 (3): 1169–1177.

Lu, X., Mcelroy, M.B., Peng, W. et al. (2016). Challenges faced by China compared with the U.S. In developing wind power. *Nature Energy* 1 (6): 16061.

Lv, C., Xu, J., Xie, H. et al. (2016). Equilibrium strategy based coal blending method for combined carbon and PM_{10} emissions reductions. *Applied Energy* 183: 1035–1052.

Mao, X.Q., Zeng, A., Hu, T. et al. (2014). Co-control of local air pollutants and CO_2 from the Chinese coal-fired power industry. *Journal of Cleaner Production* 67 (67): 220–227.

Panda, A., Tripathy, M., Barisal, A.K., and Prakash, T. (2017). A modified bacteria foraging based optimal power flow framework for hydro–thermal–wind generation system in the presence of STATCOM. *Energy* 124: 720–740.

Qing, W. (2007). Hongjiadu hydropower station No. 2 unit vibration and dynamic balance test. Proceedings of Symposium on Stability of Hydraulic Turbines.

Schill, W.-P., Pahle, M., and Gambardella, C. (2017). Start-up costs of thermal power plants in markets with increasing shares of variable renewable generation. *Nature Energy* 2: 17050.

Seguin, S., Fleten, S.E., Pichler, A., and Audet, C. (2016). Stochastic short-term hydropower planning with inflow scenario trees. *European Journal of Operational Research* 259: 1156–1168.

Shi, R.J., Fan, X.C., and He, Y. (2017). Comprehensive evaluation index system for wind power utilization levels in wind farms in China. *Renewable and Sustainable Energy Reviews* 69: 461–471.

Singh, V.K. and Singal, S.K. (2017). Operation of hydro power plants-a review. *Renewable and Sustainable Energy Reviews* 69: 610–619.

Skorek-Osikowska, A., Bartela, ?., and Kotowicz, J. (2017). Thermodynamic and ecological assessment of selected coal-fired power plants integrated with carbon dioxide capture. *Applied Energy* 200: 73–88.

Squalli, J. (2017). Renewable energy, coal as a baseload power source, and greenhouse gas emissions: evidence from U.S. state-level data. *Energy* 127: 479–488.

Sun, Y., Dong, J., and Ding, L. (2017). Optimal day-ahead wind-thermal unit commitment considering statistical and predicted features of wind speeds. *Energy Conversion and Management* 142: 347–356.

Tapia, J.F.D., Lee, J.Y., Ooi, R.E.H. et al. (2016). Planning and scheduling of CO_2 capture, utilization and storage (CCUS) operations as a strip packing problem. *Process Safety and Environmental Protection* 104: 358–372.

Tsai, S.C. and Chen, S.T. (2016). A simulation-based multi-objective optimization framework: a case study on inventory management. *Omega* 70: 148–159.

Wang, Q., Guan, Y., and Wang, J. (2012). A chance-constrained two-stage stochastic program for unit commitment with uncertain wind power output. *IEEE Transactions on Power Systems* 27 (1): 206–215.

Wang, L., Li, Q., Ding, R. et al. (2017). Integrated scheduling of energy supply and demand in microgrids under uncertainty: a robust multi-objective optimization approach. *Energy* 130: 1–14.

Wang, C., Liu, F., Wei, W. et al. (2016). Robust unit commitment considering strategic wind generation curtailment. *Power and Energy Society General Meeting*, pp. 1–5.

Xie, K., Dong, J., Singh, C., and Hu, B. (2016). Optimal capacity and type planning of generating units in a bundled wind–thermal generation system. *Applied Energy* 164: 200–210.

Xiong, W., Wang, Y., Mathiesen, B.V., and Zhang, X. (2016). Case study of the constraints and potential contributions regarding wind curtailment in Northeast China. *Energy* 110: 55–64.

Xu, J. and Zhou, X. (2011). *Fuzzy-Like Multiple Objective Decision Making*. Berlin, Heidelberg: Springer-Verlag.

Xu, C.X., Liu, Y., Zhang, F. et al. (2017). Clean coal technologies in China based on methanol platform. *Catalysis Today*. 298: 61–68.

Zeng, Z., Xu, J., Wu, S., and Shen, M. (2014). Antithetic method based particle swarm optimization for a queuing network problem with fuzzy data in concrete transportation systems. *Computer-Aided Civil and Infrastructure Engineering* 29 (10): 771–800.

Zhang, N., Lu, X., Mcelroy, M.B. et al. (2016). Reducing curtailment of wind electricity in China by employing electric boilers for heat and pumped hydro for energy storage. *Applied Energy* 184: 987–994.

Zhou, B., Geng, G., and Jiang, Q. (2016). Hydro-thermal–wind coordination in day-ahead unit commitment. *IEEE Transactions on Power Systems* 31 (6): 4626–4637.

11

Economic-Environmental Equilibrium-Based Wind–Solar–Thermal Power Generation System

Due to the continued population growth, economic development, and living standards improvements, demand for electricity supply has been dramatically increased in the past decades [Chen et al., 2016]. However, as BP stated, 38.1%, 23.2%, and 3.5% of the world's total electricity generation in 2017 were provided by coal, natural gas, and oil. The world's remaining reserves would only last 134, 52.6, and 50.2 years if they continued to be mined at the 2017 level. What's more, the combustion of fossil fuels has caused serious environmental problems, such as acid rain, greenhouse effect, and photochemical smog pollution [Liu et al., 2017]. It is found that the electric power industry contributed to more than 23%, 45%, and 64% of particle material emission, sulfur dioxide emission, and nitrogen oxide emission as well as 44% of the carbon emissions [Yuan et al., 2018]. But coal thermal power generation continues to be the most reliable sector in electricity generation, and it is expected that its proportion of electricity generation will remain high in the following decades. Therefore, the construction of environmental-friendly power generation systems is becoming more and more urgent.

11.1 Background Introduction

Developing generation systems characterized with high efficiency and low emissions is an efficient way to achieve environmental-friendly power generation. In literature, Liu et al. [2018] tried to remove sulfur dioxide and nitric oxide from flue gas using vacuum ultraviolet light/heat/peroxymonosulfate of the vacuum ultraviolet light spraying reactor and received the efficiencies of 100% and 91.3%, respectively. Ryzhkov et al. Ryzhkov et al. [2018] studied the basic methods for raising the efficiency of air-blown integrated gasification-combined cycles and proposed a scheme that enabled the approximation of cycle efficiency in a natural gas-combined cycle. Cai et al. [2017] proposed a new power and heat generation system to better exploit the heat content of flue gases, their simulation results achieved up to 82.7% and 28.8% of the thermal and energy efficiencies, which were 8.9% and 3.0% higher than the reference system. Generally, these methods are very effective in dealing with environment and efficiency issues of thermal power generation, but the immature technology, the capital costs, and the additional operating fees are still solid barrier for companies to deploy these equipments on large scale.

Innovative Approaches towards Ecological Coal Mining and Utilization, First Edition.
Jiuping Xu, Heping Xie, and Chengwei Lv.

Applying management tools to balance the economic profits and environmental protection trade-offs are therefore becoming increasingly important. For example, Roque et al. [2017] proposed a meta-heuristic approach to solve the economical-environmental unit commitment problem. Felipe and Das Feijoo and Das [2014] developed a two-layer mathematical–statistical model for the optimal design of carbon dioxide cap-and-trade policies, and Ji et al. Ji et al. [2017] proposed a systematic Pareto optimal method to deal with the allocation of emission permits problems. Singh et al. [2016] introduced a synergic predator-prey optimization algorithm to solve the economic load dispatch problem of thermal units. Mason et al. [2017] examined the performance of a number of particle swarm optimization variants to the dynamic economic dispatch problem. These explorations have made significant contributions in dealing with the equilibrium between economic profits and environment influences of power supply process. But with higher proportion of renewable energy integrated, traditional methods are facing new challenges [Du et al., 2018].

To integrate environmental-friendly renewable sources into traditional electricity generation system, the economic efficiency is an inevitable problem that needs wide discussion. In previous research, Barros et al. [2016] compared the life cycle cost of renewable and non-renewable power plants and found that conventional power plants were still the most competitive options. Vithayasrichareon et al. [2017] tested the cost and technical impacts associated with thermal units' ramp rates, minimum operating levels, and startup costs and found it is less important compared with synchronous generation. Sunet et al. compared the total social costs integration ratio of renewable sources of the large-scale and long-distance transmission scheme and the local consumption scheme [Sun et al., 2017a]. Hirth et al. [2015] analyzed and compared the integration costs of wind and solar power, including forecast storage costs, long-distance transmission costs, and deviation costs, etc. It must be said that the high capital investment in renewable energy is an important factor affecting its expansion, but the sustainable advantages of these resources cannot be neglected, for example, they are fuel-free and very clean during operation.

Establishing hybrid systems to integrate solar and wind sources into traditional power networks has attract global researchers' attention. For instance, Suresh and Sreejith [2017] proposed a generation dispatch method for the solar–thermal integrated generation systems, which apply the dragonfly algorithm to solve a economic dispatch problem, and efficiently validated the method in South Indian test systems. Sun et al. [2017b] proposed the generation scheduling method for the wind–thermal integrated generation systems which take the full consideration of the statistical and predicted features of wind speed. Reddy [2017] proposed a optimal scheduling strategy for the hybrid thermal–wind–solar generation system, in which the uncertainties of wind, solar photovoltaic, and load demand are simultaneously considered, and their simulation results indicate that cost increase in day ahead schedule will lead to substantial reduction of real-time cost. These researches have made significant contributions in wind and solar resource integration, however, to simultaneously consider the trade-off between economic profits and environmental protection, while including the seasonal fluctuations in wind speed and the weather-driven characteristics of solar power is still a difficult problem.

Based on the discussion above, we tries to propose an equilibrium strategy-based optimization method for the hybrid wind–solar–thermal generation system in this chapter, in which the environmental impacts, the economic benefits, the seasonal fluctuation of wind speed, and the weather-driven characteristic of solar irradiation will be fully considered. The generation sources inclosed in the hybrid system are fully constructed, located near, and connected to the same main grid. The uncertainty broughtby wind and solar sources will be properly handled by establishing 12 typical scheduling scenarios, and the trade-off of economic profits and environment protection will be achieved by applying multi-objective method. The proposed optimization method in this chapter is composed of three main parts, including building the physical model, building the mathematical model, and applying the method to real case, and the method details are shown in Figure 11.1. The remainder of this chapter is organized as follows. The key problems associated with hybrid wind–solar–thermal systems are discussed in Section 11.2 and the associated mathematical model is built in Section 11.3. In Sections 11.4 and 11.5, a real-world case is applied and discussed. Section 11.6 gives the conclusions and future study directions.

11.2 Key Problem Statement

Before developing the multi-objective model and outlining the scheduling strategies, methods to handle uncertainty and trade-offs of economic profits environmental protection need detailed descriptions. A related figure of the key problem this chapter dealt with is shown in Figure 11.2.

The uncertainties involved in wind and solar resources increased the difficulty of designing reasonable generation scheduling. This is because average wind speed varies across the four seasons due to the influence of geographical location and weather conditions [Feng and Shen, 2017], and the solar power output is strongly dependent on solar radiation conditions [Reddy and Momoh, 2015]. These imitations of wind and solar power indicate that they cannot be scheduled in the same way as conventional thermal units [Reddy, 2017]. Shao et al. [2017] used a flexible uncertainty set to deal with the variable wind power to propose the robust security-constrained unit commitment model for wind–thermal system. Suresh and Sreejith [2017] applied β distribution function to handle the uncertain solar farm output, so as to develop the generation dispatch for combined solar thermal systems. However, few research has considered the seasonal fluctuation of wind speed and weather-driven characteristic of solar power. Therefore, in this chapter, 12 typical scheduling scenarios are established based on the wind and solar power uncertainties. The scheduling scenarios are generated by interacting wind speed scenario set and weather condition scenario set. The wind speed set includes four elements, representing typical daily wind speed data in the four different seasons. The solar irradiation set includes three weather conditions: sunny, cloudy, and rainy. Details of these scenarios are shown in Figure 11.2.

Economical power generation and environmental protection are two important goals that need to be considered simultaneously. In practical, more than 90% of SO_x and NO_x can be removed during the coal combustion process [Liu et al., 2018].

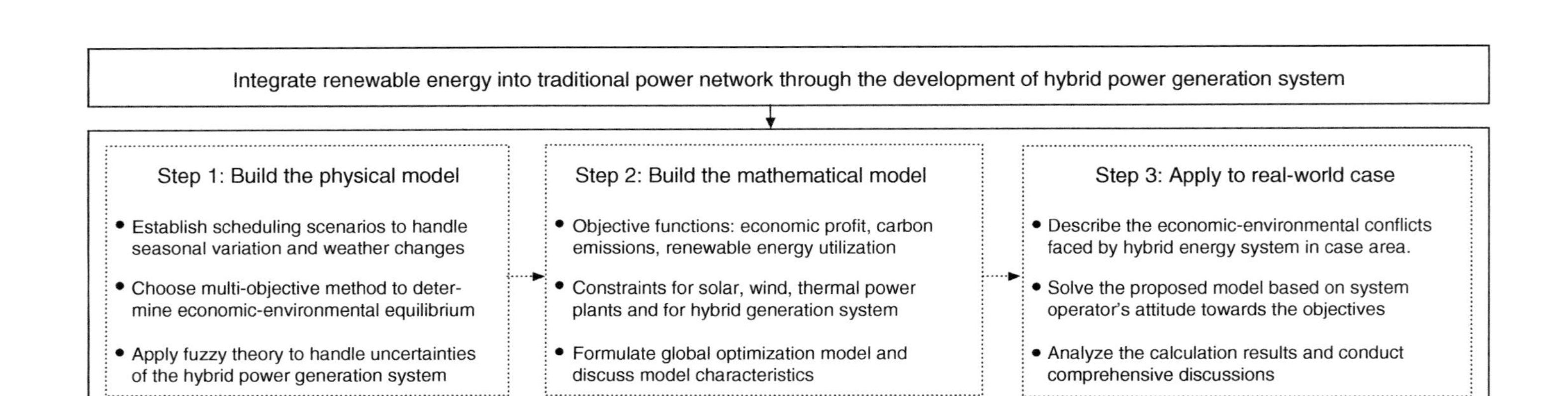

Figure 11.1 Flowchart of the equilibrium strategy-based optimization method proposed in this chapter.

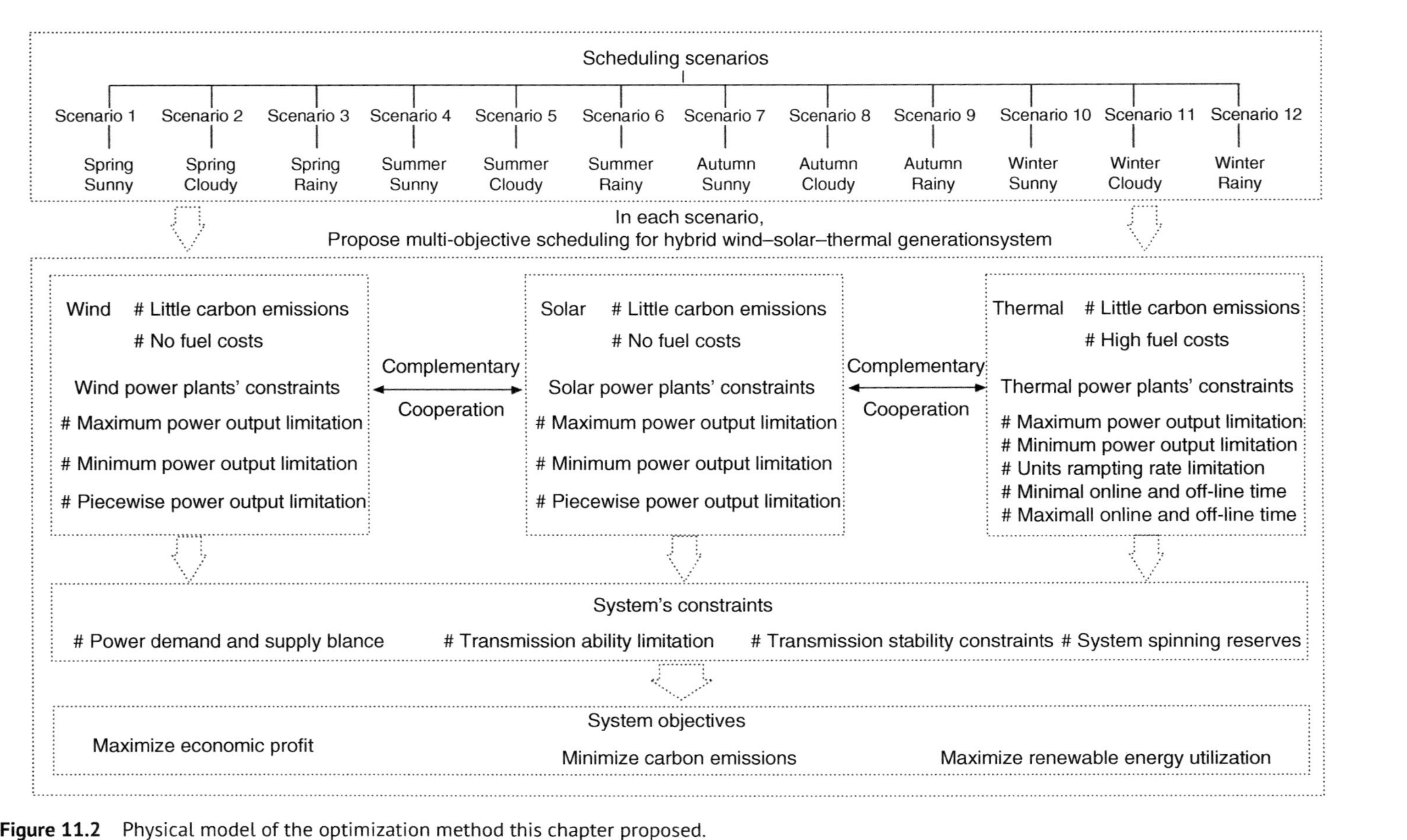

Figure 11.2 Physical model of the optimization method this chapter proposed.

But to remove carbon emissions is not so easy and highly efficient. Because the technology to reduce carbon emissions during generation process is so complex and expensive, the significant carbon emissions are still being released. Therefore, the renewable energy sources are integrated, so that the system operators can pursue an environmental-friendly generation by scheduling less proportions of thermal power output. On the other hand, due to the subsidy policy of renewable sources, the on-gird price of solar and wind power is much higher than that of thermal power. Therefore, if the system operator pursues better financial revenue to ensure steady company development, a higher proportion of renewable energy access would be necessary. Based on the above discussion, it is easy to find that increasing renewable power utilization is vital for system operators to pursue higher economic profits and promote environmental protection. Further, we also found that the hybrid wind–solar–thermal scheduling problem is multi-objective in nature. Therefore, multi-objective programming, which has been found to be very effective in solving the trade-offs between different objectives, is applied in this chapter to optimize equilibrium for the environmental impact and economic benefits.

Even in each typical scheduling scenario, uncertainties of wind and solar power also exist and influence the designing of a reasonable scheduling. This is because the wind speed and direction change even in every minute or second, on the other hand, as photovoltaic (PV) technology immediately changes solar irradiations into solar power, even a cloud shadow can change solar power production. Generally, the daily scheduling is done a day before or several hours before, to accurately predict the wind speed and solar irradiations will do great help, but it is still difficult to provide a precise value. Reddy [2017] applied Weibull Probability Density Function to simulate the wind speed and used bimodel distribution to simulate solar irradiance. While Zubo et al. [2018] built scenario trees to model solar irradiance and wind speed-associated uncertainties. Their efforts made important contributions in dealing with renewable sources uncertainties but require somewhat heavy calculations. In this chapter, the fuzzy numbers are adopted to describe the daily wind and solar uncertainties. Fuzzy number theory was first proposed by Dubois and Zimmermann in 1991 and has explored extensive applications in describing vague values [Zimmermann, 2011], such as uncertain conversion and emission parameters of thermal power units, uncertainties in water resources allocation [Xu et al., 2012], and so on.

11.3 Modeling

In this section, the mathematical model for the above-described equilibrium strategy-based hybrid wind–solar–thermal generation system is given.

11.3.1 Notations

The following notations will be used in the proposed model.

11.3.2 Objectives

11.3.2.1 Economic Profits

The income of power supplier mainly includes electricity sales income and renewable energy subsidies, while the coal purchase cost is the main expenditure.

Generally, power supplier needs to maximum their economic profits to ensure steady company development. To represent this situation, here we let a_1 and a_2 be the subsidies of 1 mW wind/solar power that the government pay to power supplier, b_{lt} be the power output of wind farm l at time period t, and c_{kt} be the power output of solar photovoltaic park k at time period t. Then the total government subsidy is $\sum_t \sum_l b_{lt}a_1 + \sum_t \sum_k c_{kt}a_2$. Let a_3 represent price of each tonnes of coal and x_{ijt} represent the coal to be burned by thermal power plant i unit j at time t, then the total fuel cost is $\sum_t \sum_i \sum_j x_{ijt}a_3$. Let e_t represent the total power generated by hybrid wind–solar–thermal system at time t, and a_4 represent the on-grid price of power generated by hybrid system, then the income of selling electricity is $\sum_t e_t a_4$. Based on the description above, maximizing the economic profits can be represented as follows.

11.3.2.2 Carbon Emissions

Power industry emission reduction is significantly important for global sustainable development. Therefore, reducing carbon emissions while generating electricity is a necessary condition for both enterprise survival and social development. Here in the hybrid wind–solar–thermal power generation system, coal combustion is the only carbon emission source. We let g_{ijt} represent the on/off status of thermal unit j of thermal power plant i at time t, x_{ijt} represent the coal to be burned by thermal power plant i unit j at time t, and $\widetilde{h}_{ij}$ represent the coal to carbon emission parameter. Then the power supplier needs to minimize the carbon emission of the total scheduling period, and this situation can be represented as follows.

$$\min f_2 = \sum_i \sum_i \sum_j x_{ijt} g_{ijt} E(\widetilde{h}_{ij}) \tag{11.1}$$

Here $E(\widetilde{h}_{ij})$ is the expected value of the uncertain parameter $\widetilde{h_{ij}}$ calculated by the method proposed by Xu and Zhou [2013]. Due to the lack of historical data and immature technology, the exact value of the coal to carbon parameter is very difficult to obtain; however, through field research with practical engineers, we found that the value of this parameter always belongs to an interval, which is very similar to the character of fuzzy number. Therefore, to ensure a more accurate and more scientific value, $\widetilde{h}_{ij}$ is represented by a triangular fuzzy number. We let L_h and R_h represent the left and right bound of the uncertain parameter $\widetilde{h}_{ij}$, and $\alpha(h)$ represent the most likely value. By applying the expected value calculation method, the expected value of parameter $\widetilde{h}_{ij}$ is as follows, in which $\beta, (0 \leq \beta \leq 1)$ is positive and negative attitude parameter of power generator toward parameter $\widetilde{h}_{ij}$.

$$E(\widetilde{h}_{ij}) = \frac{(1-\beta)L_h + \alpha(h) + \beta R_h}{2} \tag{11.2}$$

$$\max f_1 = \sum_t e_t a_4 + \sum_t \sum_l b_{lt} a_1 + \sum_t \sum_k c_{kt} a_2 - \sum_t \sum_i \sum_j x_{ijt} a_3 \tag{11.3}$$

11.3.2.3 Renewable Energy Utilization

Promoting renewable energy utilization is a global urgent problem. Although the global installed capacity of wind and solar power are substantially improving, the utilization efficiency of these facilities needs further attention, and this problem is even more prominent in China. For example, although the net wind and solar-installed capacity of China were 129.3 and 43.2 GW in 2017, the wind and solar curtailment rate in northwest China were as high as 24.6% and 14.1%, which is a double waste of resources and capital. As in the wind–solar–thermal generation system, these power plants are already well constructed and connected to the main grid, and higher proportion of renewable energy utilization is conditional and can be achieved. Here we let b_{lt} represent the power output of wind farm l at time period t, and c_{kt} represent the power output of solar photovoltaic park k at time period t, to maximize renewable energy utilization and can be represented as follows.

$$\max f_3 = \sum_t \sum_l b_{lt} + \sum_t \sum_k c_{kt} \tag{11.4}$$

Indices

i = Index for coal-fired power plants.
j = Index for thermal units.
k = Index for solar power plants.
l = Index for wind farms.
t = Index for time period.

Decision variables

g_{ijt} = on/off status of thermal unit j of thermal power plant i at time t.
x_{ijt} = coal to be burned by thermal power plant i unit j at time t.
y_{lt} = wind turbines to be operated at wind farm l at time t.
z_{kt} = photovoltaic arrays to be operated at solar power plant k at farm t.

Certain parameters

a = system load spinning reserve rate.
a_1 = subsidies of 1 mW wind power that the government pay to power supplier.
a_2 = subsidies of 1 mW solar power that the government pay to power supplier.
a_3 = price of each tonnes of coal.
a_4 = on-grid price of power generated by hybrid system.
$a_{ij}^{\max}$ = maximum technical power output of unit j of thermal power plant i.
b_{lt} = the power output of wind farm l at time period t.
c_{kt} = the power output of solar photovoltaic park k at time period t.
C = carbon emissions of previous production period.
e_t = total power generated by hybrid wind–solar–thermal system at time t.
$E(\widetilde{d}_t)$ = expected value of the uncertain parameter $\widetilde{d}_t$.
$E(\widetilde{h}_{ij})$ = expected value of the uncertain parameter $\widetilde{h}_{ij}$.
$E(\widetilde{o}_{ij})$ = expected value of the uncertain parameter $\widetilde{o}_{ij}$.
L_h = left bound of the uncertain parameter $\widetilde{h}_{ij}$.
L_n = the lower bound of the designed transmission ability of transmission line.
m_{it} = power output of thermal power plant i at time t.
$\sum_i m_{it}^{\max}$ = the maximum outputs from the thermal power plants.

$\sum_i m_{it}^{\min}$ = the minimum outputs from the thermal power plants.
m^- = the lower bound of power variation that the transmission line can withstand.
m^+ = the upper bound of power variation that the transmission line can withstand.
N = renewable energy utilization of the last production period.
p_r^w = rated output of the wind turbines.
p_r^s = rated output of the photovoltaic arrays.
P_{ij}^- = lower bound of thermal unit ramp rate limitations.
P_{ij}^+ = upper bound of thermal unit ramp rate limitations.
$q_{w-l}^{\min}$ = minimum power output at wind farm l.
$q_{w-l}^{\min}$ = maximum power output at wind farm l.
$q_{s-k}^{\min}$ = minimum power output at photovoltaic park k.
$q_{s-k}^{\max}$ = maximum power output at photovoltaic park k.
R_0 = certain irradiation point, set as $150\,\mathrm{W/m^2}$.
R_h = right bound of the uncertain parameter $\widetilde{h}_{ij}$.
R_n = the upper bound of the designed transmission ability of transmission line.
v_{in} = cut-in speed of the wind turbines.
v_r = rated speed of the wind turbines.
v_{out} = cut-out speed of the wind turbines.
U_0 = solar irradiation in the standard environment, set as $1000\,\mathrm{W/m^2}$.
$\alpha(h)$ = the most likely value of uncertain parameter $\widetilde{h}_{ij}$.
β = positive and negative attitude parameter of power generator toward parameter $\widetilde{h}_{ij}$.
ε = system operator's attitude toward renewable energy utilization.
μ = system operator's attitude toward carbon emission reduction.

Uncertain parameters

$\widetilde{d}_t$ = power demand of time t.
$\widetilde{h}_{ij}$ = coal to carbon emission parameter of unit j at power plant i.
$\widetilde{O}_{ij}$ = coal to power parameter of unit j at power plant i.
$\widetilde{U}_{kt}$ = solar irradiation forecast for solar power plant k at time t.

11.3.3 Constraints

In this section, constraints for the equilibrium strategy-based scheduling of hybrid wind–solar–thermal generation system are given as follows.

11.3.3.1 Constraints of Hybrid System

11.3.3.1.1 Power Balance

Here, we let $\sum_i m_{it}$ represent total thermal power output at time t, in which m_{it} is the power output of thermal power plant i at time t; $\sum_l b_{lt}$ represent the total wind power output at time t and $\sum_k c_{kt}$ represent the total solar power output at time t, then the total power supply of time t is $\sum_l b_{lt} + \sum_k c_{kt} + \sum_i m_{it}$. Here we let $\tilde{d}_t$ represent power demand of time t, which is also an uncertain parameter, then the total power demand must be supplied.

$$\sum_l b_{lt} + \sum_k c_{kt} + \sum_i m_{it} - E\left(\tilde{d}_t\right) = 0 \tag{11.5}$$

11.3.3.1.2 Transmission Ability

The transmission line has a designed transmission capability with an upper and lower bound. Here the electricity power to be transmitted cannot be greater or lower than the designed value. We let L_n and R_n represent the lower and upper bound of the designed transmission ability of transmission line, then this can be represented as follows.

$$L_n \leq \sum_l b_{lt} + \sum_k c_{kt} + \sum_i m_{it} \leq R_n \tag{11.6}$$

11.3.3.1.3 Transmission Stability

Due to the high fluctuation involved in renewable energy sources, to ensure stable amount of transmitted electricity must be considered [Bhandari et al., 2014]. Here we let m^- represent the lower bound power variation that the transmission line can withstand, and m^+ is the upper bound of variation, then this constraint can be written as:

$$m^- \leq \left(\sum_l b_{lt} + \sum_k c_{kt} + \sum_i m_{it}\right) - \left(\sum_l b_{l(t-1)} + \sum_k c_{k(t-1)} + \sum_i m_{i(t-1)}\right) \leq m^+ \tag{11.7}$$

11.3.3.1.4 Spinning Reserve

The stochastic characteristic of wind and solar power indicates that thermal power needs to assign spinning reserve for the hybrid generation system. Here we let $\sum_i m_{it}^{\max}$ and $\sum_i m_{it}^{\min}$ represent the maximum and minimum outputs from the thermal power plants, let a represent the system load spinning reserve rate for the hybrid generation system. Then the spinning reserve of this system can be written as:

$$\begin{aligned} &\sum_i m_{it}^{\max} + \sum_l b_{lt} + \sum_k c_{kt} \geq E\left(\tilde{d}_t\right)(1 + \mathrm{a}\%) \\ &\sum_i m_{it}^{\min} + \sum_l b_{lt} + \sum_k c_{kt} \leq E\left(\tilde{d}_t\right) \end{aligned} \tag{11.8}$$

11.3.3.2 Constraints of Thermal Power Plant

As we discussed above, here we use x_{ijt} to represent the coal to be burned by thermal power plant i unit j at time t, $\tilde{o}_{ij}$ represents the coal to the power parameter of unit j at power plant i, g_{ijt} represents the on/off status of thermal unit j in thermal power plant i at time t, then the total power output of thermal power plant i at time t can be written as:

$$m_{it} = \sum_i x_{ijt} g_{ijt} E\left(\tilde{o}_{ij}\right) \tag{11.9}$$

11.3.3.2.1 Output Limitation

For each thermal unit, the maximum technical power output limitation should be taken into consideration. Let $a_{ij}^{\max}$ represents the maximum technical power output of unit j of thermal power plant i, then, this constraint can be written as:

$$x_{ijt}E\left(\widetilde{o}_{ij}\right) \leq a_{ij}^{\max} g_{ijt} \tag{11.10}$$

11.3.3.2.2 Unit Ramping Rate

To maintain a sustainable and stable utilization at the coal-fired power plants, a unit ramp rate must be considered while shaving the peak. Let P_{ij}^{-} and P_{ij}^{+} be the lower and the upper ramp rate limitations; therefore, this constraint can be written as:

$$\begin{aligned} &x_{ijt}E\left(\widetilde{o}_{ij}\right) - x_{ij(t-1)}E\left(\widetilde{o}_{ij}\right) \leq P_{ij}^{+}g_{ijt} + a_{ij}^{\max}(1 - g_{ij(t-1)}) \\ &x_{ij(t-1)}E\left(\widetilde{o}_{ij}\right) - x_{ijt}E\left(\widetilde{o}_{ij}\right) \leq P_{ij}^{-}g_{ijt} + a_{ij}^{\max}(1 - g_{ijt}) \end{aligned} \tag{11.11}$$

11.3.3.2.3 On- and Off-Line Time

There exists upper bound of thermal units' working hours; similarly, they also exist the minimum turn-off time of thermal units. Here, we use the following mathematical form to represent the thermal unit's minimal online time:

$$\begin{aligned} &\sum_{t=1}^{R_{ij}^{on}} (1 - g_{ijt}) = 0 \\ &\sum_{n=t}^{t+S_{ij}^{on}-1} g_{ijn} \geq S_{ij}^{on}\left(g_{ijt} - g_{ij(t-1)}\right), \quad \forall t = R_{ij}^{on} + 1, \ldots, T - S_{ij}^{on} + 1 \end{aligned} \tag{11.12}$$

in which, $R_{ij}^{on} = \min\left[T, (S_{ij}^{on} - S_{ij}^{on}(0))g_{ij}(0)\right]$ represents the time period thermal unit k must be running at the beginning of the scheduling period and is related to the initial operating condition, S_{ij}^{on} represents the minimum time thermal unit j must be running, $S_{ij}^{on}(0)$ represents the time thermal unit j has been running at the beginning of a scheduling period, and $g_{ij}(0)$ represents the initial status of thermal unit k, with 1 representing on and 0 representing off.

The minimal off-line time constraint is

$$\begin{aligned} &\sum_{t=1}^{R_{ij}^{off}} (1 - g_{ijt}) = 0 \\ &\sum_{n=t}^{t+S_{ij}^{off}-1} g_{ijn} \geq S_{ij}^{off}\left(g_{ijt} - g_{ij(t-1)}\right), \quad \forall t = R_{ij}^{off} + 1, \ldots, T - S_{ij}^{off} + 1 \end{aligned} \tag{11.13}$$

in which, $[R_{ij}^{off} = \min\left[T, (S_{ij}^{off} - S_{ij}^{off}(0))\, g_{ij}(0)\right]$ represents the time period thermal unit k must be off-line at the beginning of the scheduling period, which is related to the initial operating condition, S_{ij}^{off} represents the minimum off-line time for thermal unit k, and $S_{ij}^{off}(0)$ represents the time that thermal unit k has been off-line at the beginning of the scheduling time.

11.3.3.3 Constraints of Wind Power Plant

Wind turbine power output curve is characterized by cut-in, rated, and cut-out speeds [Aliari and Haghani, 2016]. Here we let b_{lt} be the power output of wind farm j at time t, y_{lt} be the wind turbines to be operated at wind farm l at time t, v_{in} be the cut-in speed of the wind turbines, v_r be the rated speed of the wind turbines, v_{out} be the cut-out speed of the wind turbines, p_r^w be the rated output of the wind turbines, and $\widetilde{v_{lt}}$ be the wind speed at wind farm l at time t, then wind power output can be written as:

$$b_{lt} = \begin{cases} 0, & E\left(\widetilde{v_{lt}}\right) < v_{in} \\ y_{lt}p_r^w \frac{E(\widetilde{v_{lt}})-v_{in}}{v_r - v_{in}}, & v_{in} \leq E\left(\widetilde{v_{lt}}\right) < v_r \\ y_{lt}p_r^w, & v_r \leq E\left(\widetilde{v_{lt}}\right) < v_{out} \\ 0, & v_{out} \leq E\left(\widetilde{v_{lt}}\right) \end{cases} \tag{11.14}$$

11.3.3.3.1 Power Output

Wind power plant has the technical power output limitation. Here, we let $q_{w-l}^{\min}$ represent the minimum power output at wind farm l, and $q_{w-l}^{\min}$ represent the maximum power output at wind farm l; then, this constraint can be written as:

$$q_{w-l}^{\min} \leq b_{lt} \leq q_{w-l}^{\max} \tag{11.15}$$

11.3.3.4 Constraints of Solar Power Plant

Solar power output curve is a piecewise function characterized by a certain irradiation point [Biswas et al., 2017]. Here we use c_{kt} to represent the power output at solar plant k at time t, z_{kt} to represent the photovoltaic arrays to be operated at solar power plant k at farm t, $\widetilde{U_{kt}}$ represent the solar irradiation forecast for solar power plant k at time t, which is a triangular fuzzy number, U_0 is the solar irradiation in the standard environment, set as $1000\,\mathrm{W/m^2}$, and R_0 is the certain irradiation point, set as $150\,\mathrm{W/m^2}$, p_r^s be the rated output of the photovoltaic arrays. Therefore, the solar irradiation to energy conversion function for the PV generator is as follows:

$$c_{kt} = \begin{cases} z_{kt}p_s^r \frac{E(\widetilde{U_{kt}})^2}{U_0 R_0}, & 0 < E\left(\widetilde{U_{kt}}\right) \leq R_0 \\ z_{kt}p_s^r \frac{E(\widetilde{U_{kt}})}{U_0}, & E\left(\widetilde{U_{kt}}\right) > R_0 \end{cases} \tag{11.16}$$

11.3.3.4.1 Power Output

Solar photovoltaic power plant has the technical power output limitation. Here, we use $q_{s-k}^{\min}$ to represent the minimum power output at photovoltaic park k and $q_{s-k}^{\max}$ to represent the maximum power output at photovoltaic park k, then this constraint can be written as:

$$q_{s-k}^{\min} \leq c_{kt} \leq q_{s-k}^{\max} \tag{11.17}$$

11.3.4 Global Model

From the above process, the scheduling model of hybrid wind–solar–thermal power plant is built, and this model takes full consideration of economic benefits, environmental impacts, and renewable energy utilization of electricity generation process.

The system operator seeks to maximize electricity generation profit, minimize carbon emissions, and increase the proportion of renewable energy utilization at the same time, but these objectives subject to complex constraints, including constraints for the hybrid wind–solar–thermal generation system are shown in Eqs. (11.5)–(11.8), and constraints for the thermal, wind, solar power plants are shown in Eqs. (11.9)–(11.13), Eqs. (11.14)–(11.15), and Eqs. (11.16) and (11.17). Generally, this optimal model is proposed from a regional perspective; however, as there exist three different kind of power sources in the hybrid system, which need related cooperation, the decision variables are set directly for each power plant to facilitate the operations. The multi-objective optimal scheduling model for the hybrid wind–solar–thermal generation system is formulated in Eq. (11.18).

$$
\begin{aligned}
&\max f_1 = \sum_t e_t a_4 + \sum_t \sum_l b_{lt} a_1 + \sum_t \sum_k c_{kt} a_2 - \sum_t \sum_i \sum_j x_{ijt} a_3 \\
&\min f_2 = \sum_i \sum_i \sum_j x_{ijt} g_{ijt} E(\widetilde{h}_{ij}) \\
&\max f_3 = \sum_t \sum_l b_{lt} + \sum_t \sum_k c_{kt} \\
&\begin{cases}
\sum_l b_{lt} + \sum_k c_{kt} + \sum_i m_{it} - E\left(\widetilde{d}_t\right) = 0 \\
L_n \le \sum_l b_{lt} + \sum_k c_{kt} + \sum_i m_{it} \le R_n \\
m^- \le \left(\sum_l b_{lt} + \sum_k c_{kt} + \sum_i m_{it}\right) - \left(\sum_l b_{l(t-1)} + \sum_k c_{k(t-1)} + \sum_i m_{i(t-1)}\right) \le m^+ \\
\sum_i m_{it}^{\max} + \sum_l b_{lt} + \sum_k c_{kt} \ge E\left(\widetilde{d}_t\right)(1 + \mathrm{a}\%) \\
\sum_i m_{it}^{\min} + \sum_l b_{lt} + \sum_k c_{kt} \le E\left(\widetilde{d}_t\right) \\
m_{it} = \sum_i x_{ijt} g_{ijt} E\left(\widetilde{o}_{ij}\right) \\
x_{ijt} E\left(\widetilde{o}_{ij}\right) \le a_{ij}^{\max} g_{ijt} \\
x_{ijt} E\left(\widetilde{o}_{ij}\right) - x_{ij(t-1)} E\left(\widetilde{o}_{ij}\right) \le P_{ij}^{+} g_{ijt} + a_{ij}^{\max}(1 - g_{ij(t-1)}) \\
x_{ij(t-1)} E\left(\widetilde{o}_{ij}\right) - x_{ijt} E\left(\widetilde{o}_{ij}\right) \le P_{ij}^{-} g_{ijt} + a_{ij}^{\max}(1 - g_{ijt}) \\
\sum_{t=1}^{R_{ij}^{on}} (1 - g_{ijt}) = 0, \quad \sum_{n=t}^{t+S_{ij}^{on}-1} g_{ijn} \ge S_{ij}^{on}\left(g_{ijt} - g_{ij(t-1)}\right), \quad \forall t = R_{ij}^{on} + 1, \ldots, T - S_{ij}^{on} + 1 \\
\sum_{t=1}^{R_{ij}^{off}} (1 - g_{ijt}) = 0, \quad \sum_{n=t}^{t+S_{ij}^{off}-1} g_{ijn} \ge S_{ij}^{off}\left(g_{ijt} - g_{ij(t-1)}\right), \quad \forall t = R_{ij}^{off} + 1, \ldots, T - S_{ij}^{off} + 1 \\
b_{lt} = \begin{cases}
0, & E\left(\widetilde{v}_{lt}\right) < v_{in} \\
y_{lt} p_r^w \frac{E(\widetilde{v}_{lt}) - v_{in}}{v_r - v_{in}}, & v_{in} \le E\left(\widetilde{v}_{lt}\right) < v_r \\
y_{lt} p_r^w, & v_r \le E\left(\widetilde{v}_{lt}\right) < v_{out} \\
0, & v_{out} \le E\left(\widetilde{v}_{lt}\right)
\end{cases} \\
q_{w-l}^{\min} \le b_{lt} \le q_{w-l}^{\max} \\
c_{kt} = \begin{cases}
z_{kt} p_s^r \frac{E(\widetilde{U}_{kt})^2}{U_0 R_0}, & 0 < E\left(\widetilde{U}_{kt}\right) \le R_0 \\
z_{kt} p_s^r \frac{E(\widetilde{U}_{kt})}{U_0}, & E\left(\widetilde{U}_{kt}\right) > R_0
\end{cases} \\
q_{s-k}^{\min} \le c_{kt} \le q_{s-k}^{\max}
\end{cases}
\end{aligned}
\tag{11.18}
$$

In above model, we take wind, solar, and thermal power plants into consideration to avoid a single generation disadvantage. The objectives of pursue electricity generation profits, carbon emission reduction, and renewable energy utilization are simultaneously considered by applying multi-objective method; therefore, system operator can obtain an equilibrium scheduling strategy. Thirdly, two methods are applied to handle the uncertainty of renewable sources, to deal with the hourly fluctuation of wind speed and solar irradiation, and we applied fuzzy numbers to represent these values, because the average wind speed and solar irradiation are always interval values; as for the seasonal fluctuation of wind speed and weather-driven character of solar irradiation, we choose to establish 12 typical scheduling scenarios to handle them. Therefore, the multi-objective model-based scheduling strategy this chapter proposed has a more comprehensive, systematic structure.

11.4 Case Study

To testify the feasibility and efficiency of the suggested programming approach, this section gives an applied mixed wind–solar–thermal generation example from Hami, Xinjiang, China.

11.4.1 Case Description

Hami, a city in the east of Xinjiang Province, stores about 5.7 billion tonnes of coal resources, the most in Xinjiang Province and an eighth of total reserves in China [China Energy News, 2016]. In June 2014, according to the Energy Development Strategy Action Plan (2014–2020) published by the General Office of the State Council, Hami became one of the nine large-scale coal and electricity bases with a tens of millions kilowatt level. Historical data show that in Hami, there generated 25 billion kWh of electricity by thermal power in 2016, which covers 73.48% of total power generation in Xinjiang Province. Even though the coal in Hami is always regarded as low sulfur, low phosphorus, low ash powder, and high calorific value, it is still a main source of carbon dioxide emissions because of the large-scale coal burning. Consequently, it is quite essential to find a way to reduce carbon emission in power generation in order to protect the environment and achieve the sustainable development. Also, in the last several years, coal resources in Hami have been left behind by the rapid development of economy. In 2016, the annual added value for thermal power production and supply industry was 4.23 billion CNY, which was 46.73% of total industrial added value at 9.05 billion CNY. This implies that the coal and electricity industry has become a significant supporting industry for Hami's economic growth. On condition that Hami's economy needs to keep growing, the coal and electricity industry also have to keep developing. Then, it is of great significance to balance the conflict between carbon emissions reduction and economic growth in Hami.

Renewable energy has been integrated into the traditional power generation networks to reduce the contradiction between environmental protection and economic development, which has been recognized by more and more industries. Hami has clear advantages on renewable energy. 62.9% of the total needs of Xinjiang Province can be supplied by Hami's exploitable wind energy resources with certain technology. It is the reason why there are three major wind zones of the total seven planned in China. With annual sunshine hours of 3170–3380, which is an advantage of solar power generation, Hami is considered to have the best solar energy resources in Xinjiang Province. It is reported from The Hami Power Dispatching Center that the installed wind and solar power capacities in Hami were 9750 and 2400 MW, respectively, which is 64.8% of the total installed power capacity [Xinhua News Agency, 2017]. Whereas, Xinjiang has a tough problem of wind and solar curtailment due to the insufficient local electricity adoptions. To deal with this, the government launched the Hami South-Zhengzhou Ultra-High-Voltage Network (Ha-Zheng Line) to transmit the electricity generated in Xinjiang Province to end users in Zhengzhou Province, and this project has totally transmitted 90 billion kWh of electricity.

What makes the Ha-Zheng Line unmatched is that the Line used the wind–solar–thermal power bundling mode, which makes it more competitive and efficient than other projects. And it also helps the western region in raising the consumption of renewable energy. The supporting power supply of Ha-Zheng Line is constituted of three parts: 8000 MW of wind power, 1250 MW of solar power, and four thermal power plants with a total installed capacity of 6600 MW. As a key project of Hami Energy Base, the wind–solar–thermal three-in-one energy output is applied to search for new applications of energy and to promote the sustainable development of China's power generation industry. Whereas, it is impossible to arrange wind and solar in the same as thermal units as they are more reliant on natural environment, which makes them it hard to control; to deal with, what we should do is to explore a proper dispatching method. Consequently, this project could be considered as a typical example, the practical model of wind–solar–thermal power bundling mode used in Hami Energy Base is shown in Figure 11.3.

11.4.2 Model Transformation

When scheduling electricity generation, the company's system operator needs to ensure enterprise survival first, so that the economic profits objective is significantly important. Further, as Hami is still a developing region in China, the economy is growing, to ensure stable and reliable electricity is very important for social development. On the other hand, to comply with national environmental protection and resource conservation policies and make reasonable use of the Ha-Zheng Line Transmission Network, carbon emissions must be minimized and renewable energy utilization must be maximized within a certain range. Encouraged by Zeng et al.'s

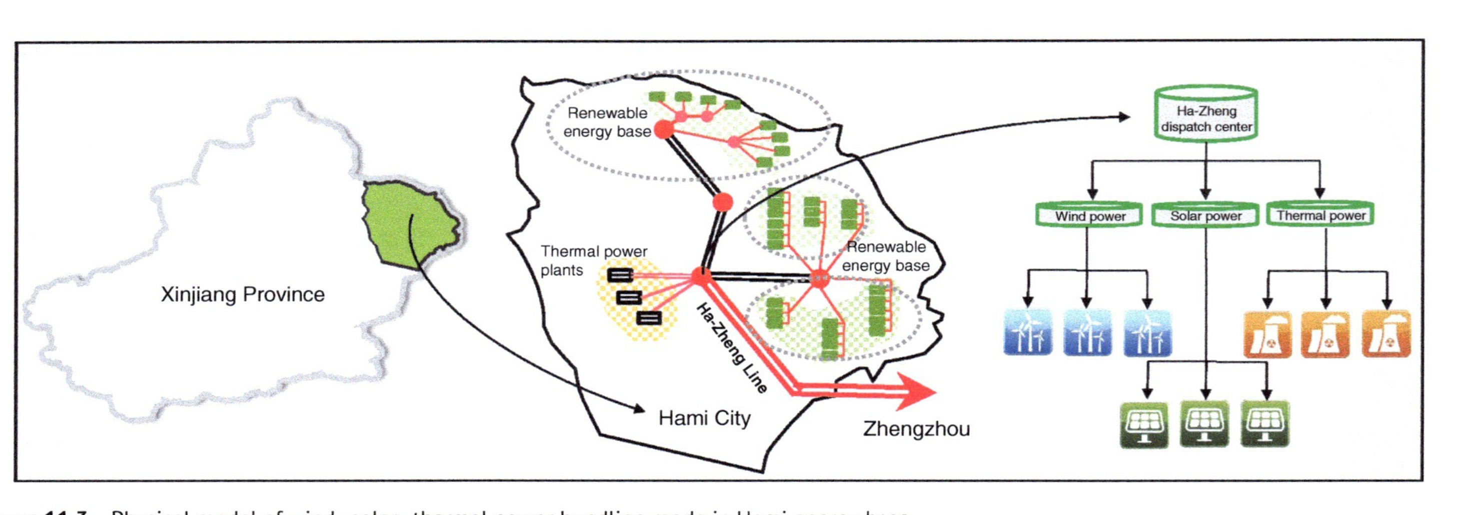

Figure 11.3 Physical model of wind–solar–thermal power bundling mode in Hami energy base.

method proposed in Zeng et al. [2014], which has been efficiently used by Lv et al. [2016] and Xu et al. [2018], two system operator attitude parameters for carbon emissions reductions and renewable energy utilization are introduced to transform model 11.18 into its equivalent single-objective form. Parameter μ means that the carbon emissions in this scheduling period is μ times less than or equal to that of the previous production period, and parameter ε means that the wind and solar power utilization in this scheduling period is ε times more than or equal to that of the previous production period. C and N, respectively, represent the carbon emissions and the renewable energy utilization of the previous production period. Therefore, the multi-objective model in Eq. (11.18) can be transformed in its equivalent form as Eq. (11.19).

By introducing the system operators' attitude parameter toward carbon emission reduction and renewable energy utilization, the global model is transformed into its equivalent single form (Eq. (11.19)) with integer variables and piecewise functions. In our model (Eq. (11.19)), the decision variable y_{lt}, which indicates the number of wind turbines to be operated at wind farm l during time t, and z_{kt}, which indicates the solar PV arrays to be operated in solar power plant k at time t, are both integer variables. LINGO software, which has been used around the world, is very efficient in solving single-objective model mixed with integer variable. For example, Jabir et al. Jabir et al. [2017] used Lingo solver to successfully solve an integer linear programming model for a capacitated multi-depot green vehicle routing problem, and Ware et al. Ware et al. [2014] used Lingo software to solve a dynamic supplier selection problem modeled as a mixed-integer nonlinear program. Encouraged by these success, this software was chosen to solve the single-objective model (11.19) in this chapter.

11.4.3 Data Collection

Before application, the relevant data were collected. According to the *Benchmark on-gird Price for Wind Power ([2009] No.1906)* published by Chinese National Development and Reform Commission (CNDRC) [Chinese National Development and Reform Commission, 2009], China's onshore wind energy resources were divided into four categories. Under this sort of classification, Hami, China belongs to the third type area, of which, the on-grid price is determined as 0.58 CNY/kWh (data of 2009, it has been renewed as 0.56 CNY/kWh for wind farms connected to the grid in 2014). Similarly, in the *Benchmark on-gird Price for PV Power ([2013] No.1638)* [Chinese National Development and Reform Commission, 2013], China's solar PV resources were divided into three categories, and Hami belongs to the first type area, with an on-grid price of 0.90 CNY/kWh. The latest documents from the Xinjiang Uygur Autonomous Regional Development and Reform Commission declared that the on-grid price for Ha-Zheng Line at the Hami Energy Base was 0.21 CNY/kWh [Uygur, 2014]. For wind and solar power, the price within coal thermal power benchmark price was paid by the power grid, while the higher part was provided by the National Renewable Energy Development Fund.

$$
\max f = \sum_t e_t a_4 + \sum_t \sum_l b_{lt} a_1 + \sum_t \sum_k c_{kt} a_2 - \sum_t \sum_i \sum_j x_{ijt} a_3
$$

$$
\begin{cases}
\sum_i \sum_i \sum_j x_{ijt} g_{ijt} E(\widetilde{h}_{ij}) \le \mu C \\
\sum_t \sum_l b_{lt} + \sum_t \sum_k c_{kt} \ge \varepsilon N \\
\sum_l b_{lt} + \sum_k c_{kt} + \sum_i m_{it} - E\left(\widetilde{d}_t\right) = 0 \\
L_n \le \sum_l b_{lt} + \sum_k c_{kt} + \sum_i m_{it} \le R_n \\
m^- \le \left(\sum_l b_{lt} + \sum_k c_{kt} + \sum_i m_{it}\right) - \left(\sum_l b_{l(t-1)} + \sum_k c_{k(t-1)} + \sum_i m_{i(t-1)}\right) \le m^+ \\
\sum_i m_{it}^{\max} + \sum_l b_{lt} + \sum_k c_{kt} \ge E\left(\widetilde{d}_t\right)(1 + \mathrm{a}\%) \\
\sum_i m_{it}^{\min} + \sum_l b_{lt} + \sum_k c_{kt} \le E\left(\widetilde{d}_t\right) \\
m_{it} = \sum_i x_{ijt} g_{ijt} E\left(\widetilde{o}_{ij}\right) \\
x_{ijt} E\left(\widetilde{o}_{ij}\right) \le a_{ij}^{\max} g_{ijt} \\
x_{ijt} E\left(\widetilde{o}_{ij}\right) - x_{ij(t-1)} E\left(\widetilde{o}_{ij}\right) \le P_{ij}^{+} g_{ijt} + a_{ij}^{\max}(1 - g_{ij(t-1)}) \\
x_{ij(t-1)} E\left(\widetilde{o}_{ij}\right) - x_{ijt} E\left(\widetilde{o}_{ij}\right) \le P_{ij}^{-} g_{ijt} + a_{ij}^{\max}(1 - g_{ijt}) \\
\sum_{t=1}^{R_{ij}^{on}} (1 - g_{ijt}) = 0, \quad \sum_{n=t}^{t+S_{ij}^{on}-1} g_{ijn} \ge S_{ij}^{on}\left(g_{ijt} - g_{ij(t-1)}\right), \quad \forall t = R_{ij}^{on} + 1, \ldots, T - S_{ij}^{on} + 1 \\
\sum_{t=1}^{R_{ij}^{off}} (1 - g_{ijt}) = 0, \quad \sum_{n=t}^{t+S_{ij}^{off}-1} g_{ijn} \ge S_{ij}^{off}\left(g_{ijt} - g_{ij(t-1)}\right), \quad \forall t = R_{ij}^{off} + 1, \ldots, T - S_{ij}^{off} + 1 \\
b_{lt} = \begin{cases} 0, & E\left(\widetilde{v}_{lt}\right) < v_{in} \\ y_{lt} p_r^w \frac{E(\widetilde{v}_{lt}) - v_{in}}{v_r - v_{in}}, & v_{in} \le E\left(\widetilde{v}_{lt}\right) < v_r \\ y_{lt} p_r^w, & v_r \le E\left(\widetilde{v}_{lt}\right) < v_{out} \\ 0, & v_{out} \le E\left(\widetilde{v}_{lt}\right) \end{cases} \\
q_{w-l}^{\min} \le b_{lt} \le q_{w-l}^{\max} \\
c_{kt} = \begin{cases} z_{kt} p_s^r \frac{E(\widetilde{U}_{kt})^2}{U_0 R_0}, & 0 < E\left(\widetilde{U}_{kt}\right) \le R_0 \\ z_{kt} p_s^r \frac{E(\widetilde{U}_{kt})}{U_0}, & E\left(\widetilde{U}_{kt}\right) > R_0 \end{cases} \\
q_{s-k}^{\min} \le c_{kt} \le q_{s-k}^{\max}
\end{cases}
\tag{11.19}
$$

Certain data for the wind–solar–thermal power plants are presented in Table 11.1. Coal price was collected from the website. The transmission capacity and the transmission ramping bound for the Ha-Zheng Line were collected from the research chapter in Yuan et al. [2017]. The minimum and maximum wind and solar power plant outputs were collected from the Hami Government website. The data for the uncertain parameters: the coal to power parameter, the coal to carbon parameter (Table 11.2), as well as the wind speed data (Table 11.3), solar irradiation data (Table 11.4), and power demand (Table 11.5) were collected from field research

Table 11.1 Certain parameters for the case.

Parameters	Values	Parameters	Values
Power on grid price (CNY/MW)	210	Min. output of wind and solar power plants (MW)	0
Allowance of wind power (CNY/MW)	370	Max. output of wind power plant 1, 2, 3 (MW)	2400, 2800, 2300
Allowance of solar power (CNY/MW)	690	Max. output of solar power plant 1, 2, 3 (MW)	150, 300, 500
Price of coal (CNY/MW)	170	Min. output of thermal units (MW)	264
Transmission capacity (MW)	8000	Max. output of thermal units (MW)	660
Ramping upper bound (MW)	1000	Lower bound of ramping rate (MW/h)	−396
Ramping lower bound (MW)	−1000	Upper bound of ramping rate (MW/h)	396

Table 11.2 Coal to electricity parameter and coal to carbon emission parameter.

Thermal units	Coal to power (MW/tonne)	Coal to carbon (tonne/tonne)	Thermal units	Coal to power (MW/tonne)	Coal to carbon (tonne/tonne)
Unit 1	(2.21, 2.23, 2.25)	(1.96, 1.97, 1.98)	Unit 5	(2.28, 2.30, 2.33)	(1.95, 1.97, 1.98)
Unit 2	(2.22, 2.25, 2.28)	(1.97, 1.98, 1.99)	Unit 6	(2.20, 2.24, 2.27)	(1.95, 1.96, 1.98)
Unit 3	(2.21, 2.26, 2.29)	(1.98, 1.99, 2.00)	Unit 7	(2.22, 2.23, 2.24)	(1.96, 1.97, 1.99)
Unit 4	(2.08, 2.11, 2.15)	(1.98, 1.99, 2.01)	Unit 8	(2.18, 2.19, 2.20)	(1.95, 1.97, 1.98)

through interviews with experts and engineers as well as historical data, as follows. (i) Interviews were conducted with experts and engineers from each power plant associated with the hybrid wind–solar–thermal generation system, in which they were asked to give a range for each uncertain parameter; some first-hand measurement data were also obtained. (ii) Then, the initial collected data were analyzed and extreme values were eliminated. (iii) The minimum value for the remaining data was set as the lower bound for each uncertain parameter and the maximum value was set as the upper bound. (iv) Without a loss of generality, the values with the highest frequency were assumed to be the most possible range for each uncertain parameter.

11.4.4 Results and Analysis

In this chapter, three aspects of the calculation results for the 12 scheduling scenarios are analyzed; the output ratio, the hourly output, and an objective values comparison. Results under different scenarios are shown in Tables 11.6 to 11.10.

Table 11.3 Wind speed data in fuzzy numbers (m/s).

Time period	Wind speed data			Time period	Wind speed data		
	Farm 1	Farm 2	Farm 3		Farm 1	Farm 2	Farm 3
1	(2.70 4.10 11.0)	(2.90 7.10 14.0)	(5.20 9.50 12.6)	13	(3.70 7.30 17.0)	(3.10 5.50 9.90)	(5.00 8.20 11.5)
2	(2.00 5.00 10.0)	(2.20 5.50 11.0)	(3.60 9.00 14.3)	14	(3.70 12.0 16.0)	(1.90 3.10 15.0)	(3.00 5.00 12.0)
3	(3.10 3.80 8.00)	(5.70 8.50 13.0)	(5.20 6.30 13.0)	15	(2.80 4.00 10.9)	(0.60 5.30 8.10)	(2.00 7.20 11.0)
4	(4.40 4.70 8.00)	(5.50 8.10 9.80)	(4.30 8.00 12.0)	16	(1.00 6.00 15.0)	(1.00 6.40 9.50)	(2.80 6.00 9.00)
5	(5.70 8.00 12.0)	(5.60 8.70 10.2)	(5.70 8.90 13.5)	17	(2.50 7.00 16.5)	(2.70 6.00 16.5)	(3.00 12.0 13.2)
6	(4.30 6.20 9.00)	(5.00 8.00 14.0)	(4.40 13.0 16.0)	18	(0.10 5.00 12.0)	(2.30 10.7 11.2)	(4.00 8.00 15.7)
7	(5.20 6.30 13.0)	(4.80 5.80 9.50)	(3.10 7.00 9.80)	19	(0.70 8.20 14.0)	(3.80 8.70 10.3)	(3.70 7.10 9.00)
8	(5.50 8.00 9.00)	(5.40 8.10 9.40)	(5.30 8.60 13.2)	20	(1.20 9.00 13.0)	(4.60 8.30 9.70)	(1.20 3.60 13.6)
9	(5.20 6.50 11.4)	(3.00 57.50 16.0)	(2.70 7.80 12.9)	21	(1.80 5.00 7.00)	(3.50 6.00 15.0)	(3.70 4.40 12.0)
10	(4.80 5.60 17.3)	(1.30 8.00 11.0)	(2.80 9.00 16.0)	22	(1.30 4.00 12.0)	(3.80 7.40 11.0)	(2.90 8.30 12.4)
11	(3.00 9.00 13.5)	(1.10 7.00 8.40)	(4.00 8.00 15.0)	23	(2.30 8.00 13.5)	(4.20 7.60 10.8)	(4.20 8.70 14.3)
12	(2.90 8.90 15.8)	(1.80 6.10 11.0)	(1.30 6.70 9.60)	24	(2.80 6.00 15.2)	(4.00 8.60 9.70)	(5.30 8.80 12.8)

Table 11.4 Solar irradiation data in fuzzy numbers (W/m^2).

Time period	Solar irradiation data			Time period	Solar irradiation data		
	Solar plant 1	Solar plant 2	Solar plant 3		Solar plant 1	Solar plant 2	Solar plant 3
1	(0, 0, 0)	(0, 0, 0)	(0, 0, 0)	13	(756, 778, 798)	(722, 755, 788)	(756, 777, 789)
2	(0, 0, 0)	(0, 0, 0)	(0, 0, 0)	14	(777, 789, 799)	(720, 768, 798)	(777, 785, 799)
3	(0, 0, 0)	(0, 0, 0)	(0, 0, 0)	15	(688, 697, 725)	(678, 702, 732)	(657, 678, 699)
4	(0, 0, 0)	(0, 0, 0)	(0, 0, 0)	16	(633, 642, 656)	(630, 653, 670)	(623, 655, 678)
5	(0, 0, 0)	(0, 0, 0)	(0, 0, 0)	17	(547, 577, 589)	(534, 566, 587)	(554, 566, 598)
6	(0, 0, 0)	(0, 0, 0)	(0, 0, 0)	18	(388, 400, 427)	(401, 423, 456)	(340, 288, 400)
7	(105, 110, 115)	(78, 100, 109)	(100, 109, 120)	19	(230, 250, 275)	(234, 245, 261)	(234, 250, 277)
8	(203, 215, 220)	(198, 208, 224)	(210, 222, 230)	20	(88, 110, 130)	(98, 102, 118)	(88, 99, 120)
9	(350, 377, 390)	(320, 355, 370)	(367, 376, 399)	21	(0, 0, 0)	(0, 0, 0)	(0, 0, 0)
10	(480, 499, 507)	(457, 478, 499)	(521, 543, 567)	22	(0, 0, 0)	(0, 0, 0)	(0, 0, 0)
11	(587, 596, 618)	(566, 589, 600)	(589, 605, 624)	23	(0, 0, 0)	(0, 0, 0)	(0, 0, 0)
12	(650, 699, 725)	(650, 678, 699)	(678, 700, 712)	24	(0, 0, 0)	(0, 0, 0)	(0, 0, 0)

Table 11.5 Power demand data of Ha-Zheng Line (MW).

Time	Power demand	Time	Power demand	Time	Power demand	Time	Power demand
1	(3000, 3480, 3670)	7	(5698, 5718, 5820)	13	(6930, 6960, 7003)	19	(5888, 5967, 6001)
2	(3500, 3729, 3832)	8	(5877, 5967, 6002)	14	(6302, 6462, 6503)	20	(6503, 6624, 6702)
3	(4010, 4224, 4322)	9	(6200, 6462, 6567)	15	(5879, 5958, 6005)	21	(6362, 6462, 6503)
4	(4701, 4711, 4756)	10	(6670, 6790, 6809)	16	(5100, 5220, 5300)	22	(5401, 5469, 5569)
5	(4891, 4971, 5002)	11	(6987, 7074, 7129)	17	(4781, 4971, 5004)	23	(4120, 4473, 4564)
6	(5400, 5469, 5567)	12	(7100, 7224, 7324)	18	(5300, 5469, 5556)	24	(3898, 3978, 4023)

When focusing on the power output ratio analysis, the wind, solar, and thermal power output of the 12 scenarios are shown in Figure 11.4. It is worth noting that base case is also presented in this figure to compare the results of these scenarios. In the base case, power output of wind, solar, and thermal resources are 39%, 2%, and 59%, respectively. For the 12 subfigures in the right part, the horizontal subgraph shows the output of each power source in the four seasons under a fixed weather condition; taking sunny weather in the four seasons (scenarios 1, 4, 7, 10) as examples, solar power output in spring and summer are relatively higher than that of autumn and winter. The vertical subgraph shows the output of each energy source under different weather conditions in a certain season, taking the scenarios 1, 2, and 3 as examples, when weather conditions change from sunny to cloudy, and from cloudy to rainy, the output of solar power decreased accordingly and the output of wind and thermal power increased accordingly.

The hourly power output analysis is another interesting topic and as shown in Figure 11.5, the hourly output of wind, solar, and thermal power, respectively. The horizontal subgraph shows the output of the three types of energy under three weather conditions in a certain season, and the vertical subgraph shows the output of the three types of energy in different seasons. From these subfigures, it can be easily seen that variation of thermal and wind power is very obvious, but the output of solar power is relatively fluent. What's more, in each season, when solar power output changes due to the weather condition, the corresponding load is supplied by the thermal power rather than the wind power, and this indicates that in the hybrid system, the power more reliable source has priority in terms of power output. Another obvious condition in this figure is that wind power satisfied 38.1–51.8% load of the demand, indicating that hybrid mode is a very sustainable options for future energy use.

The analysis of trade-off among the conflict objectives in the proposed model will be given in this section. Objective values of our proposed model, including economic profit, carbon emissions, and renewable energy utilization, are presented in Figure 11.6. From Figure 11.5a, it is obvious that economic profits of the 12 scenarios are higher than that of the base case, especially in the spring season, and this is because more renewable energy is exported and the corresponding on-grid

Table 11.6 Power output of 12 scheduling scenarios (MW).

Season		Spring			Summer	
Weather	Sunny	Cloudy	Rainy	Sunny	Cloudy	Rainy
Scenarios	Scenario 1	Scenario 2	Scenario 3	Scenario 4	Scenario 5	Scenario 6
Thermal power output	58 601	59 195	60 197	70 993	72 042	73 866
Wind power output	67 527	67 705	67 990	59 059	59 185	59 302
Solar PV power output	5046	4275	2987	7681	6506	4566
Season		Autumn			Winter	
Weather	Sunny	Cloudy	Rainy	Sunny	Cloudy	Rainy
Scenarios	Scenario 7	Scenario 8	Scenario 9	Scenario 10	Scenario 11	Scenario 12
Thermal power output	73 324	73 957	75 050	75 770	76 270	77 114
Wind power output	48 034	48 060	48 077	60 067	60 067	60 067
Solar PV power output	4570	3910	2801	3208	2709	1865

Table 11.7 Hourly output of thermal power, wind power, and solar power when weather is sunny (MW).

	Scenario 1 Spring-Sunny			Scenario 4 Summer-Sunny			Scenario 7 Autumn-Sunny			Scenario 10 Winter-Sunny		
Time	Thermal	Wind	Solar	Thermal	Wind	Solar	Thermal	Wind	Solar	Thermal	Wind	Solar
1	2112	2388	0	2112	2613	0	3089	1231	0	2112	2658	0
2	2190	2400	0	2459	2360	0	3154	1252	0	2112	2753	0
3	2112	2778	0	2882	2253	0	2573	2122	0	2264	2919	0
4	2141	2858	0	2581	2668	0	2371	2428	0	2658	2641	0
5	2112	3121	0	2935	2560	0	2254	2770	0	2581	2966	0
6	2802	2409	0	2647	2744	81	2132	2870	0	3530	1994	0
7	2289	3087	57	2678	2754	273	2412	2758	46	3381	2378	0
8	2112	3401	165	3546	2011	405	2112	3211	128	3151	2868	0
9	2478	3142	280	4324	1400	470	2581	2881	201	3658	2500	95
10	3127	2659	392	3980	1973	534	3168	2485	278	3824	2463	262
11	3188	2679	455	3841	2207	590	2418	3228	423	4131	2230	340
12	2712	3262	527	3259	2902	663	3349	2401	490	4439	2003	448
13	2460	3077	585	2815	2894	719	3196	2104	577	3259	2695	535
14	2304	3004	591	3239	2244	710	2809	2253	601	4174	1590	488
15	2112	2932	523	2374	2819	652	2782	2032	531	3602	1870	429
16	2112	2714	496	2775	2232	580	2280	2224	605	3595	1735	311
17	3921	747	431	2112	2723	520	2564	1981	351	3939	1291	177
18	2814	2214	304	2112	3021	466	4132	770	217	2303	3225	124
19	2573	2909	189	2560	3019	375	4292	1027	125	3673	2338	0
20	2293	3454	52	2823	2953	314	3937	1631	0	3440	2708	0
21	2300	3267	0	4331	1284	230	3948	1396	0	2955	2946	0
22	2112	3229	0	3764	1747	98	3905	1223	0	2767	2895	0
23	2112	3010	0	2256	3123	0	4043	874	0	2112	3317	0
24	2112	2788	0	2588	2557	0	3822	882	0	2112	3082	0

Table 11.8 Hourly output of thermal power, wind power, and solar power when weather is cloudy (MW).

	Scenario 1 Spring-Cloudy			Scenario 5 Summer-Cloudy			Scenario 8 Autumn-Cloudy			Scenario 11 Winter-Cloudy		
Time	Thermal	Wind	Solar	Thermal	Wind	Solar	Thermal	Wind	Solar	Thermal	Wind	Solar
1	2112	2388	0	2112	2613	0	3089	1231	0	2112	2658	0
2	2190	2400	0	2459	2360	0	3154	1252	0	2112	2753	0
3	2112	2778	0	2882	2253	0	2573	2122	0	2264	2919	0
4	2141	2858	0	2581	2668	0	2371	2428	0	2658	2641	0
5	2112	3121	0	2935	2560	0	2254	2770	0	2581	2966	0
6	2802	2409	0	2669	2744	58	2132	2870	0	3530	1994	0
7	2305	3087	41	2719	2754	232	2425	2758	33	3381	2378	0
8	2112	3426	140	3607	2011	344	2112	3237	102	3151	2868	0
9	2520	3142	238	4394	1400	400	2611	2881	171	3684	2500	69
10	3185	2659	333	4060	1973	454	3210	2485	236	3863	2463	222
11	3257	2679	387	3929	2207	502	2481	3228	360	4182	2230	289
12	2791	3262	448	3359	2902	564	3423	2401	416	4506	2003	381
13	2548	3077	497	2923	2894	611	3283	2104	490	3339	2695	454
14	2393	3004	502	3346	2244	604	2900	2253	511	4247	1590	415
15	2112	3010	445	2472	2819	554	2861	2032	451	3666	1870	365
16	2112	2789	421	2863	2232	493	2326	2224	559	3641	1735	265
17	3986	747	367	2134	2779	442	2617	1981	298	3965	1291	150
18	2859	2214	259	2112	3091	396	4164	770	184	2327	3225	100
19	2601	2909	160	2617	3019	319	4317	1027	99	3673	2338	0
20	2308	3454	38	2870	2953	267	3937	1631	0	3440	2708	0
21	2300	3267	0	4365	1284	196	3948	1396	0	2955	2946	0
22	2112	3229	0	3791	1747	70	3905	1223	0	2767	2895	0
23	2112	3010	0	2256	3123	0	4043	874	0	2112	3317	0
24	2112	2788	0	2588	2557	0	3822	882	0	2112	3082	0

Table 11.9 Hourly output of thermal power, wind power, and solar power when weather is rainy (MW).

Time	Scenario 3 Spring-Rainy			Scenario 6 Summer-Rainy			Scenario 9 Autumn-Rainy			Scenario 12 Winter-Rainy		
	Thermal	Wind	Solar	Thermal	Wind	Solar	Thermal	Wind	Solar	Thermal	Wind	Solar
1	2112	2388	0	2112	2613	0	3089	1231	0	2112	2658	0
2	2190	2400	0	2459	2360	0	3154	1252	0	2112	2753	0
3	2112	2778	0	2882	2253	0	2573	2122	0	2264	2919	0
4	2141	2858	0	2581	2668	0	2371	2428	0	2658	2641	0
5	2112	3121	0	2935	2560	0	2254	2770	0	2581	2966	0
6	2802	2409	0	2698	2744	29	2132	2870	0	3530	1994	0
7	2326	3087	21	2787	2754	164	2442	2758	16	3381	2378	0
8	2136	3457	86	3708	2011	243	2145	3254	52	3151	2868	0
9	2590	3142	168	4512	1400	282	2661	2881	121	3718	2500	34
10	3283	2659	235	4193	1973	320	3279	2485	167	3928	2463	157
11	3370	2679	273	4077	2207	354	2587	3228	254	4267	2230	204
12	2922	3262	316	3525	2902	398	3545	2401	294	4618	2003	269
13	2694	3077	351	3103	2894	431	3427	2104	346	3473	2695	321
14	2541	3004	354	3524	2244	426	3050	2253	361	4369	1590	293
15	2112	3141	314	2635	2819	391	2994	2032	319	3773	1870	257
16	2112	2913	297	3008	2232	348	2402	2224	483	3719	1735	187
17	4094	747	259	2264	2779	312	2705	1981	211	4022	1291	93
18	2935	2214	183	2112	3207	280	4218	770	130	2377	3225	50
19	2649	2909	113	2711	3019	225	4367	1027	50	3673	2338	0
20	2327	3454	19	2949	2953	189	3937	1631	0	3440	2708	0
21	2300	3267	0	4423	1284	138	3948	1396	0	2955	2946	0
22	2112	3229	0	3826	1747	35	3905	1223	0	2767	2895	0
23	2112	3010	0	2256	3123	0	4043	874	0	2112	3317	0
24	2112	2788	0	2588	2557	0	3822	882	0	2112	3082	0

Table 11.10 Objective values of the 12 scheduling scenarios (MW).

Season		Spring				Summer			
Weather		Sunny	Cloudy	Rainy		Sunny	Cloudy	Rainy	
Scenarios		Scenario 1	Scenario 2	Scenario 3	Base case	Scenario 4	Scenario 5	Scenario 6	Base case
Economic profit	CNY	56 010	55 550	54 760	4828	56 070	55 310	54 010	5070
Carbon emissions	tonnes	52 213	52 709	53 563	68 948	63 213	64 120	65 816	72 367
Renewable energy	MW	72 574	71 980	70 978	53 782	66 741	65 692	63 868	56 471
Thermal power	MW	58 601	59 195	60 197	77 393	70 993	72 042	73 866	81 262
Wind power	MW	67 527	67 705	67 990	51 158	59 059	59 185	59 302	53 716
Solar PV power	MW	5046	4275	2987	2123	7681	6506	4566	2755

Season		Autumn				Winter			
Weather		Sunny	Cloudy	Rainy		Sunny	Cloudy	Rainy	
Scenarios		Scenario 7	Scenario 8	Scenario 9	Base case	Scenario 10	Scenario 11	Scenario 12	Base case
Economic profit	CNY	47 370	46 930	46 170	4597	53 640	53 290	52 710	5118
Carbon emissions	*tonnes*	65 245	65 797	66 884	67 016	67 523	67 877	68 684	73 121
Renewable energy	MW	52 606	51 971	50 878	50 595	63 275	62 776	61 932	57 009
Thermal power	MW	73 324	73 957	75 050	74 297	75 770	76 270	77 114	82 036
Wind power	MW	48 034	48 060	48 077	49 111	60 067	60 067	60 067	54 228
Solar PV power	MW	4570	3910	2801	2518	3208	2709	1865	2781

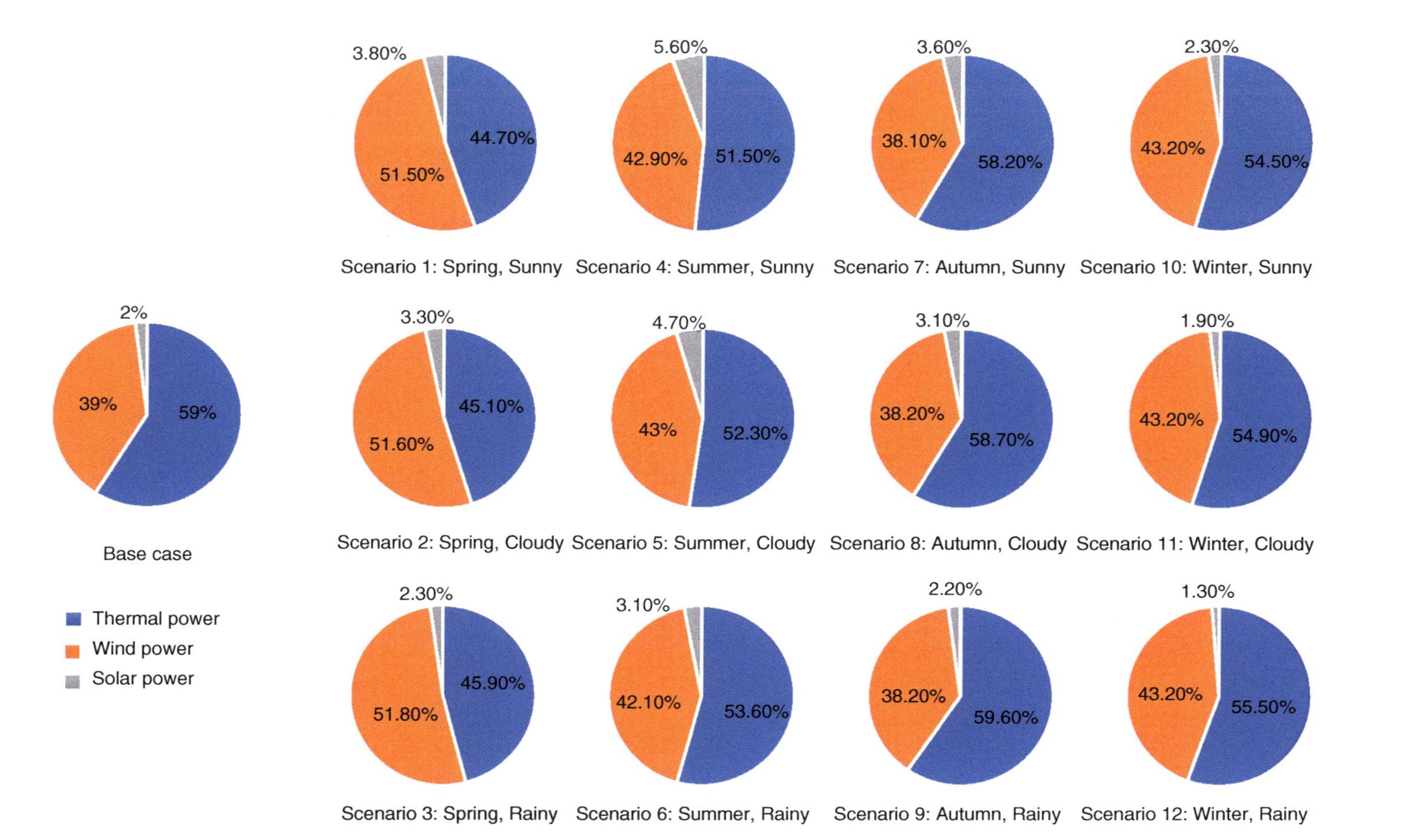

Figure 11.4 Power plant output ratios under different scenarios.

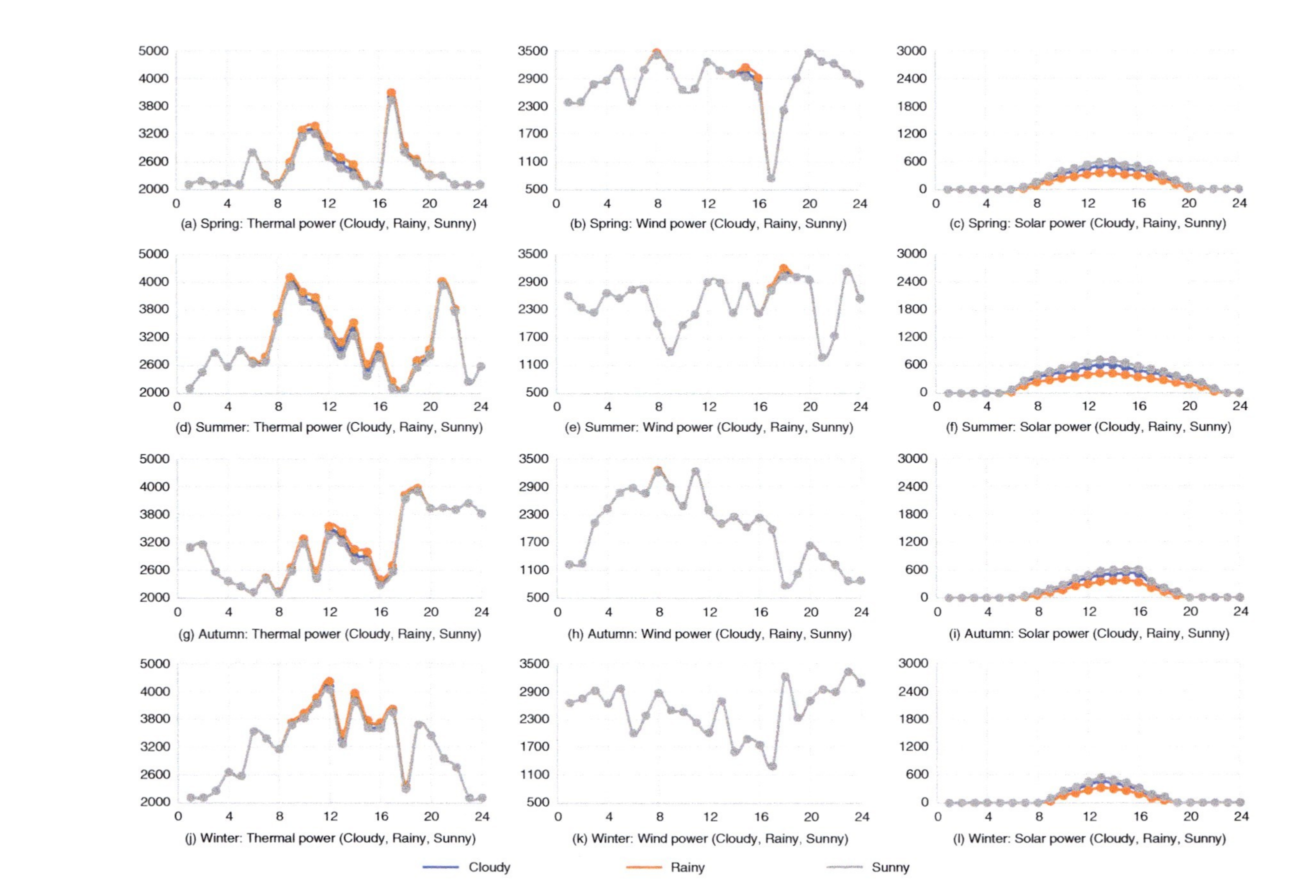

Figure 11.5 Hourly outputs of thermal power, wind power, and solar energy in the different scenarios (MW).

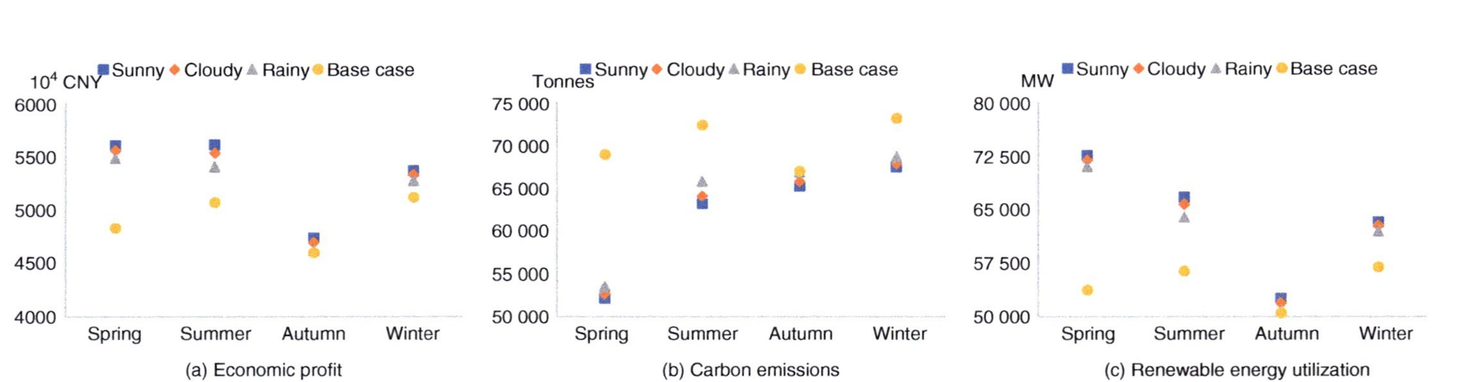

Figure 11.6 Objective values for the 12 scheduling scenarios and base cases.

price is higher. As can be seen in Figure 11.5b, the carbon emissions of the four seasons are all decreased when compared with the base case, especially in the spring. What's more, in summer, the weather condition changes influenced the carbon emissions. Accordingly, in Figure 11.5c, the renewable energy utilization of all the 12 scenarios increased, but increase in autumn is relatively lower than that in other seasons.

11.5 Discussion

A comprehensive discussion from the calculation results of the 12 scheduling scenarios is given in this section, including related propositions, comparative study, and management recommendations.

11.5.1 Propositions and Analysis

Based on the analysis above, three core conclusions can be obtained. Firstly, the proposed optimal scheduling strategy can effectively reduce carbon emissions, increase profits, and increase the renewable energy utilization for the wind–solar–thermal hybrid system. This can be seen in Figure 11.4, in the base case, the thermal, wind, solar, and power outputs were fixed at 59%, 39%, and 2%, respectively, but in the 12 scheduling scenario with the adoption of an optimal scheduling method, the wind and solar power utilization increases. For example, in scenario 1, daily power output ratios for the wind and solar power plants are 51.5%, and 3.8%, showing a significant increase in the proportion of renewable power. Further, the carbon emissions in this scenario also decreased and the profit increased. The same conventional power decrease and renewable energy increase can also be seen in the other scenarios. Secondly, when solar power output decreases, the reliable thermal power has an advantage in terms of the connection to the main grid than the unreliable wind power sources. Taking scenarios 1 and 3 in Figure 11.4 as an example, when the weather conditions change from sunny to rainy, the solar PV power output ratio decreases from 3.8% to 2.3%, and the wind power output ratio changes from 51.5% to 51.8%; however, the thermal power output ratio changes from 44.7% to 45.9%, an increase of 1.2%, almost four times that of the wind power. A similar situation is also seen in the other scenarios. Thirdly, when the weather conditions were fixed, the cooperative effect of the wind power and thermal power was very obvious. Taking scenarios 1, 4, 7, and 9 in Figure 11.4 as examples, in which there were sunny weather conditions. In spring and autumn, the share of wind power is 51.5% and 38.1%, which are the highest and the lowest in the four scenarios; however, the thermal power share in spring and autumn is 44.7% and 58.2%, which are the lowest and the highest in the four scenarios, which proves the obvious complementary capacity of wind and thermal power generation.

Compared with other studies, the study of Azizipanah-Abarghooee developed a multi-objective framework to handle a combined heat and power economic load dispatch that considered the stochastic characteristics of the wind and photovoltaic power outputs and the electrical and heat load demand [Azizipanah-Abarghooee et al., 2015]. This chapter was conducted from a different optimization perspective, that is, this chapter proposed an optimal scheduling strategy from the perspective of economic-environmental equilibrium, which is therefore more applicable to developing countries such as China that have an urgent need for economic development and an important responsibility to protect the environment. While compared with Peng et al. Peng et al. [2016]'s research, in which they established a flexible robust optimization dispatch model for hybrid wind–photovoltaic–hydro–thermal power systems. This chapter applied different optimization technology, that is, this chapter established 12 typical scheduling scenarios to handle the seasonal uncertainties in wind power and the weather-driven characteristics of solar power, greatly reduced the computational burden of scheduling due to processing uncertainty, and improved the practical convenience of the scheduling strategy.

To summarize, when comparing optimization perspectives, this chapter proposed an optimal scheduling strategy from the perspective of economic-environmental equilibrium, which is therefore more applicable to developing countries such as China that have an urgent need for economic development and an important responsibility to protect the environment. For the optimization technology, this chapter established 12 typical scheduling scenarios to handle the seasonal uncertainties in wind power and the weather-driven characteristics of solar power, greatly reduced the computational burden of scheduling due to processing uncertainty, and improved the practical convenience of the scheduling strategy.

11.5.2 Management Recommendations

Based on the above analysis and discussion, three management recommendations are given in this section to better elaborate how this optimal strategy could be further used in real application.

Firstly, the multi-objective optimal scheduling strategy proposed in this chapter can achieve equilibrium between profits and emissions, as well as increase the proportion of renewable energy utilization. Therefore, for cleaner and more economical electricity generation, it is necessary for electricity generators to apply the equilibrium-based optimal strategy. Secondly, government should develop focused policies and regulations to encourage a wider recognition establishment of hybrid generation systems. This is because although hybrid generation systems have the ability to overcome the weaknesses of one source with the strengths of another, increasing the utilization of renewable sources and reducing fossil fuel utilization, cooperation between the various power supply sources is still inefficient and ineffective. Thirdly, the cooperative effect of wind and thermal power proves the complementary capacity of wind–solar–thermal hybrid generation systems; however, the effect between solar and thermal was not found to be as strong as that of wind and thermal power, as a result of huge difference in installed capacity. Therefore, improvements in solar power's installed capacity and new technology need to be considered to explore a better complementary performance.

References

Aliari, Y. and Haghani, A. (2016). Planning for integration of wind power capacity in power generation using stochastic optimization. *Renewable and Sustainable Energy Reviews* 59: 907–919.

Azizipanah-Abarghooee, R., Niknam, T., Bina, M.A., and Zare, M. (2015). Coordination of combined heat and power–thermal–wind–photovoltaic units in economic load dispatch using chance-constrained and jointly distributed random variables methods. *Energy* 79: 50–67.

Barros, J.J.C., Coira, M.L., de la Cruz López, M.P., and del Ca no Gochi, A. (2016). Probabilistic life-cycle cost analysis for renewable and non-renewable power plants. *Energy* 112: 774–787.

Bhandari, B., Lee, K.-T., Lee, C.S. et al. (2014). A novel off-grid hybrid power system comprised of solar photovoltaic, wind, and hydro energy sources. *Applied Energy* 133: 236–242.

Biswas, P.P., Suganthan, P.N., and Amaratunga, G.A.J. (2017). Optimal power flow solutions incorporating stochastic wind and solar power. *Energy Conversion and Management* 148: 1194–1207.

Cai, B., Li, H., Hu, Y. et al. (2017). Theoretical and experimental study of combined heat and power (CHP) system integrated with ground source heat pump (GSHP). *Applied Thermal Engineering* 127: 16–27.

Chen, Y., He, L., Li, J. et al. (2016). An inexact bi-level simulation-optimization model for conjunctive regional renewable energy planning and air pollution control for electric power generation systems. *Applied Energy* 183: 969–983.

China Energy News (2016). HAMI: by 2020, the capacity of 12 million kW of coal-fired power and renewable energy electricity was reached to 16 million kilowatts. http://news.bjx.com.cn/html/20160914/772601.shtml (accessed 29 May 2021).

Chinese National Development and Reform Commission (2009). Notice of the national development and reform commission on perfecting the policy of wind power on-grid price. http://www.ndrc.gov.cn/zcfb/zcfbtz/200907/t20090727_292827.html.

Chinese National Development and Reform Commission (2013). Notice of CNDRC on exploiting price leverage to promote healthy development of PV industry. http://www.ndrc.gov.cn/zcfb/zcfbtz/201308/t20130830_556000.html.

Du, E., Zhang, N., Hodge, B.-M. et al. (2018). Economic justification of concentrating solar power in high renewable energy penetrated power systems. *Applied Energy* 222: 649–661.

Feijoo, F. and Das, T.K. (2014). Design of Pareto optimal CO_2 cap-and-trade policies for deregulated electricity networks. *Applied Energy* 119: 371–383.

Feng, J. and Shen, W.Z. (2017). Wind farm power production in the changing wind: Robustness quantification and layout optimization. *Energy Conversion and Management* 148: 905–914.

Hirth, L., Ueckerdt, F., and Edenhofer, O. (2015). Integration costs revisited: an economic framework for wind and solar variability. *Renewable Energy* 74: 925–939.

Jabir, E., Panicker, V.V., and Sridharan, R. (2017). Design and development of a hybrid ant colony-variable neighbourhood search algorithm for a multi-depot green vehicle routing problem. *Transportation Research Part D: Transport and Environment* 57: 422–457.

Ji, X., Li, G., and Wang, Z. (2017). Allocation of emission permits for China's power plants: a systemic Pareto optimal method. *Applied Energy* 204: 607–619.

Liu, Y., He, L., and Shen, J. (2017). Optimization-based provincial hybrid renewable and non-renewable energy planning: a case study of Shanxi, China. *Energy* 128: 839–856.

Liu, Y., Wang, Y., Wang, Q. et al. (2018). Simultaneous removal of NO and SO_2 using vacuum ultraviolet light (VUV)/heat/peroxymonosulfate (PMS). *Chemosphere* 190: 431–441.

Lv, C., Xu, J., Xie, H. et al. (2016). Equilibrium strategy based coal blending method for combined carbon and PM_{10} emissions reductions. *Applied Energy* 183: 1035–1052.

Mason, K., Duggan, J., and Howley, E. (2017). Multi-objective dynamic economic emission dispatch using particle swarm optimisation variants. *Neurocomputing* 270: 188–197.

Peng, C., Xie, P., Pan, L., and Yu, R. (2016). Flexible robust optimization dispatch for hybrid wind/photovoltaic/hydro/thermal power system. *IEEE Transactions on Smart Grid* 7 (2): 751–762.

Reddy, S.S. (2017). Optimal scheduling of thermal–wind–solar power system with storage. *Renewable Energy* 101: 1357–1368.

Reddy, S.S. and Momoh, J.A. (2015). Realistic and transparent optimum scheduling strategy for hybrid power system. *IEEE Transactions on Smart Grid* 6 (6): 3114–3125.

Roque, L., Fontes, D., and Fontes, F. (2017). A metaheuristic approach to the multi-objective unit commitment problem combining economic and environmental criteria. *Energies* 10 (12): 2029.

Ryzhkov, A., Bogatova, T., and Gordeev, S. (2018). Technological solutions for an advanced IGCC plant. *Fuel* 214: 63–72.

Shao, C., Wang, X., Shahidehpour, M. et al. (2017). Security-constrained unit commitment with flexible uncertainty set for variable wind power. *IEEE Transactions on Sustainable Energy* 8 (3): 1237–1246.

Singh, N.J., Dhillon, J.S., and Kothari, D.P. (2016). Synergic predator-prey optimization for economic thermal power dispatch problem. *Applied Soft Computing* 43: 298–311.

Sun, B., Yu, Y., and Qin, C. (2017a). Should China focus on the distributed development of wind and solar photovoltaic power generation? A comparative study. *Applied Energy* 185: 421–439.

Sun, Y., Dong, J., and Ding, L. (2017b). Optimal day-ahead wind-thermal unit commitment considering statistical and predicted features of wind speeds. *Energy Conversion and Management* 142: 347–356.

Suresh, V. and Sreejith, S. (2017). Generation dispatch of combined solar thermal systems using dragonfly algorithm. *Computing* 99 (1): 59–80.

Uygur, X. (2014). Autonomous Region Development and Reform Commission. Electricity price adjustment scheme of Hazheng line. http://www.xjdrc.gov.cn/info/10509/10981.htm.

Vithayasrichareon, P., Riesz, J., and MacGill, I. (2017). Operational flexibility of future generation portfolios with high renewables. *Applied Energy* 206: 32–41.

Ware, N.R., Singh, S.P., and Banwet, D.K. (2014). A mixed-integer non-linear program to model dynamic supplier selection problem. *Expert Systems with Applications* 41 (2): 671–678.

Xinhua News Agency (2017). The new energy installed capacity in Hami, Xinjiang Province accounts for over 60% of the total installed capacity. http://news.bjx.com.cn/html/20170703/834557.shtml (accessed 29 May 2021).

Xu, J. and Zhou, X. (2013). Approximation based fuzzy multi-objective models with expected objectives and chance constraints: application to earth-rock work allocation. *Information Sciences* 238: 75–95.

Xu, J., Tu, Y., and Zeng, Z. (2012). Bilevel optimization of regional water resources allocation problem under fuzzy random environment. *Journal of Water Resources Planning and Management* 139 (3): 246–264.

Xu, J., Wang, F., Lv, C., and Xie, H. (2018). Carbon emission reduction and reliable power supply equilibrium based daily scheduling towards hydro–thermal–wind generation system: a perspective from China. *Energy Conversion and Management* 164: 1–14.

Yuan, T., Ma, T., Sun, Y. et al. (2017). Game-based generation scheduling optimization for power plants considering long-distance consumption of wind–solar–thermal hybrid systems. *Energies* 10 (9): 1260.

Yuan, J., Na, C., Lei, Q. et al. (2018). Coal use for power generation in China. *Resources, Conservation and Recycling* 129: 443–453.

Zeng, Z., Xu, J., Wu, S., and Shen, M. (2014). Antithetic method-based particle swarm optimization for a queuing network problem with fuzzy data in concrete transportation systems. *Computer-Aided Civil and Infrastructure Engineering* 29 (10): 771–800.

Zimmermann, H.-J. (2011). *Fuzzy Set Theory and Its Applications*. Springer Science & Business Media.

Zubo, R.H.A., Mokryani, G., and Abd-Alhameed, R. (2018). Optimal operation of distribution networks with high penetration of wind and solar power within a joint active and reactive distribution market environment. *Applied Energy* 220: 713–722.

12

Carbon Emissions Reductions-Oriented Equilibrium Strategy for Municipal Solid Waste with Coal Co-combustion

The 2017 BP statistical review of World Energy reported that coal's share of global primary energy consumption was still around 27.6% and ranked second after oil [BP Global, 2018]. The world has heavily relied on coal for power generation for over a century and will possibly continue to do for the foreseeable future due to its relative abundance and low price [Qi et al., 2016]. In 2011, coal-fired power generation accounted for 40.6% of total power generation, and even though there was a 1.2% annual increase from 2016 to 2022, coal's share of the power mix is expected to fall to 36% by 2022 [Jin et al., 2013, International Energy Agency, 2017]. However, the massive annual carbon emissions from coal-fired power plants have contributed significantly to climate change. Taking China as an example, the carbon emissions from coal combustion constituted as much as 80% of total emissions between 2000 and 2013 [Liu et al., 2015]. The United Nations World Meteorological Organization (WMO) announced in the 2017 Greenhouse Gas Bulletin that the growth rate in global carbon dioxide concentrations had reached its highest level in millions of years; therefore, reducing the carbon emissions from coal-fired power plants has become extremely urgent and is a matter of public concern [World Meteorological Organization, 2018].

12.1 Background Introduction

There have been many advanced hard and soft technologies proposed to reduce the carbon emissions from coal-fired power plants, with the hard technologies focusing on clean technological improvements and the soft technologies focusing on effective optimization methods or emissions reduction policies. In regard to hard technology, Mao et al. outlined some mature carbon emissions reduction measures for the power industry such as coal washing, pulverized coal low-power igniting techniques, flue gas denitrification from front-end controls, in-the-process controls, and end-of-pipe controls [Mao et al., 2014]. Service suggested that more carbon capture and storage (CCS) projects needed to be applied to power plants and gave an example of Petra Nova, where the world's largest CCS system has been installed that can capture 1.6 million tonnes of carbon dioxide per year [Service, 2017]. Although these technologies can significantly reduce carbon emissions, at

Innovative Approaches towards Ecological Coal Mining and Utilization, First Edition.
Jiuping Xu, Heping Xie, and Chengwei Lv.

present they are very expensive; therefore, in many countries and particularly in developing countries, soft technology approaches such as carbon taxes, carbon trading schemes, and co-combustion methods are cheaper and easier to implement. Olsen et al. established a bilevel planning model to determine an optimal tax rate and considered unit commitments based on cyclic representative days to achieve emissions reductions targets [Olsen et al., 2018]. Fang et al. developed a carbon trading optimization model based on a novel Chinese energy-saving and emissions reduction system and found that carbon trading was able to effectively control carbon emissions [Fang et al., 2018]. Of all the technologies, municipal solid waste (MSW)/coal co-combustion has been found to have a positive effect on both the environment and the economy [Dong et al., 2002, Mcphail et al., 2013].

MSW, commonly known as rubbish or garbage, consists of various materials such as food waste, wood and yard trimmings, cotton, leather, and many others [Cheng and Hu, 2010a]. Because of the significant increase in urbanization and industrialization, MSW increased to 1.3 billion metric tonnes in 2011 and is expected to increase to 2.2 billion metric tonnes by 2025 [Levis et al., 2013]. United Nations Environment Programme (UNEP) claimed that MSW has posed a serious threat to the environment and human health [United Nations Environment Programme, 2016]. In China, MSW increased to 179.36 million tonnes in 2011 and has been growing by 8–10% each year, becoming a serious barrier to sustainable development [Zheng et al., 2014, Hoornweg et al., 2005, Sipra et al., 2018]. However, turning waste into wealth is an option. As MSW incineration has the advantages of carbon emissions reduction, energy extraction, and air pollutant control, MSW could be seen as a valuable renewable energy that could partly replace coal for power generation in a carbon-constrained world [Jin et al., 2013, Muthuraman et al., 2010, Alkhatib et al., 2010].

Based on the above discussion, to achieve carbon emissions reductions, an MSW/coal co-combustion method could be more suitable. MSW/coal co-combustion has been attracting increased research attention. Desroches et al. investigated the air pollutant emissions during the co-combustion process and found lower N_2O and SO_2 emissions and an improvement in the combustion rate [Desroches-Ducarne et al., 1998]. Wei et al. investigated the influences of temperature and the co-combustion ratio on reactivity and synergy and concluded that a higher reactivity was exhibited at higher MSW char proportions and temperatures [Wei et al., 2017]. McPhail et al. found that using MSW and coal as co-fuels offered potential energy recovery, reduced air emissions, and offered overall cost-savings to the electricity producer [Mcphail et al., 2013]. Cheng and Hu examined MSW incineration and found that it emitted a high level of dioxins and recommended that the dioxin emissions standard be lowered, fly ash management strengthened, and regulation enforcement improved [Cheng and Hu, 2010b]. However, these studies have mainly only considered a single decision-maker, and have only focused on carbon emissions reductions or dioxin emissions [Yang et al., 2012, Wang et al., 2015, Warren and El-Halwagi, 1996]. In reality, there are multiple decision-makers involved in carbon emissions reductions, especially the authority and the municipal solid waste-added coal power plants (MSWACPPs), which usually have conflicting objectives; the authority attaches great importance to overall economic conditions and environmental protection, while the MSWACPPs pursue the greatest profits. In addition, more extensive environmental protection needs to be considered; when seeking to reduce carbon emissions, it is also important to ensure that dioxin emissions are also below the emissions standard. Therefore, although previous studies have made significant contributions, the realistic situation is more complicated and further research is needed.

This chapter proposes a bilevel multi-objective carbon emissions reduction model that fully considers the relationship between the authority and the MSWACPPs and seeks economic development and environmental protection equilibrium by resolving the conflicts. First, because of limited resources, there is a contradiction between environmental protection and economic development, especially in the developing countries. Therefore, this chapter employs multi-objective programming to determine the trade-off between the economy and the environment. Second, due to conflicting objectives, the relationship between the authority and the MSWACPPs is difficult to reconcile; the authority, as the representative of the people, has the right to make the first decision, and the MSWACPP, as the lower-level stakeholder, must abide by the authority's decisions. This chapter employs bilevel programming to realize a harmonious relationship between the authority and the MSWACPPs. In this chapter, fuzzy theory is also used to deal with the uncertainties such as fuel quality and carbon emissions per unit of fuel.

12.2 Key Problem Statement

An equilibrium strategy-based bilevel multi-objective carbon emissions reduction model is developed in this chapter. Before establishing the proposed model, some basic background is given.

12.2.1 Conflict and Cooperation Between the Decision-Makers

There are many decision-makers involved in carbon emissions reduction, in particular, the authority and the MSWACPPs. However, objective conflicts exist between the authority and the MSWACPPs; the authority seeks maximal financial revenues and minimal carbon emissions while the MSWACPPs seek maximal profits. In the decision-making system, the authority first makes an initial decision about the carbon emissions quota allocation, following which the MSWACPPs make their own decisions with full knowledge of the authority's initial decision, which is then sent back to the authority. As the MSWACPPs' decisions may influence the authority's objectives, the authority adjusts its initial decision and then gives the updated strategy to the MSWACPPs. The MSWACPPs then adjust their decisions again and provide feedback to the authority. After several interactions, a satisfactory solution for both decision-makers is agreed to. Therefore, a Stackelberg–Nash equilibrium that reflects the global satisfactory solution is reached and can be transformed into the corresponding mathematical model

[Nash, 1951, Sinha et al., 2018]. Bilevel programming, a powerful tool for solving decentralized decision-making problems involving two decision-makers, has been widely used in many fields [Zhang et al., 2015]. Feng et al. developed a bilevel multistage programming method with multiple objective optimization to examine an integrated production–distribution–construction system that involved a construction department and material suppliers, and Gao and You proposed a game-theory-based two-stage stochastic bilevel model for optimizing decentralized supply chains under uncertainty based on one leader and multiple followers [Feng et al., 2018, Gao and You, 2018]. Motivated by this research, bilevel programming is employed in this chapter to resolve the conflicts and enhance the cooperation between the authority and the MSWACPPs in which the authority on the upper level has the first-mover advantage while the MSWACPPs on the lower level are the followers, as shown in Figure 12.1.

12.2.2 Trade-Off Between the Economy and the Environment

In the decision-making system, there is a conflict between economic development and environmental protection. Pollution and greenhouse gas emissions are the major environmental problems associated with coal-powered generation. While nitride oxide and sulfur dioxide emissions are easily and cheaply dealt with, carbon emissions are difficult and costly to control [Liu et al., 2017]. As carbon emissions are a major contributor to climate change and the focus of low-carbon economy policies, in this chapter, carbon emissions reduction is seen as equivalent to environmental protection. However, to ensure socioeconomic growth, the authority needs to guarantee financial revenue, with the higher the financial revenue, the more prosperous the economy. Therefore, in this chapter, increase in financial revenue is seen as equivalent to economic development. However, as economic development and environmental protection have a negative correlation, it is not possible for both to be concurrently realized. However, multi-objective programming has been shown to be effective in resolving multiple and conflicting objectives. Simab et al. established a multi-objective model to minimize operational costs and emissions in a short-term hydrothermal scheduling problem that included pumped-storage units [Simab et al., 2018]. Xie et al. presented a novel multi-objective active distribution network planning model to minimize total financial costs and the minimum spanning tree [Xie et al., 2018]. Because multi-objective programming is suitable for these types of situations, it is introduced in this chapter to achieve a trade-off between economic development and environmental protection, as shown in Figure 12.1.

12.2.3 Problem Analysis for MSW/Coal Co-combustion

The MSW/coal co-combustion methods to reduce carbon emissions have been widely researched on the theory and practice [Suksankraisorn, 2004, Xie et al., 2017, Pires et al., 2011]. However, there have been two crucial barriers in utilizing MSW/coal to generate electricity. First, the MSW heat value is too low to achieve

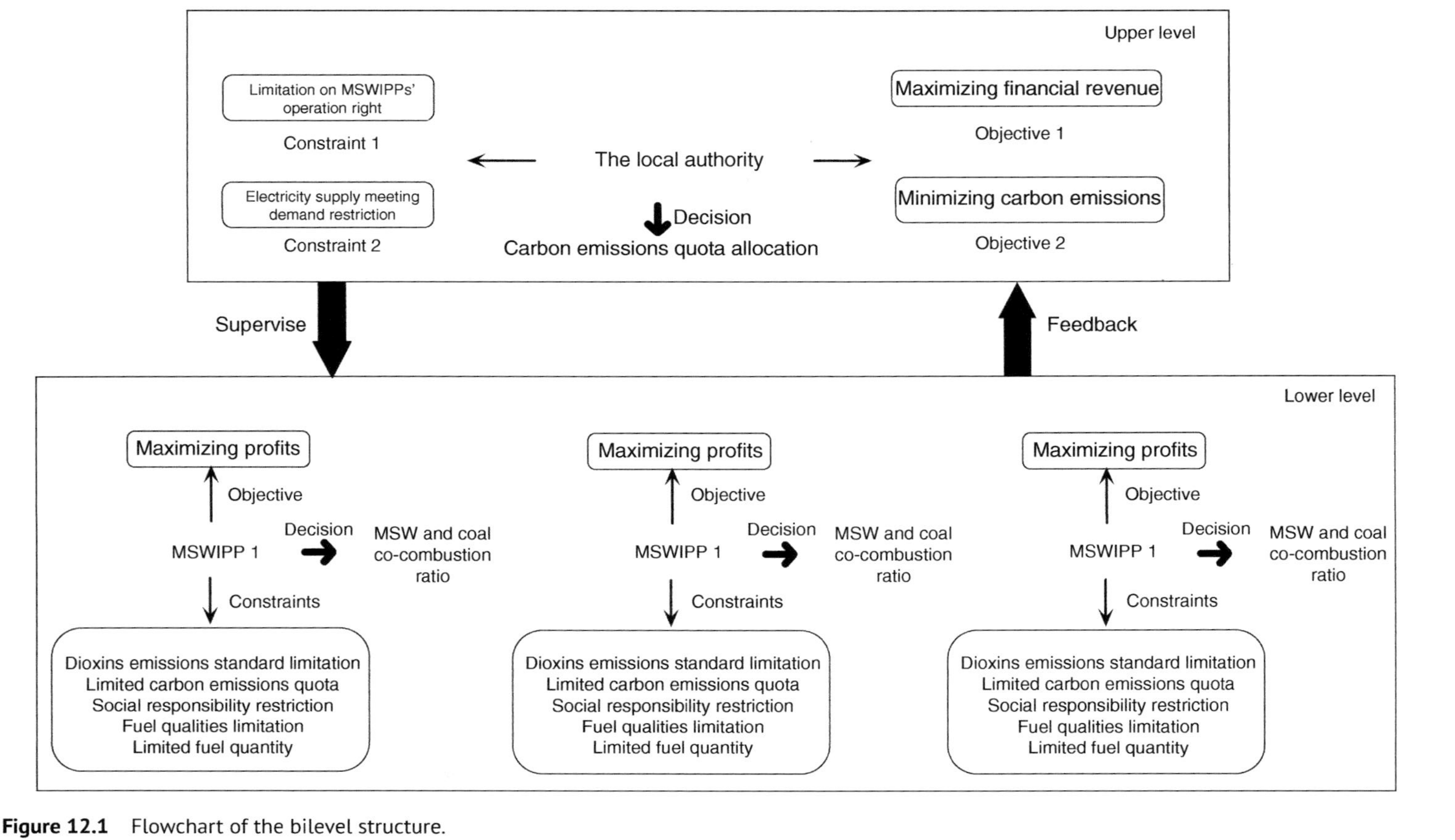

Figure 12.1 Flowchart of the bilevel structure.

combustion efficiency. Specifically, for the lack of strict MSW classification, the MSW moisture and ash content are low, which influences the heat value; therefore, coal needs to be added to the furnace to enhance the total heat value [Zhou et al., 2014, Yang et al., 2012]. The second barrier is associated with the dioxins emitted from MSW incineration. Dioxins, refer to polychlorinated dibenzo-*p*-dioxins (PCDD) and polychlorinated dibenzofurans (PCDF), are ubiquitous environmental pollutants [Huang et al., 2016]. Therefore, because of these dioxin emissions, the public perception often had a so-called not in my back yard (NIMBY) attitude. However, Gullett and Xu found in trials that the high concentrations of sulfur dioxide released through coal combustion can significantly inhibit dioxins production and that dioxin emissions could be controlled below the authority's emissions standard [Gullett et al., 1992, Xu, 2001]. Given these research results, in this chapter, a practical and green model that considers both heat value and dioxins emission is proposed.

From what have been discussed above, this chapter proposes an MSW/coal co-combustion model that integrates bilevel and multi-objective programming to seek authority-MSWACPPs equilibrium and economic-environmental equilibrium. The modeling details are given in Section 12.3.

12.3 Modeling

12.3.1 Assumptions

(1) All MSWACPPs are entitled to subsidies from the authority.
(2) The value-added tax that the MSWACPPs pay to the authority is the same.
(3) This is a single period production decision problem with the single period being one year.
(4) The penalties of dioxins emissions beyond the emissions standard are not considered.

12.3.2 Notations

The following notations will be employed in this chapter.

12.3.3 Allocation Scheme for the Authority

12.3.3.1 Maximizing Financial Revenue

The authority raises financial revenue from the society by various taxes to implement public policies and provide public services. In this chapter, value-added tax, carbon emissions quota fees, and MSW usage subsidies are considered as financial revenue. Let $\widetilde{H}_{ij}$ be the conversion parameter for a unit of fuel j to electricity at MSWACPP i, and φ be value-added tax; therefore, the total value-added tax from all MSWACPPs is $\varphi \sum_{i=1}^{I} \sum_{j=1}^{J} \widetilde{H}_{ij} x_{ij}$. Set λ as the price of a unit of the carbon emissions quota; therefore, the total carbon emissions quota fees on are $\lambda \sum_{i=1}^{I} y_i$. The authority encourages the

MSWACPPs to use as much MSW as possible and in return gives them a certain MSW subsidy [Saini et al., 2012]. Let γ be the MSW-electricity subsidies; therefore, the subsidy expense is $\gamma \sum_{i=1}^{I} \sum_{j=1}^{N} \widetilde{H}_{ij} x_{ij}$. Therefore, the financial revenue of the authority can be mathematically formulated as:

$$\max F_1 = \varphi \sum_{i=1}^{I} \sum_{j=1}^{J} \widetilde{H}_{ij} x_{ij} + \lambda \sum_{i=1}^{I} y_i - \gamma \sum_{i=1}^{I} \sum_{j=1}^{N} \widetilde{H}_{ij} x_{ij} \tag{12.1}$$

12.3.3.2 Minimizing Carbon Emissions

As well as economic development, the authority also attaches great importance to environmental protection. Since large amount of carbon emissions from coal-fired power plants have been caused and already stimulating global warming, the authority seeks to minimize carbon emissions. Let $\sum_{i=1}^{I} y_i$ be the gross carbon emissions quotas which should be as small as possible. Therefore, this objective of minimizing carbon emissions can be formulated as:

$$\min F_2 = \sum_{i=1}^{I} y_i \tag{12.2}$$

12.3.3.3 Electricity Supply Meeting Demand

In modern industries, agriculture, and other sectors of the national economy, electricity is the main source of energy. Without electricity it will bring a lot of inconvenience and cause huge losses. As a result, ensuring that provisioning can steadily meet demand is vital. Let $\widetilde{P}_i$ represent the conversion parameter for a unit of carbon emissions to electricity at MSWACPP i, so $\sum_{i=1}^{I} \widetilde{P}_i y_i$ means the total electricity supply, which must exceed electricity demand U in a region. Thus, the mathematical formula of this constraint can be:

$$\sum_{i=1}^{I} \widetilde{P}_i y_i \geq U \tag{12.3}$$

Indices

i = MSWACPP index, $i \in \delta = 1, 2, \ldots, I$.
j = Fuel index, $j \in \sigma = 1, 2, \ldots, N, \ldots, J$, where $1 - N$ represent MSW and $(N+1) - J$
t = Fuel quality index, $t \in \psi = 1, 2, \ldots T$.
r = Pollutant index, $r \in \xi = 1, 2 \ldots R$.

Certain parameters

φ = Value-added tax.
ω = Unit electricity price.
λ = Price of the unit of the carbon emissions quota.
U = Electricity demand in a region.

(Continued)

(Continued)

Q_i^L = Lower bound for the carbon emissions at MSWACPP i.
Q_i^U = Upper bound for the carbon emissions at MSWACPP i.
C_j = Procurement costs for a unit of fuel j.
U_i = Electricity that MSWACPP i must generate.
O_{ij}^U = Maximum available quantity of fuel j at MSWACPP i.
η_i = Inhibition efficiency of coal to dioxins at MSWACPP i.
κ_i = Residual ratio of dioxins at MSWACPP i.
ϖ = Dioxin emissions limit that is regulated by the authority.
F_{ir} = Pollutant r treatment costs at MSWACPP i.
γ = Subsidies for MSW usage granted to the MSWACPPs by the authority.
L = Actual carbon emissions in the last production period.

Uncertain parameters

$\widetilde{H_{ij}}$ = Conversion parameter from a unit of fuel j to electricity at MSWACPP i.
$\widetilde{D_{ij}}$ = Dioxins emissions from burning a unit of fuel j at MSWACPP i.
$\widetilde{R_{ij}}$ = Flue gas from burning a unit of fuel j at MSWACPP i.
$\widetilde{P_i}$ = Conversion parameter for a unit of carbon emissions to electricity at MSWACPP i.
$\widetilde{S_{ijt}}$ = The t-th quality of fuel j at MSWACPP i.
$\widetilde{Q_{ij}}$ = Carbon emissions from burning per unit of fuel j at MSWACPP i.
$\widetilde{N_{ijr}}$ = Quantity of pollutant r from burning per unit of fuel j.
$\widetilde{S_{it}^L}$ = Lower bounds of the t-th quality at MSWACPP i.
$\widetilde{S_{it}^U}$ = Upper bounds of the t-th quality at MSWACPP i.

Control parameters

α = Attitude toward dioxin emissions below the emissions standard.
β = Authority's attitude toward carbon emissions reduction.

Decision variables

y_i = Carbon emissions quota that the authority allocates to MSWACPP i.
x_{ij} = Fuel j burnt by MSWACPP i.

12.3.3.4 Requirements for the MSWACPPs' Operating Rights

The enterprise has a basic legal right to operate and make a profit. Therefore, when the authority makes decisions on the carbon emissions quotas, it must consider the MSWACPP's rights and allocate carbon emissions quotas that each MSWACPP can carry [Lv et al., 2016]. Q_i^L indicates the smallest carbon emissions under which each MSWACPP can maintain basic operations. Q_i^U represents the largest carbon emissions that each MSWACPP emits under full-load power generation. Therefore, the carbon emissions quotas that the authority allocates should be higher than or equal to the lower bound Q_i^L but lower than or equal to the upper bound Q_i^U. This limitation can be described as:

$$Q_i^L \le y_i \le Q_i^U \tag{12.4}$$

12.3.4 Production Strategy for MSWACPPs

12.3.4.1 Pursuing Maximum Profits

The fundamental purpose of an enterprise is to pursue the maximum profits under the carbon emissions quotas allocated by the authority. In this case, each MSWACPP's profits are equal to operating revenue plus the authority's subsidy minus procurement costs and taxes. Let ω denote the unit electricity price, and φ denote value-added tax. Let $\widetilde{H_{ij}}$ be the conversion parameter for a unit of fuel j to electricity; therefore, $\sum_{j=1}^{J} \widetilde{H_{ij}} x_{ij}$ is the total electricity. Then after-tax income is $(\omega - \varphi) \sum_{j=1}^{J} \widetilde{H_{ij}} x_{ij}$. C_j is procurement cost of a unit of fuel j, and the fuel cost is $\sum_{j=1}^{J} C_j x_{ij}$. F_{ir} and $\widetilde{N_{ijr}}$, respectively, denote pollutant r treatment costs and the amount of pollutant r generated from burning a unit of fuel j; therefore, the total pollutant treatment cost is $\sum_{j=1}^{J} \sum_{r=1}^{R} F_{ir} \widetilde{N_{ijr}} x_{ij}$. The MSWACPPs also pay fees λy_i for the carbon emissions quotas and receive subsidy $\gamma \sum_{j=1}^{N} \widetilde{H_{ij}} x_{ij}$. Therefore, the profits for each MSWACPP are formulated as:

$$\max f = (\omega - \varphi) \sum_{j=1}^{J} \widetilde{H_{ij}} x_{ij} - \sum_{j=1}^{J} C_j x_{ij} - \sum_{j=1}^{J} \sum_{r=1}^{R} F_{ir} \widetilde{N_{ijr}} x_{ij} - \lambda y_i + \gamma \sum_{j=1}^{N} \widetilde{H_{ij}} x_{ij} \tag{12.5}$$

12.3.4.2 Coal's Inhibitory Effect on Dioxin Emissions

The MSWACPPs usually use coal as a supplementary fuel as the addition of coal increases the MSW heat value and also has an inhibitory effect on the MSW dioxin emissions. Test results showed that the inhibitory effect of coal on dioxin emissions was related to the proportion of coal in the mixed fuel [Gullett et al., 1998, Raghunathan et al., 1997, Gullett et al., 1992, Xu, 2001]. Let η_i denote the inhibition efficiency of coal to dioxins. In Briank Gullet's research, the inhibitory effect is given as:

$$\eta_i = \begin{cases} 0, & \sum_{j=N+1}^{J} x_{ij} \le 0.04 \sum_{j=1}^{J} x_{ij} \\ 9.1447 \dfrac{\sum_{j=N+1}^{J} x_{ij}}{\sum_{j=1}^{J} x_{ij}} - 0.0372, & 0.04 \sum_{j=1}^{J} x_{ij} < \sum_{j=N+1}^{J} x_{ij} < 0.08 \sum_{j=1}^{J} x_{ij} \\ 0.7, & \sum_{j=N+1}^{J} x_{ij} \ge 0.08 \sum_{j=1}^{J} x_{ij} \end{cases} \tag{12.6}$$

12.3.4.3 Dioxin Emissions Risk Control

The MSWACPPs have a social responsibility to protect the environment and follow the authority's environmental protection policy. ϖ represents the dioxin emissions standard set by the authority. $\sum_{j=1}^{N}(1-\eta_i)\widetilde{D_{ij}}x_{ij}$ denotes the dioxin emissions from MSW incineration, and $\sum_{j=N+1}^{J}\widetilde{D_{ij}}x_{ij}$ denotes the dioxin emissions produced by coal; therefore, $\sum_{j=1}^{N}(1-\eta_i)\widetilde{D_{ij}}x_{ij}+\sum_{j=N+1}^{J}\widetilde{D_{ij}}x_{ij}$ is dioxins produced by MSW and coal. Let κ_i be residual ratio at MSWACPP i; therefore, the final dioxin emissions of the combustion process is $\kappa_i[\sum_{j=1}^{N}(1-\eta_i)\widetilde{D_{ij}}x_{ij}+\sum_{j=N+1}^{J}\widetilde{D_{ij}}x_{ij}]$. $\widetilde{R_{ij}}$ is flue gas from burning a unit of fuel i; therefore, $\sum_{j=1}^{J}\widetilde{R_{ij}}x_{ij}$ is the total flue gas, and the dioxin concentration is $\frac{\kappa_i[\sum_{j=1}^{N}(1-\eta_i)\widetilde{D_{ij}}x_{ij}+\sum_{j=N+1}^{J}\widetilde{D_{ij}}x_{ij}]}{\sum_{j=1}^{J}\widetilde{R_{ij}}x_{ij}}$. However, because of the fuel quality and combustion process uncertainty, there is an excessive dioxin emissions risk even though the MSWACPPs attempt to keep the dioxin emissions below the emissions standard ϖ as far as possible. Therefore, the fuzzy chance constraint method is introduced to tackle with the problem [Xu and Zhou, 2011]. Let α be attitude toward dioxin emissions below the emissions standard; therefore, this constraint can be written as:

$$Pos\left\{\frac{\kappa_i\left[\sum_{j=1}^{N}(1-\eta_i)\widetilde{D_{ij}}x_{ij}+\sum_{j=N+1}^{J}\widetilde{D_{ij}}x_{ij}\right]}{\sum_{j=1}^{J}\widetilde{R_{ij}}x_{ij}}\le\varpi\right\}\ge\alpha \tag{12.7}$$

where Pos is the possibility measure proposed by Dubois and Prade Dubois and Prade [1983].

12.3.4.4 Limited Carbon Emissions Quota

As upper-level decision-maker, the authority first decides on the carbon emissions quota allocations. As the MSWACPPs are in the subordinate position, they must abide by the authority's decisions. Due to the special relationship between the authority and the MSWACPPs, the actual carbon emissions that the MSWACPPs produce cannot exceed the carbon emissions quotas the authority allocates. In this chapter, let $\widetilde{Q_{ij}}$ be the carbon emissions from burning per unit of fuel j; therefore, $\sum_{j=1}^{J}\widetilde{Q_{ij}}x_{ij}$ is the actual carbon emissions. This constraint is as follows:

$$\sum_{j=1}^{J}\widetilde{Q_{ij}}x_{ij}\le y_i \tag{12.8}$$

12.3.4.5 Social Responsibility

Since electricity is of great importance, every MSWACPP, being one of the members of the society, should undertake social responsibility to ensure electricity supply. $\widetilde{H_{ij}}$ is the power produced from burning a unit of fuel j. The actual power generated is $\sum_{j=1}^{J}\widetilde{H_{ij}}x_{ij}$, which should not be lower than the social responsibility U_i.

$$\sum_{j=1}^{J}\widetilde{H_{ij}}x_{ij}\ge U_i \tag{12.9}$$

12.3.4.6 Fuel Quality Required by the Incinerators

The quality of the mixed fuels should satisfy the different incinerator standards. $\widetilde{S_{ijt}}$ is the t-th quality of fuel j at MSWACPP i; when $t = 1, 2, 3, 4$, and 5, respectively, $\widetilde{S_{ijt}}$, respectively, represents the heat rate, volatile matter content, moisture content, ash content, and sulfur content [Tsumura et al., 2003]. $\widetilde{S_{it}^L}$ and $\widetilde{S_{it}^U}$, respectively, represent lower and upper bounds of t-th quality. Therefore, the constraint can be expressed as:

$$\widetilde{S_{it}^L} \leq \frac{\sum_{j=1}^{J} \widetilde{S_{ijt}} x_{ij}}{\sum_{j=1}^{J} x_{ij}} \leq \widetilde{S_{it}^U} \tag{12.10}$$

12.3.4.7 Limited Fuel Quantity

In fact, the fuel quantity that every MSWACPP can use is limited. There is maximum quantity that MSWACPPs can be supplied. O_{ij}^U indicates maximum available quantity; therefore, the fuel j quantity ranges from zero to O_{ij}^U. This constraint can be mathematically modeled as:

$$0 \leq x_{ij} \leq O_{ij}^U \tag{12.11}$$

12.3.5 Global Model

A bilevel multi-objective model is established to resolve the inevitable conflicts and obtain a desirable satisfactory solution. In this model, there is a leader–follower relationship between the two decision-makers, with the authority being the upper-level decision-maker and the MSWACPPs being the lower-level decision-makers. The two decision-makers not only consider their own objectives and constraints but also the other decision-maker's decisions. In the decision-making system, the authority allocates the initial carbon emissions quotas to each MSWACPP on the basis of its objectives and constraints. Under the initial allocation decision, the MSWACPPs make decisions and then feed back to the superior authority, after which the authority reallocates the updated carbon emissions quotas, which compels the MSWACPPs to redetermine their decisions. This process is repeated several times until the results are acceptable to both decision-makers. The authority, however, has its own conflicting objectives; it attempts to gain maximal financial revenue, as shown in Eq. (12.1), while also emphasizing carbon emissions reductions, as shown in Eq. (12.2). These objectives are restricted by the constraints described in Eqs. (12.3) and (12.4). The MSWACPPs pursue maximum profits, as shown in Eq. (12.6) under the constraints related to the standard dioxin emissions limitations, the limited carbon emissions quota, social responsibility commitments, fuel quality limitations, and limited fuel quantity, as shown in Eqs. (12.6)–(12.11). Therefore, by integrating Eqs. (12.1)–(12.11), the global model for the MSWACPP carbon emissions reductions can be formulated, as shown in Eq. (12.12).

Compared with previous MSW/coal co-combustion models, this model has three advantages. First, it is more practical and more appropriate for actual production as it considers multiple decision-makers and multiple conflicts. Using the proposed model, the authority and the MSWACPPs can jointly cooperate to reduce carbon emissions, and a successful trade-off between economic development and the environmental protection can be achieved. Second, this model comprehensively considers both carbon emissions and dioxin emissions, thereby more fully protecting the environment. This model minimizes carbon emissions and ensures that the dioxin emissions are maintained below a certain concentration to prevent any adverse effects on human health and the ecosystem. Third, this model can be widely applied by adjusting some parameters or adding or removing some objectives and constraints. In the proposed model, α represents the attitude toward dioxin emissions below the emissions standards, and β represents the authority's attitude toward carbon emissions reduction, both of which make the model flexible and robust. Further, some objectives and constraints can be changed, added or removed to make the model more pervasive.

$$
\begin{aligned}
&\max F_1 = \varphi \sum_{i=1}^{I}\sum_{j=1}^{J}\widetilde{H_{ij}}x_{ij} + \lambda\sum_{i=1}^{I}y_i - \gamma\sum_{i=1}^{I}\sum_{j=1}^{N}\widetilde{H_{ij}}x_{ij}\\
&\min F_2 = \sum_{i=1}^{I}y_i\\
&\text{s.t.}\begin{cases}
\sum_{i=1}^{I}\widetilde{P_i}y_i \geq U\\
Q_i^L \leq y_i \leq Q_i^U, \quad \forall i \in \delta\\
\max f = (\omega - \varphi)\sum_{j=1}^{J}\widetilde{H_{ij}}x_{ij} - \sum_{j=1}^{J}C_j x_{ij} - \sum_{j=1}^{J}\sum_{r=1}^{R}F_{ir}\widetilde{N_{ijr}}x_{ij} - \lambda y_i + \gamma\sum_{j=1}^{N}\widetilde{H_{ij}}x_{ij}\\
\text{s.t.}\begin{cases}
\eta_i = \begin{cases}
0, & \sum_{j=N+1}^{J}x_{ij} \leq 0.04\sum_{j=1}^{J}x_{ij}\\
9.1447\frac{\sum_{j=N+1}^{J}x_{ij}}{\sum_{j=1}^{J}x_{ij}} - 0.0372, & 0.04\sum_{j=1}^{J}x_{ij} \leq \sum_{j=N+1}^{J}x_{ij} \leq 0.08\sum_{j=1}^{J}x_{ij}\\
0.7, & \sum_{j=N+1}^{J}x_{ij} \geq 0.08\sum_{j=1}^{J}x_{ij}
\end{cases}\\
Pos\left\{\frac{\kappa_i\left[\sum_{j=1}^{N}(1-\eta_i)\widetilde{D_{ij}}x_{ij} + \sum_{j=N+1}^{J}\widetilde{D_{ij}}x_{ij}\right]}{\sum_{j=1}^{J}\widetilde{R_{ij}}x_{ij}} \leq \varpi\right\} \geq \alpha, \quad \forall i \in \delta\\
\sum_{j=1}^{J}\widetilde{Q_{ij}}x_{ij} \leq y_i, \quad \forall i \in \delta\\
\sum_{j=1}^{J}\widetilde{H_{ij}}x_{ij} \geq U_i, \quad \forall i \in \delta\\
\widetilde{S_{it}^L} \leq \frac{\sum_{j=1}^{J}\widetilde{S_{ijt}}x_{ij}}{\sum_{j=1}^{J}x_{ij}} \leq \widetilde{S_{it}^U}, \quad \forall i \in \delta, \quad \forall t \in \psi\\
0 \leq x_{ij} \leq O_{ij}^U, \quad \forall i \in \delta, \quad \forall j \in \sigma
\end{cases}
\end{cases}
\end{aligned}
\tag{12.12}
$$

12.4 Case Study

In this section, the proposed model is applied to Sichuan Province to demonstrate its efficiency and practicality.

12.4.1 Case Description

Sichuan Province, located in southwest China, has the sixth largest economy and the fourth largest population in China. As MSW in Sichuan Province increased to 8.87 million tonnes in 2016, and faced with an energy shortage, Sichuan Province established around 20 MSWACPPs and further 36 MSWACPPs are under construction [Sichuan Provincial Bureau of Statistics, 2017]. Incineration is being widely employed for MSW disposal and the MSWACPPs have become important sources of power. However, there are increasing concerns in relation to the MSWACPPs. In 2017, the Sichuan authority issued Sichuan Province's Work Plan to Control Greenhouse Gas Emissions, which has a target of reducing carbon emissions by 2020 by 19.5% from 2015 levels [Sichuan Provincial Bureau of Statistics, 2017]. Moreover, because of the potential negative health and environmental effects related to dioxin toxicity, many residents do not want MSWACPPs constructed near where they live. Therefore, measures need be taken to reduce the carbon emissions from the MSWACPPs and keep dioxin emissions below the regulated standard, both of which are important indicators of a MSWACPP's operations. Therefore, to reduce carbon emissions and ensure low dioxin emissions, the MSWACPPs need to achieve sustainable development and guarantee the harmony of the economy-society-ecology system.

To decrease the calculation pressure, only three MSWACPPs in Sichuan Province were selected for this case: Jiujiang MSWACPP, Weiao MSWACPP, and Zigong MSWACPP. In Sichuan Province, the heat value of MSW is generally low, but there are abundant coal resources; as a result, the coal used in the combustion process usually exceeds 8% of the total combustion fuels [Lv et al., 2016].

12.4.2 Model Transformation and Solution Approach

The many uncertain parameters in the proposed model are estimated using a value range, with the most likely value being the smaller range. For instance, $\widetilde{H}_{ij}$ ranges from r_{11} to r_{14}, and the most likely value is between r_{12} and r_{13}. Therefore, we denote $\widetilde{H}_{ij} = (r_{11}, r_{12}, r_{13}, r_{14})(r_{11} \leq r_{12} \leq r_{13} \leq r_{14})$. In this chapter, all uncertain parameters are typical trapezoidal fuzzy variables and are difficult to calculate directly. In Xu's research, the expected value operator method was employed to determine the exact values of trapezoidal fuzzy variables [Xu and Zhou, 2011]. Using Xu's method, $\widetilde{H_{ij}}$ can be represented as:

$$\widetilde{H_{ij}} \rightarrow E[\widetilde{H_{ij}}] = \frac{1-\partial}{2}(r_{11}+r_{12}) + \frac{\partial}{2}(r_{13}+r_{14}) \tag{12.13}$$

Fuzzy theory is introduced in this chapter to measure the excessive dioxin emissions risk as shown in Eq. (12.7). In the fuzzy chance constraint, $\widetilde{D_{ij}}$ is a trapezoidal fuzzy number, denoted $\widetilde{D_{ij}} = (r_{21}, r_{22}, r_{23}, r_{24})$. As this constraint cannot be directly calculated, based on the research of Liu and and Iwamura Liu and Iwamura [1998], the computing method of fuzzy chance constraint is to convert it into it equivalent form; therefore, the certain equivalence of this fuzzy chance constraint is

$$\begin{aligned}&(1-\alpha)(r_{21}+r_{24})\left[\frac{\sum_{j=1}^{N}\kappa_i(1-\eta_i)x_{ij}+\sum_{j=N+1}^{J}\kappa_i x_{ij}}{\sum_{j=1}^{J}\widetilde{R_{ij}}x_{ij}}\right]\\&+\alpha(r_{22}+r_{23})\left[\frac{\sum_{j=1}^{N}\kappa_i(1-\eta_i)x_{ij}+\sum_{j=N+1}^{J}\kappa_i x_{ij}}{\sum_{j=1}^{J}\widetilde{R_{ij}}x_{ij}}\right]-\varpi\leq 0\end{aligned} \tag{12.14}$$

As Sichuan Province is still a developing region, economic development remains an important goal but environmental protection is also important, which means that the authority places first priority on financial revenue, and second priority on carbon emissions reductions. Therefore, in the proposed model, minimizing the carbon emissions objective can be transformed into a constraint and controlled within an acceptable range. Let β be the attitude of the authority toward carbon emissions reductions. Based on Zeng et al's research Zeng et al. [2014], Eq. (12.12) can be transformed as Eq. (12.15).

In the proposed bilevel model, the decisions on each level influence the decisions on the other decision-maker level. To tackle this complicated hierarchical optimization problem, an interactive evolutionary mechanism was developed in Matlab to determine the optimal solution [Xu et al., 2018]. First, all the objectives and constraints of the upper-level model were input into Matlab to build the feasible region of y_i for the upper-level model, after which an initial value of y_i^1 was randomly chosen in the feasible region. Then y_i^1 was transmitted into the lower-level and the lower-level model transformed into a single-level optimization model. The solution x_{ij}^1 was then calculated in Matlab, after which the lower-level solution x_{ij}^1 was transmitted into the upper-level model, at which time the upper-level model was also transformed into a single-level optimization model, and the new solution y_i^2 was calculated in Matlab. This was the initial global solution (x_{ij}^1, y_i^2); if $\frac{|y_i^2 - y_i^1|}{y_i^1} \leq 1\%$ was satisfied, solution (x_{ij}^1, y_i^2) was chosen as the satisfactory solution, otherwise y_i^2 was transmitted to the lower-level model again and calculate x_{ij}^2 which was then transmitted to the upper-level model to calculate y_i^3; therefore, the solution was updated. This process was repeated until the termination $\frac{|y_i^q - y_i^{q-1}|}{y_i^{q-1}} \leq 1\%$ was satisfied. Finally, the satisfactory solution (x_{ij}^{q-1}, y_i^q) was obtained. This interactive evolutionary mechanism is graphically represented in Figure 12.2.

$$
\max F_1 = \varphi \sum_{i=1}^{I}\sum_{j=1}^{J} E[\widetilde{H_{ij}}]x_{ij} + \lambda \sum_{i=1}^{I} y_i - \gamma \sum_{i=1}^{I}\sum_{j=1}^{N} E[\widetilde{H_{ij}}]x_{ij}
$$

$$
\text{s.t.}\begin{cases}
\sum_{i=1}^{I} y_i \le \beta L \\
\sum_{i=1}^{I} E[\widetilde{P}_i]y_i \ge U \\
Q_i^L \le y_i \le Q_i^U, \quad \forall i \in \delta \\
\max f = (\omega - \varphi)\sum_{j=1}^{J} E[\widetilde{H_{ij}}]x_{ij} - \sum_{j=1}^{J} C_j x_{ij} - \sum_{j=1}^{J}\sum_{r=1}^{R} F_{ir} E[\widetilde{N_{ijr}}]x_{ij} - \lambda y_i + \gamma \sum_{j=1}^{N} E[\widetilde{H_{ij}}]x_{ij} \\
\text{s.t.}\begin{cases}
\eta_i = \begin{cases}
0, & \sum_{j=N+1}^{J} x_{ij} \le 0.04 \sum_{j=1}^{J} x_{ij} \\
9.1447 \frac{\sum_{j=N+1}^{J} x_{ij}}{\sum_{j=1}^{J} x_{ij}} - 0.0372, & 0.04 \sum_{j=1}^{J} x_{ij} \le \sum_{j=N+1}^{J} x_{ij} \le 0.08 \sum_{j=1}^{J} x_{ij} \\
0.7, & \sum_{j=N+1}^{J} x_{ij} \ge 0.08 \sum_{j=1}^{J} x_{ij}
\end{cases} \\
(1-\alpha)(r_{21}+r_{24})\left[\frac{\sum_{j=1}^{N} \kappa_i (1-\eta_i) x_{ij} + \sum_{j=N+1}^{J} \kappa_i x_{ij}}{\sum_{j=1}^{J} E[\widetilde{R_{ij}}]x_{ij}}\right] \\
\quad + \alpha(r_{22}+r_{23})\left[\frac{\sum_{j=1}^{N} \kappa_i (1-\eta_i) x_{ij} + \sum_{j=N+1}^{J} \kappa_i x_{ij}}{\sum_{j=1}^{J} E[\widetilde{R_{ij}}]x_{ij}}\right] - \varpi \le 0, \quad \forall i \in \delta \\
\sum_{j=1}^{J} E[\widetilde{Q_{ij}}]x_{ij} \le y_i, \quad \forall i \in \delta \\
\sum_{j=1}^{J} E[\widetilde{H_{ij}}]x_{ij} \ge U_i, \quad \forall i \in \delta \\
E[\widetilde{S_{it}^L}] \le \frac{\sum_{j=1}^{J} E[\widetilde{S_{ijt}}]x_{ij}}{\sum_{j=1}^{J} x_{ij}} \le E[\widetilde{S_{it}^U}], \quad \forall i \in \delta,\ \forall t \in \psi \\
0 \le x_{ij} \le O_{ij}^U, \quad \forall i \in \delta,\ \forall j \in \sigma
\end{cases}
\end{cases}
\tag{12.15}
$$

12.4.3 Data Collection

The data needed for the proposed model are shown in Tables 12.1–12.4 and are divided into uncertain data and certain data. The uncertain data shown in Tables 12.1 and 12.2 were collected from the MSWACPPs official websites, the MSWACPPs environmental impact report of municipal waste incineration power generation project, and the Sichuan Provincial Environment and Ecology Department. As the uncertain parameters were typical trapezoidal fuzzy variables in this

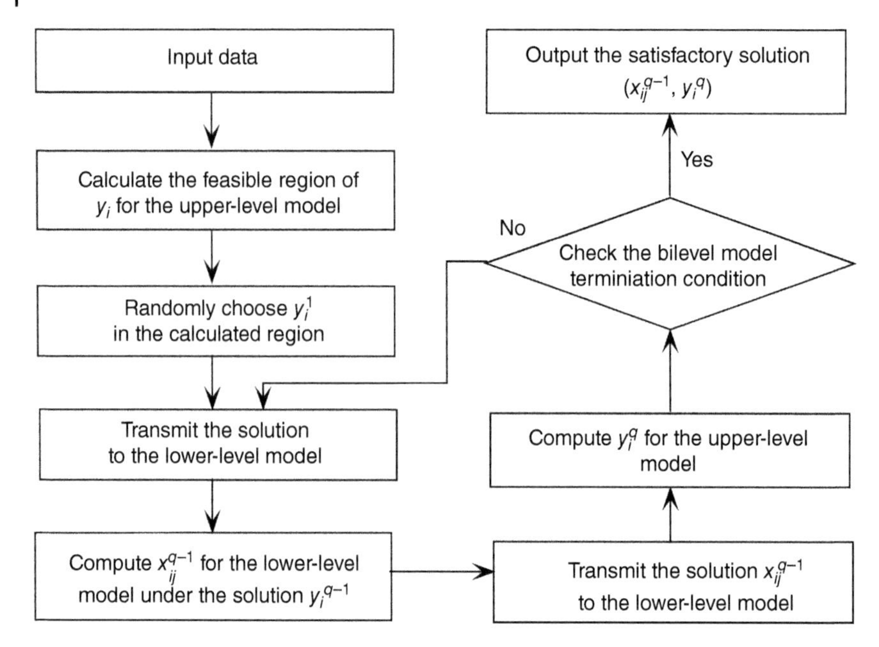

Figure 12.2 Interactive evolution mechanism flowchart.

chapter, they were collected in the fuzzy form of $(r_{i1}, r_{i2}, r_{i3}, r_{i4})$. Table 12.1 shows the fuel characteristics and carbon emissions to power parameters at the MSWACPPs. Table 12.2 shows the lower and upper bounds for the fuel characteristics at each MSWACPP. Tables 12.3 and 12.4 show the certain data collected from MSWACPP official websites and Sichuan Province Statistical Yearbook. Table 12.3 shows the power demand, procurement costs, and so on at the MSWACPPs. Table 12.4 shows the other parameters needed in the proposed model.

12.4.4 Results Under Different Scenarios

In this section, results under various scenarios are given. β and α are the control parameters. β represents the attitude of the authority toward carbon emissions reduction; the lower the β, the stricter the attitude. As total power demand must be met as shown in Eq. (12.3), based on the collected data, β can be adjusted to its minimum value 0.9. α represents the attitude toward dioxin emissions below the emissions standard; the higher the α, the stricter the attitude. Due to power demand restrictions, incinerator requirements, and fuel quantity constraints, as shown in Eqs. (12.9)–(12.11), α can be adjusted to its minimum value 0.85. From what has been discussed above, β can vary from 1 to 0.9 while α can vary from 1 to 0.85. Therefore, four scenarios are shown in Tables 12.5–12.7. Scenario 1 is shown in Table 12.5, in which α is set at 1 and β changes from 1 to 0.9. Scenario 2 is shown in Table 12.6, in which α is set at 0.95 and β changes from 1 to 0.9. Scenario 3 is shown in Table 12.7, in which α is set at 0.9 and β changes from 1

Table 12.1 Uncertain parameters of each fuel and carbon to power parameter at MSWACPPs.

	Jiujiang MSWACPP		Weiao MSWACPP		Zigong MSWACPP	
	MSW	**Coal**	**MSW**	**Coal**	**MSW**	**Coal**
Fuel to power $\widetilde{H}_{ij}$ (kWh/tonne)	(1300,1365,1435, 1500)	(2350,2450,2550, 2650)	(1180,1255,1340, 1425)	(2070,2150,2245, 2335)	(1105,1170,1235, 1290)	(1880,1965,2035, 2120)
Carbon emissions $\widetilde{Q}_{ij}$ (kg/tonne)	(590,660,735,815)	(2080,1925,2000, 2065)	(510,575,630,685)	(1850,1925,2000, 2065)	(400,475,525,600)	(1810,1880,1950, 2040)
Pollutants emissions $\widetilde{N}_{ijr}$ (kg/tonne)						
For NO_x ($r = 1$)	(2.1,2.6,3.3,4)	(5.9,6.6,7.3,8.2)	(2.5,3.2,4,4.7)	(6.5,7.3,8,8.6)	(3,3.7,4.3,5)	(7,7.9,9,10.1)
For SO_2 ($r = 2$)	(4,4.7,5.3,6)	(7.1,8,9,9.9)	(4.5,5,5.5,6.2)	(8,9.2,10.3,11.3)	(4.8,5.6,6.4,7.2)	(8.8,9.5,10.4,11.3)
Flue gas emissions $\widetilde{R}_{ij}$($\times 10^5$ Nm^3/tonne)	(1.2,1.6,2,2.4)	(4.2,4.3,4.4,4.5)	(1,1.4,1.7,1.9)	(3.5,3.8,4.3,4.8)	(0.9,1.2,1.6,1.9)	(3.2,4,4.8,5.6)
Dioxins emissions $\widetilde{D}_{ij}$ (kg/tonne)	(0.45,0.49,0.553, 0.57)	(0.15,0.18,0.22, 0.25)	(0.45,0.5,0.55, 0.62)	(0.14,0.18,0.24, 0.28)	(0.48,0.53,0.59, 0.64)	(0.15,0.21,0.27, 0.33)
Fuel quality $\widetilde{S}_{ijt}$						
Heat value ($t = 1$) (Gj/tonne)	(16.8,17.7,18.4, 19.1)	(24.2,24.8,25.3, 25.7)	(15,15.7,16.3, 17)	(24,24.6,25.5, 25.9)	(13.7,14.6,15.5, 16.2)	(23.9,24.7,25.4, 26)
Volatile matter ($t = 2$) (wt%)	(28.5,29.3,30.5, 31.7)	(22.2,22.7,23.2, 23.9)	(26.1,26.8,27.3, 27.8)	(22.4,22.8,23.1, 23.7)	(31,31.7,32.4, 32.9)	(22.5,22.8,23.1, 23.6)
Moisture content ($t = 3$) (wt%)	(13.8,14.6,15.4, 16.2)	(2.9,3.6,4.4,5.1)	(13,13.7,14.4, 14.9)	(3.1,3.8,4.3,4.8)	(12.2,12.7,13.4, 13.7)	(3.3,3.7,4.2,4.8)
Ash content ($t = 4$) (wt%)	(13.8,14.6,15.4, 16.2)	(2.9,3.6,4.4,5.1)	(13,13.7,14.4, 14.9)	(3.1,3.8,4.3,4.8)	(12.2,12.7,13.4, 13.7)	(3.3,3.7,4.2,4.8)
Sulfur content ($t = 5$) (wt%)	(0.14,0.18,0.22, 0.26)	(0.3,0.36,0.43, 0.51)	(0.08,0.09,0.11, 0.12)	(0.32,0.38,0.43, 0.47)	(0.07,0.09,0.11, 0.13)	(0.28,0.35,0.44, 0.53)
Carbon to power $\widetilde{P}_i$(kWh/tonne)	(940,1020,1100,1180)		(910,1000,1070,1140)		(870,965,1040,1125)	

Table 12.2 Lower and upper bounds of fuel qualities at MSWACPPs.

	Jiujiang MSWACPP		Weiao MSWACPP		Zigong MSWACPP	
Fuel quality	**Lower bound:** $\widetilde{S_{it}^{L}}$	**Upper bound:** $\widetilde{S_{it}^{U}}$	**Lower bound:** $\widetilde{S_{it}^{L}}$	**Upper bound:** $\widetilde{S_{it}^{U}}$	**Lower bound:** $\widetilde{S_{it}^{L}}$	**Upper bound:** $\widetilde{S_{it}^{U}}$
Heat value ($t = 1$) (Gj/tonne)	(20.8,21.6,22.4, 23.2)	—	(19.7,20.5,21.4, 22.4,)	—	(18.7,19.5,20.4, 21.4)	—
Volatile matter ($t = 2$) (wt%)	(4.8,5.7,6.4,7.1)	(25.7,26.5,27.4, 28.4)	(6.1,6.7,7.3,7.9)	(26.7,27.6,28.5, 29.2)	(7.8,8.6,9.4,10.2)	(34.9,35.6,36.3, 37.2)
Moisture content ($t = 3$) (wt%)	—	(4,4.7,5.3,6)	—	(4.9,5.7,6.3,7.1)	—	(3.8,4.5,5.4,6.3)
Ash content ($t = 4$) (wt%)	—	(18.7,19.6,20.5, 21.2)	—	(20,20.6,21.2, 22.2)	—	(20.9,21.7,22.4, 23)
Sulfur content ($t = 5$) (wt%)	—	(0.83,0.93,1.07, 1.17)	—	(1.08,1.16,1.24, 1.32)	—	(1.39,1.46,1.54, 1.61)

Table 12.3 Certain parameters at each MSWACPP.

	Jiujiang MSWACPP	Weiao MSWACPP	Zigong MSWACPP
Dioxins residual ratio: κ_i(%)	0.2	0.18	0.17
Power demand: U_i (kWh)	8×10^9	6.3×10^9	4.45×10^9
Pollutants treatment costs (CNY/kg)			
For NO_x	15.7	15.5	14.8
For SO_2	2.5	2.3	1.8
Procurement cost: F_{ir} (CNY/tonne)			
MSW	280	295	290
Coal	610	550	510
Maximum quantity: O_{ij}^U (tonnes)			
MSW	3×10^6	2.7×10^6	2.3×10^6
Coal	4.5×10^6	3.9×10^6	3.3×10^6
Carbon emission quota (tonnes)			
Lower bound: Q_i^L	4.35×10^6	2.5×10^6	1.8×10^6
Upper bound: Q_i^U	1.19×10^7	7.9×10^6	6.45×10^6

Table 12.4 Other parameters used in the proposed model.

Value-added tax: φ (%)	0.01
Carbon emissions quota price: λ (CNY/tonne)	30
Subsidies of waste-to-power: γ (%)	0.01
Power price: ω (CNY/kWh)	0.4
Dioxins emissions limit: ϖ (ngTEQ/Nm^3)	0.1
The total power demand in a region: U (kWh)	1.875×10^{10}
Carbon emissions in the last production period: L (kWh)	2.395×10^7

to 0.9. Scenario 4 is shown in Table 12.8, in which α is set at 0.85 and β changes from 1 to 0.9.

Based on the results under the four scenarios, two obvious conclusions were reached. First, as β reduces, the total carbon emissions, financial revenue, carbon emissions quotas that the MSWACPPs receive, and MSWACPP profits all reduce. The reason for this is that with lower β there is a stricter attitude toward carbon emissions reductions, which results in less fees for carbon emissions quotas, which restricts the MSWACPPs' power generation capacity and therefore their respective profits also reduce. Second, as α reduces, the authority's financial revenue reduces while the MSWACPPs' profits increase. The reason for this is that with lower α which means a more relaxed attitude toward dioxin emissions below the regulated standard, the MSWACPPs can increase the amount of cheaper MSW and gain

Table 12.5 Results when $\alpha = 1$ and β is changing.

		Jiujiang MSWACPP	Weiao MSWACPP	Zigong MSWACPP	Authority
$\beta = 1$	MSW: x_{i1} (tonnes)	1 210 651	879 330	703 334	
	Coal: x_{i2} (tonnes)	4 015 818	3 742 648	3 134 231	
	Carbon emissions quota: y_i (tonnes)	9 682 254	7 863 189	6 404 557	2 3950 000
	Profits: f (CNY)	915 458 781	530 521 169	287 803 279	
	Financial revenue: F_1 (CNY)				963 918 322
$\beta = 0.975$	MSW: x_{i1} (tonnes)	1 144 560	874 506	700 364	
	Coal: x_{i2} (tonnes)	3 796 590	3 722 115	3 120 996	
	Carbon emissions quota: y_i (tonnes)	9 153 689	7 820 049	6 377 512	23 351 250
	Profits: f (CNY)	865 482 857	527 610 591	286 587 936	
	Financial revenue: F_1 (CNY)				939 758 686
$\beta = 0.95$	MSW: x_{i1} (tonnes)	1 080 549	868 781	696 452	
	Coal: x_{i2} (tonnes)	3 584 260	3 697 747	3 103 564	
	Carbon emissions quota: y_i (tonnes)	8 641 756	7 768 853	6 341 891	22 752 500
	Profits: f (CNY)	817 079 511	524 156 446	284 987 225	
	Financial revenue: F_1 (CNY)				915 603 213
$\beta = 0.925$	MSW: x_{i1} (tonnes)	1 024 573	858 565	689 892	
	Coal: x_{i2} (tonnes)	3 398 584	3 654 268	3 074 333	
	Carbon emissions quota: y_i (tonnes)	8 194 087	7 677 504	6 282 159	22 153 750
	Profits: f (CNY)	774 752 281	517 993 206	282 303 068	
	Financial revenue: F_1 (CNY)				891 457 651
$\beta = 0.9$	MSW: x_{i1} (tonnes)	985 380	840 098	676 696	
	Coal: x_{i2} (tonnes)	3 268 578	3 575 668	3 015 527	
	Carbon emissions quota: y_i (tonnes)	7 880 637	7 512 369	6 161 994	21 555 000
	Profits: f (CNY)	745 115 507	506 851 737	276 903 181	
	Financial revenue: F_1 (CNY)				867 339 680

Table 12.6 Results when $\alpha = 0.95$ and β is changing.

		Jiujiang MSWACPP	Weiao MSWACPP	Zigong MSWACPP	Authority
$\beta = 1$	MSW: x_{i1} (tonnes)	1 916 096	1 220 768	1 219 497	
	Coal: x_{i2} (tonnes)	3 791 358	3 638 127	2 986 372	
	Carbon emissions quota: y_i (tonnes)	9 682 254	7 863 189	6 404 557	23 950 000
	Profits: f (CNY)	1 018 448 997	566 582 148	331 083 043	
	Financial revenue: F_1 (CNY)				953 050 160
$\beta = 0.975$	MSW: x_{i1} (tonnes)	1 811 494	1 214 070	1 214 347	
	Coal: x_{i2} (tonnes)	3 584 383	3 618 167	2 973 761	
	Carbon emissions quota: y_i (tonnes)	9 153 689	7 820 049	6 377 512	23 351 250
	Profits: f (CNY)	962 850 721	563 473 730	329 684 937	
	Financial revenue: F_1 (CNY)				929 221 965
$\beta = 0.95$	MSW: x_{i1} (tonnes)	1 710 184	1 206 122	1 207 565	
	Coal: x_{i2} (tonnes)	3 383 922	3 594 480	2 957 151	
	Carbon emissions quota: y_i (tonnes)	8 641 756	7 768 853	6 341 891	22 752 500
	Profits: f (CNY)	909 001 941	559 784 798	327 843 512	
	Financial revenue: F_1 (CNY)				905 394 610
$\beta = 0.925$	MSW: x_{i1} (tonnes)	1 621 591	1 191 940	1 196 191	
	Coal: x_{i2} (tonnes)	3 208 624	3 552 214	2 929 299	
	Carbon emissions quota: y_i (tonnes)	8 194 087	7 677 504	6 282 159	22 153 750
	Profits: f (CNY)	861 912 847	553 202 625	324 755 712	
	Financial revenue: F_1 (CNY)				881 562 795
$\beta = 0.9$	MSW: x_{i1} (tonnes)	1 559 560	1 166 303	1 173 311	
	Coal: x_{i2} (tonnes)	3 085 884	3 475 810	2 873 267	
	Carbon emissions quota: y_i (tonnes)	7 880 637	7 512 369	6 161 994	21 555 000
	Profits: f (CNY)	828 941 900	541 303 840	318 543 792	
	Financial revenue: F_1 (CNY)				857 730 264

Table 12.7 Results when $\alpha = 0.9$ and β is changing.

		Jiujiang MSWACPP	Weiao MSWACPP	Zigong MSWACPP	Authority
$\beta = 1$	MSW: x_{i1} (tonnes)	2 664 841	1 905 750	1 800 222	
	Coal: x_{i2} (tonnes)	3 553 121	3 428 438	2 820 018	
	Carbon emissions quota: y_i (tonnes)	9 682 254	7 863 189	6 404 557	23 950 000
	Profits: f (CNY)	1 127 760 571	638 926 634	379 776 220	
	Financial revenue: F_1 (CNY)				939 154 022
$\beta = 0.975$	MSW: x_{i1} (tonnes)	2 519 364	1 895 295	1 792 620	
	Coal: x_{i2} (tonnes)	3 359 152	3 409 629	2 808 110	
	Carbon emissions quota: y_i (tonnes)	9 153 689	7 820 049	6 377 512	23 351 250
	Profits: f (CNY)	1 066 194 854	635 421 315	378 172 491	
	Financial revenue: F_1 (CNY)				915 690 327
$\beta = 0.95$	MSW: x_{i1} (tonnes)	2 378 465	1 882 887	1 782 607	
	Coal: x_{i2} (tonnes)	3 171 287	3 387 307	2 792 425	
	Carbon emissions quota: y_i (tonnes)	8 641 756	7 768 853	6 341 891	22 752 500
	Profits: f (CNY)	1 006 566 408	631 261 359	376 060 242	
	Financial revenue: F_1 (CNY)				892 226 420
$\beta = 0.925$	MSW: x_{i1} (tonnes)	2 255 253	1 860 747	1 765 818	
	Coal: x_{i2} (tonnes)	3 007 004	3 347 477	2 766 125	
	Carbon emissions quota: y_i (tonnes)	8 194 087	7 677 504	6 282 159	22 153 750
	Profits: f (CNY)	954 423 175	623 838 736	372 518 313	
	Financial revenue: F_1 (CNY)				868 754 606
$\beta = 0.9$	MSW: x_{i1} (tonnes)	2 168 983	1 820 724	1 732 041	
	Coal: x_{i2} (tonnes)	2 891 977	3 275 477	2 713 214	
	Carbon emissions quota: y_i (tonnes)	7 880 637	7 512 369	6 161 994	21 555 000
	Profits: f (CNY)	917 913 409	610 420 646	365 392 791	
	Financial revenue: F_1 (CNY)				845 274 195

Table 12.8 Results when $\alpha = 0.85$ and β is changing.

		Jiujiang MSWACPP	Weiao MSWACPP	Zigong MSWACPP	Authority
$\beta = 1$	MSW: x_{i1} (tonnes)	2 664 841	2 578 095	2 171 036	
	Coal: x_{i2} (tonnes)	3 553 121	3 222 618	2 713 795	
	Carbon emissions quota: y_i (tonnes)	9 682 254	7 863 189	6 404 557	23 950 000
	Profits: f (CNY)	1 127 760 571	709 936 371	410 868 605	
	Financial revenue: F_1 (CNY)				932 501 531
$\beta = 0.975$	MSW: x_{i1} (tonnes)	2 519 364	2 563 951	2 161 868	
	Coal: x_{i2} (tonnes)	3 359 152	3 204 938	2 702 335	
	Carbon emissions quota: y_i (tonnes)	9 153 689	7 820 049	6 377 512	23 351 250
	Profits: f (CNY)	1 066 194 854	706 041 474	409 133 578	
	Financial revenue: F_1 (CNY)				909 071 649
$\beta = 0.95$	MSW: x_{i1} (tonnes)	2 378 465	2 547 165	2 149 793	
	Coal: x_{i2} (tonnes)	3 171 287	3 183 956	2 687 242	
	Carbon emissions quota: y_i (tonnes)	8 641 756	7 768 853	6 341 891	22 752 500
	Profits: f (CNY)	1 006 566 408	701 419 185	406 848 400	
	Financial revenue: F_1 (CNY)				885 649 040
$\beta = 0.925$	MSW: x_{i1} (tonnes)	2 255 253	2 517 214	2 129 546	
	Coal: x_{i2} (tonnes)	3 007 004	3 146 518	2 661 932	
	Carbon emissions quota: y_i (tonnes)	8 194 087	7 677 504	6 282 159	22 153 750
	Profits: f (CNY)	954 423 175	693 171 618	403 016 492	
	Financial revenue: F_1 (CNY)				862 249 642
$\beta = 0.9$	MSW: x_{i1} (tonnes)	2 168 983	2 463 072	2 088 812	
	Coal: x_{i2} (tonnes)	2 891 977	3 078 840	2 611 015	
	Carbon emissions quota: y_i (tonnes)	7 880 637	7 512 369	6 161 994	21 555 000
	Profits: f (CNY)	917 913 409	678 262 253	395 307 601	
	Financial revenue: F_1 (CNY)				838 904 184

greater profits by lowering the cost. As the conversion parameter from a unit of MSW to power is smaller than that from a unit of coal, less power is generated, which means there is less value-added tax and therefore less financial revenue for the authority.

12.5 Discussion

Based on the results under different scenarios shown above, in this section, some propositions will be proposed.

12.5.1 Propositions and Analysis

First, an equilibrium strategy toward MSW and coal co-combustion optimization method can contribute to carbon emissions reductions. From the results, it can be seen that the total carbon emissions reduce, which clearly shows that this method has the potential to reduce total carbon emissions compared with the carbon emissions in the last production period. In the proposed bilevel model, the carbon emissions quota allocation is the key factor that establishes the cooperative relationship between the authority and the power plants to realize carbon emissions reductions. The authority sets carbon emissions reductions targets and allocates carbon emissions quotas to the power plants, which with better carbon emissions reduction performances can be given a higher allocation. Under the limited carbon emissions quotas, each power plant makes efforts to update their strategies to maximize profits.

Second, when the carbon emissions reductions constraint is stricter, the marginal reduction rate in financial revenue is almost the same as that for the carbon emissions reductions. The total carbon emissions and financial revenue all reduce; however, the marginal carbon emissions are almost the same. It can be inferred that although relatively stricter carbon emissions reduction constraints reduce financial revenue, it also reduces carbon emissions to the same extent and improves the environment. In other words, the environmental protection is cost-effective (Figures 12.3 and 12.4).

Third, the environmental protection targets significantly affect the co-combustion ratio of MSW and coal. The stricter attitude toward dioxins emissions below the standard has a strong impact on the MSW/coal co-combustion. Taking Weiao coal-fired power plant as an example, when α is fixed at 1, the co-combustion ratio is 19%. When α reduces to 0.95, the MSW co-combustion ratio is 25%. When α is set at 0.9, the co-combustion ratio increases to 36%. And when α varies from 0.9 to 0.85, the MSW accounts for 44%. It can be seen that when α changes from 1 to 0.95, 0.9, and 0.85, the quantity of MSW increases and its co-combustion ratio also increases.When the attitude toward dioxin emissions below the emissions standard relaxes, more MSW with higher dioxin emissions and lower price can be burned and the combustion ratios increase (Figures 12.5 and 12.6).

Fourth, different coal-fired power plants play different roles in protecting the environment. It can be seen that different coal-fired power plants have different sensitivities as β changes. Taking βas an example, when β changes from 1 to 0.975, from

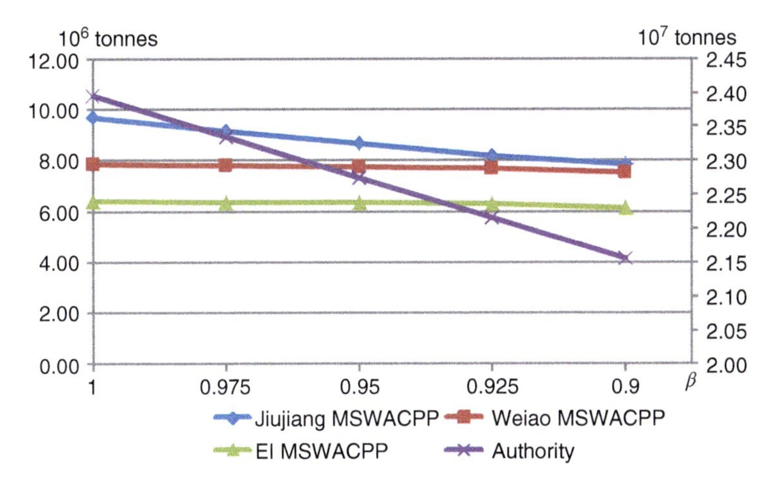

Figure 12.3 The total carbon emissions and MSWACPP carbon emissions under different β values. The total carbon emissions belong to the right vertical axis and MSWACPP carbon emissions belong to the left vertical axis.

0.975 to 0.95, from 0.95 to 0.925, and from 0.925 to 0.9, the carbon emissions quotas that Jiujiang coal-fired power plant, Weiao coal-fired power plant, and Zigong coal-fired power plant, respectively, receive reduce. However, their reduction rates are different. Jiujiang coal-fired power plant plays an important role in carbon emissions reductions and has an advantage when competing for carbon emissions quotas over the other two coal-fired power plants. The reason for this is that Jiujiang coal-fired power plant has a higher conversion parameter for a unit of carbon emissions to electricity is higher than the other two coal-fired power plants. Therefore, Jiujiang coal-fired power plant is allocated a greater carbon emissions quota by the authority.

12.5.2 Management Recommendations

Based on the above analysis and discussion, some management recommendations that include policy strategies, process procedure, and technical improvements are given in the following. First, a carbon emissions quota allocation competition mechanism should be developed. In such a mechanism, the authority allocates carbon emissions quotas to multiple coal-fired power plants, following which the coal-fired power plants decide on MSW and coal co-combustion ratio under the limited carbon emissions quotas. The authority allocates a higher carbon emissions quota to coal-fired power plants with better carbon emissions reduction performances, and the coal-fired power plants compete with each other to receive more carbon emissions quotas. It can realize carbon emissions reductions by establishing a cooperative relationship between the authority and the coal-fired power plants. Second, the authority can have a relaxed attitude in consideration of economic development. However, after a period of time, as economic developed, the authority should pay more attention to environment and therefore have a stricter attitude.

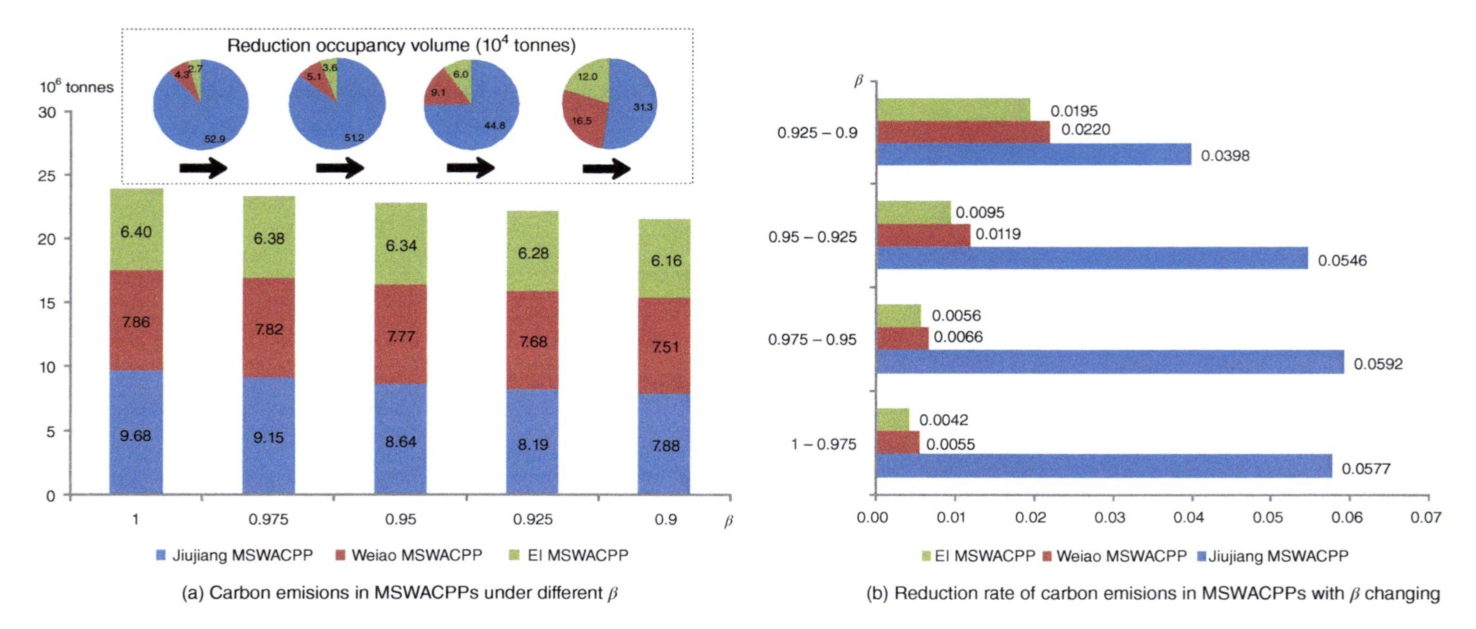

Figure 12.4 MSWACPP carbon emissions changes under different β values.

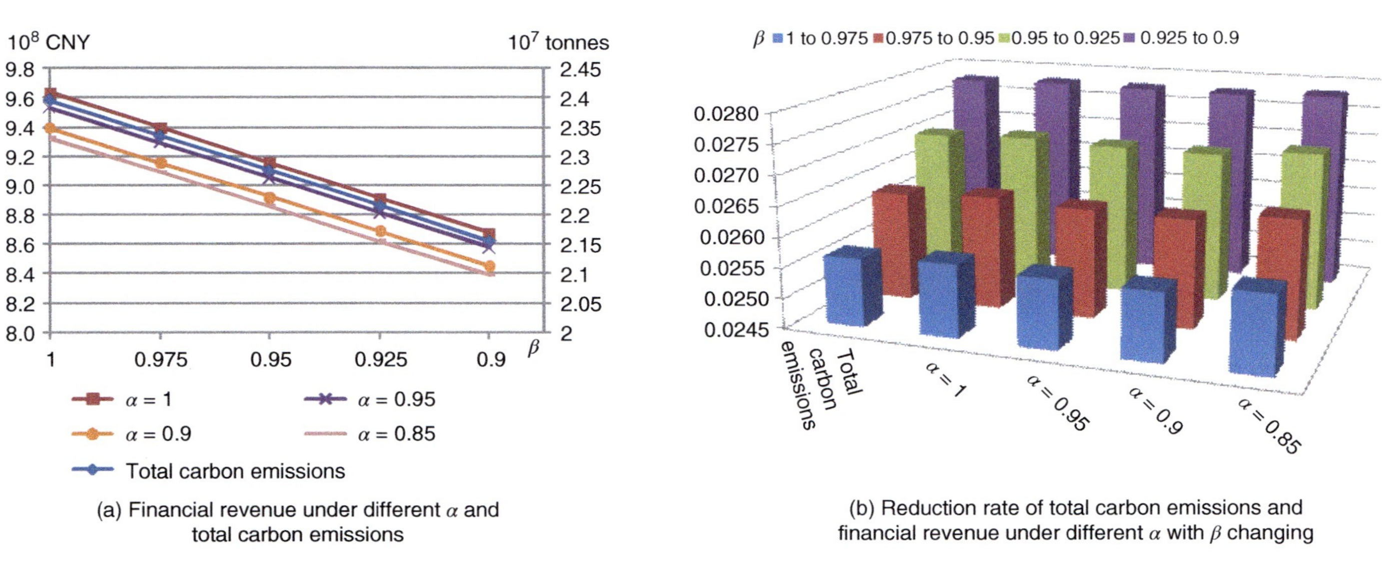

(a) Financial revenue under different α and total carbon emissions

(b) Reduction rate of total carbon emissions and financial revenue under different α with β changing

Figure 12.5 The total carbon emissions and financial revenue under different α values with β changing. In (a), the total carbon emissions belong to the right vertical axis and financial revenue belongs to the left vertical axis.

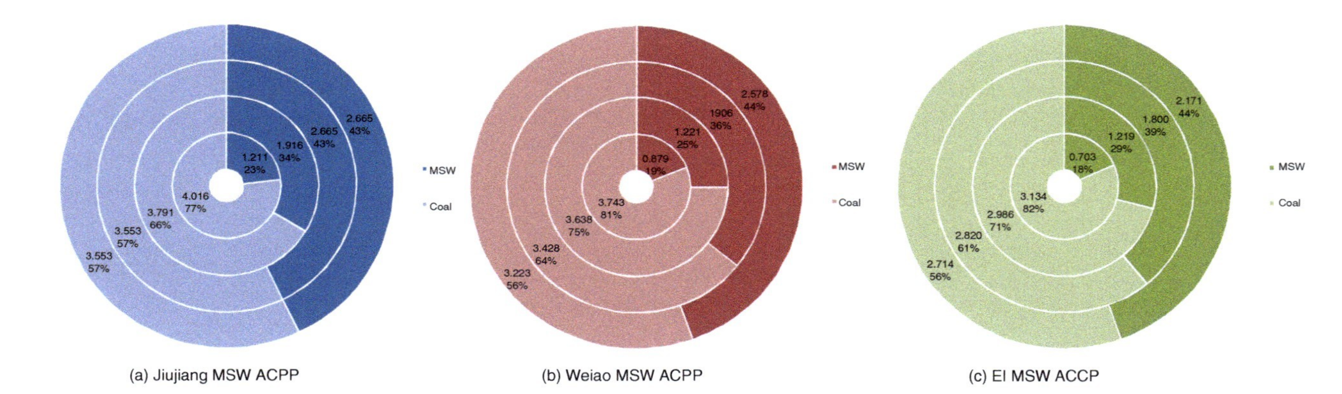

Figure 12.6 Co-combustion ratio of MSW and coal in MSWACPPs under different α values. α, respectively, equals to 1, 0.95, 0.9, and 0.85 from the inside out.

When the economy becomes better, relatively stricter environmental protection can be adopted while ensuring continuing economic development. In this way, environmental protection can be gradually adjusted to the strictest requirement. Therefore, dynamically adjusting the coordination between economic development and environmental protection can achieve sustainable development. Third, a scientifically classified MSW collection system should be adopted. Currently, Chinese MSW is collected in a mixed state. Due to lack of a scientifically classified MSW collection system, the MSW heat value is too low to burn in the combustors. A scientifically classified MSW collection system could eliminate the unburnable materials and improve the heat value. Four dioxin emissions reduction technologies should be improved. Dioxins, ubiquitous environmental pollutants, are mainly produced unintentionally by MSW incineration. The technologies used for dioxin control, such as incineration and cooling conditions control, inhibition chemicals, flue gas scrubbing, and catalytic destruction, are available in China. However, to protect the environment when widely using MSW, it is necessary to continue to improve dioxin control strategies in the incinerators and flue gas treatment systems to further reduce dioxin emissions.

References

Alkhatib, I.A., Monou, M., Zahra, A.S.F.A. et al. (2010). Solid waste characterization, quantification and management practices in developing countries: a case study: Nablus district-palestine. *Journal of Environmental Management* 91 (5): 1131–1138.

BP Global (2018). BP Statistical Review of World Energy. https://www.bp.com/content/dam/bp/en/corporate/pdf/energy-economics/statistical-review/bp-stats-review-2018-full-report.pdf (accessed 29 May 2021).

Cheng, H. and Hu, Y. (2010a). Municipal solid waste (MSW) as a renewable source of energy: current and future practices in China. *Bioresource Technology* 101 (11): 3816–3824.

Cheng, H. and Hu, Y. (2010b). Curbing dioxin emissions from municipal solid waste incineration in China: re-thinking about management policies and practices. *Environmental Pollution* 158 (9): 2809–2814.

Desroches-Ducarne, E., Marty, E., Martin, G., and Delfosse, L. (1998). Co-combustion of coal and municipal solid waste in a circulating fluidized bed. *Fuel* 77 (12): 1311–1315.

Dong, C., Jin, B., Zhong, Z., and Lan, J. (2002). Tests on co-firing of municipal solid waste and coal in a circulating fluidized bed. *Energy Conversion and Management* 43 (16): 2189–2199.

Dubois, D. and Prade, H. (1983). Ranking fuzzy numbers in the setting of possibility theory. *Information Sciences* 30 (3): 183–224.

Fang, G., Tian, L., Liu, M. et al. (2018). How to optimize the development of carbon trading in China: enlightenment from evolution rules of the EU carbon price. *Applied Energy* 211: 1039–1049.

Feng, C., Ma, Y., Zhou, G., and Ni, T. (2018). Stackelberg game optimization for integrated production-distribution-construction system in construction supply chain. *Knowledge-Based Systems*. https://doi.org/10.1016/j.knosys.2018.05.022.

Gao, J. and You, F. A stochastic game theoretic framework for decentralized optimization of multi-stakeholder supply chains under uncertainty. (2018). *Computers and Chemical Engineering*. 122: 31–46

Gullett, B.K., Bruce, K.R., and Beach, L.O. (1992). Effect of sulfur dioxide on the formation mechanism of polychlorinated dibenzodioxin and dibenzofuran in municipal waste combustors. *Environmental Science and Technology* 26 (10): 1938–1943.

Gullett, B.K., Raghunathan, K., and Dunn, J.E. (1998). The effect of cofiring high-sulfur coal with municipal waste on formation of polychlorinated dibenzodioxin and polychlorinated dibenzofuran. *Environmental Engineering Science* 15 (1): 59–70.

Hoornweg, D., Lam, P., and Chaudhry, M. (2005). Waste management in China: issues and recommendations. *Urban Development Working Papers No. 9*.

Huang, T., Jiang, W., Ling, Z. et al. (2016). Trend of cancer risk of Chinese inhabitants to dioxins due to changes in dietary patterns: 1980–2009. *Scientific Reports* 6: 21997.

International Energy Agency (2017). Coal 2017. https://www.iea.org/coal2017/ (accessed 29 May 2021).

Jin, Y., Liang, L., Ma, X. et al. (2013). Effects of blending hydrothermally treated municipal solid waste with coal on co-combustion characteristics in a lab-scale fluidized bed reactor. *Applied Energy* 102 (2): 563–570.

Levis, J.W., Barlaz, M.A., Decarolis, J.F., and Ranjithan, S.R. (2013). A generalized multistage optimization modeling framework for life cycle assessment-based integrated solid waste management. *Environmental Modelling and Software* 50 (12): 51–65.

Liu, B. and Iwamura, K. (1998). Chance constrained programming with fuzzy parameters. *Fuzzy Sets and Systems* 94 (2): 227–237.

Liu, Z., Guan, D., Wei, W. et al. (2015). Reduced carbon emission estimates from fossil fuel combustion and cement production in China. *Nature* 524 (7565): 335.

Liu, Y., Wang, Y., Wang, Q. et al. (2017). Simultaneous removal of NO and SO_2 using vacuum ultraviolet light (VUV)/heat/peroxymonosulfate (PMS). *Chemosphere* 190: 431.

Lv, C., Xu, J., Xie, H. et al. (2016). Equilibrium strategy based coal blending method for combined carbon and PM_{10} emissions reductions. *Applied Energy* 183: 1035–1052.

Mao, X.Q., Zeng, A., Hu, T. et al. (2014). Co-control of local air pollutants and CO_2 from the Chinese coal-fired power industry. *Journal of Cleaner Production* 67 (67): 220–227.

Mcphail, A., Griffin, R., Elhalwagi, M. et al. (2013). Environmental, economic, and energy assessment of the ultimate analysis and moisture content of municipal solid waste in a parallel co-combustion process. *Energy and Fuels* 28 (2): 1453–1462.

Muthuraman, M., Namioka, T., and Yoshikawa, K. (2010). Characteristics of co-combustion and kinetic study on hydrothermally treated municipal solid waste with different rank coals: a thermogravimetric analysis. *Applied Energy* 87 (1): 141–148.

Nash, J. (1951). Non-cooperative games. *Annals of Mathematics* 54 (2): 286–295.
Olsen, D.J., Dvorkin, Y., Fernandez-Blanco, R., and Ortega-Vazquez, M.A. (2018). Optimal carbon taxes for emissions targets in the electricity sector. *IEEE Transactions on Power Systems* (99): 5892–5901.
Pires, A., Martinho, G., and Chang, N.-B. (2011). Solid waste management in European countries: a review of systems analysis techniques. *Journal of Environmental Management* 92 (4): 1033–1050.
Qi, Y., Stern, N., Wu, T. et al. (2016). China's post—coal growth. *Nature Geoscience* 9: 564–566.
Raghunathan, K., Gullett, B.K., Lee, C.W. et al. (1997). Prevention of PCDD/PCDF formation by coal co-firing.
Saini, S., Rao, P., and Patil, Y. (2012). City based analysis of MSW to energy generation in India, calculation of state-wise potential and tariff comparison with EU. *Procedia-Social and Behavioral Sciences* 37 (1): 407–416.
Service, R.F. (2017). Cleaning up coal–cost-effectively. *Science* 356 (6340): 798.
Sichuan Provincial Bureau of Statistics (2017). 2017 Sichuan Provincial Statistical Yearbook. http://www.sc.stats.gov.cn/tjcbw/tjnj/2017/zk/indexch.htm (accessed 29 May 2021).
Sichuan Provincial Bureau of Statistics (2017). Sichuan provincial statistical yearbook. http://www.sc.stats.gov.cn/tjcbw/tjnj/2017/zk/indexch.htm.
Simab, M., Javadi, M.S., and Nezhad, A.E. (2018). Multi-objective programming of pumped-hydro-thermal scheduling problem using normal boundary intersection and VIKOR. *Energy* 143: 854–866.
Sinha, A., Malo, P., and Deb, K. (2018). A review on bilevel optimization: from classical to evolutionary approaches and applications. *IEEE Transactions on Evolutionary Computation* 22 (2): 276–295.
Sipra, A.T., Gao, N., and Sarwar, H. Municipal solid waste (MSW) pyrolysis for bio-fuel production: a review of effects of MSW components and catalysts. (2018). *Fuel Processing Technology* 175: 131–147.
Suksankraisorn, K. Patumsawad, S. Vallikul, P. (2004). Co-combustion of municipal solid waste and thai lignite in a fluidized bed. *Energy Conversion & Management*. 45 (6): 947–962.
Tsumura, T., Okazaki, H., Dernjatin, P., and Savolainen, K. (2003). Reducing the minimum load and NO_x emissions for lignite-fired boiler by applying a stable-flame concept. *Applied Energy* 74 (3): 415–424.
United Nations Environment Programme (2016). Guidelines for framework legislation for integrated waste management. http://wedocs.unep.org/bitstream/handle/20.500.11822/22098/UNEP?sequence=1 (accessed 29 May 2021).
Wang, Y., Yan, Y., Chen, G. et al. (2015). Effectiveness of waste-to-energy approaches in China: from the perspective of greenhouse gas emission reduction. *Journal of Cleaner Production*. 163: 99–105.
Warren, A. and El-Halwagi, M. (1996). An economic study for the co-generation of liquid fuel and hydrogen from coal and municipal solid waste. *Fuel Processing Technologyc*.

Wei, J., Guo, Q., He, Q. et al. (2017). Co-gasification of bituminous coal and hydrochar derived from municipal solid waste: reactivity and synergy. *Bioresource Technologyc*.

World Meteorological Organization (2018). WMO greenhouse gas bulletin. https://public.wmo.int/en.

Xie, S., Hu, Z., Zhou, D. et al. (2018). Multi-objective active distribution networks expansion planning by scenario-based stochastic programming considering uncertain and random weight of network. *Applied Energy* 219: 207–255.

Xie, J., Zhong, W., Shao, Y. et al. (2017). Simulation on combustion of MSW and coal in an industrial-scale CFB boiler. *Energy and Fuels*. 31: 14248–14261

Xu, X. (2001). Study on formation mechanism and emission characteristics of polychlorinated dibenzo-*p*-dioxins and polychlorinated dibenzofrans in combustion process. PhD thesis. Hangzhou: Zhejiang University.

Xu, J. and Zhou, X. (2011). *Fuzzy-Like Multiple Objective Decision Making*. Berlin, Heidelberg: Springer-Verlag.

Xu, J., Huang, Q., Lv, C. et al. (2018). Carbon emissions reductions oriented dynamic equilibrium strategy using biomass–coal co-firing. *Energy Policy* 123: 184–197.

Yang, N., Zhang, H., Chen, M. et al. (2012). Greenhouse gas emissions from MSW incineration in China: impacts of waste characteristics and energy recovery. *Waste Management* 32 (12): 2552–2560.

Zeng, Z., Xu, J., Wu, S., and Shen, M. (2014). Antithetic method-based particle swarm optimization for a queuing network problem with fuzzy data in concrete transportation systems. *Computer-Aided Civil and Infrastructure Engineering* 29 (10): 771–800.

Zhang, G., Lu, J., and Gao, Y. (2015). *Multi-Level Decision Making*. Berlin, Heidelberg: Springer-Verlag.

Zheng, L., Song, J., Li, C. et al. (2014). Preferential policies promote municipal solid waste (MSW) to energy in China: current status and prospects. *Renewable and Sustainable Energy Reviews* 36 (C): 135–148.

Zhou, H., Meng, A., Long, Y. et al. (2014). ChemInform abstract: an overview of characteristics of municipal solid waste fuel in China: physical, chemical composition and heating value. *Renewable and Sustainable Energy Reviews* 36 (30): 107–122.

Index

Innovative Approaches towards Ecological Coal Mining and Utilization, First Edition.
Jiuping Xu, Heping Xie, and Chengwei Lv.

h

i

k

l

m

n